Roman Brick and Tile

Studies in Manufacture, Distribution and Use in the Western Empire

Edited by

Alan McWhirr

B A R International Series 68

1979

B.A.R.

B.A.R., 122 Banbury Road, Oxford OX2 7BP, England

B. A. R. - S68, 1979: "Roman Brick and Tile"

ISBN 9780860540731 paperback

ISBN 9781407350028 e-book

DOI https://doi.org/10.30861/9780860540731

A catalogue record for this book is available from the British Library

This book is available at www.barpublishing.com

CONTENTS

LIST OF CONTRIBUTORS

T. F. C. Blagg, M. A.
Lecturer in Archaeology,
School of Continuing Education, University of Kent

G. Brodribb, M. A., Dip. Ed., F. S. A.
Co-Director of Excavations, Beauport Park, 1970-to present

N. Cuomo di Caprio, dott. ssa
Laboratorio di Termoluinescenza applicata all'archeologia,
Istituto di Scienze Fisiche, Università di Milano

J. C. Carter, Ph. D.
Associate Professor of Classics
Director, Metaponto Excavations
University of Texas at Austin

C. L. Cram, M. A.
Archaeologist, Reading Museum and Art Gallery

T. C. Darvill, B. A.
Department of Archaeology, University of Southampton

P. L. Everson, M. A.
Investigator, Survey of Lincolnshire
Member of Brixworth Archaeological Committee

M. G. Fulford, B. A., Ph. D., F. S. A.
Lecturer in Roman Archaeology, University of Reading

T. K. Green, Dip. Archaeol., R. S. A., M. I. Q. A.
Industrial Statistician, Solartron-Schlumberger

M. W. C. Hassall, M. A., F. S. A.
Institute of Archaeology, University of London

D. E. Johnston, M. A.
Tutor in Archaeology, Department of Adult Education,
University of Southampton

A. D. McWhirr, B. Sc., M. A., F. S. A.
School of Humanities, Leicester Polytechnic
Director of Excavations, Cirencester Excavation Committee

G. C. Morgan
Archaeological Conservator, Department of Archaeology,
University of Leicester

D. Parsons, B. A., F. S. A.
Department of Adult Education, University of Leicester
Co-ordinator, Brixworth Archaeological Research Committee

D. P. S. Peacock, B. Sc., Ph. D., F. S. A.
Lecturer in Archaeological Science, University of Southampton

A. G. Rook, B. Sc., M. Phil.
Director, Welwyn Archaeological Society
Education Officer, Lockleys Archaeological Trust

R. S. O. Tomlin, M. A., D. Phil., F. S. A.
University Lecturer in Late Roman History
Fellow of Wolfson College, Oxford

G. A. Webster, M. A., Ph. D., F. S. A., A. M. A.
Reader in Romano-British Archaeology, Department of Extramural Studies, University of Birmingham

J. D. Wilcock, M. A., B. Sc., Ph. D., C. Eng., M. I. E. E., M. B. C. S.,
Head, Research Centre for Computer Archaeology
Senior Lecturer, Department of Computing, North Staffordshire Polytechnic

D. F. Williams, B. A., Ph. D.
Research Fellow, Department of Archaeology, University of Southampton

J. J. Wilkes, B. A., Ph. D., F. S. A.
Professor of Archaeology of the Roman Provinces, Institute of Archaeology, University of London

R. J. A. Wilson, M. A., D. Phil.
Louis Claude Purser Lecturer in Classical Archaeology, Trinity College, Dublin

T. P. Wiseman, M. A., D. Phil., F. S. A.
Professor of Classics, University of Exeter

C. J. Young, M. A., D. Phil., F. S. A.
Inspector of Ancient Monuments, Department of the Environment

LIST OF FIGURES AND PLATES

FOREWORD

"Finally it may be said that the topic of Roman brick-making has not received as much attention as it deserves."

R. G. Goodchild, 1937, 90

The above quotation came at the end of Goodchild's report on his excavations at Wykehurst Farm, Cranleigh, Surrey (see page 167) and although he drew attention to the lack of research on this topic in Britain it had little effect in promoting further research or publication. It might seem strange in retrospect for Goodchild to have made such a statement when only seven years earlier Professor W. F. Grimes had published his survey of excavations at the legionary pottery and tilery at Holt (Grimes, 1930). In that work, Grimes concentrated on kilns and so perhaps Goodchild was right to point out that brick-making had still not received enough attention from those interested in Roman studies.

Once Holt had appeared, Grimes turned his attention to other archaeological matters and did not return to any further extensive studies of Roman brick and tile. A. W. G. Lowther on the other hand had a life-long interest in the subject no doubt stimulated by his involvement with excavations at Ashstead, Surrey, between 1926-8 where a large quantity of tile was found and where he first began to study relief-patterned flue-tiles in detail. This interest eventually found its way into print as a Research Paper of the Surrey Archaeological Society published in 1948 (Lowther, 1948). The situation in other provinces of the Roman Empire seems to be similar. Occasionally one finds papers on topics relating to tile and brick, but rarely does one come across such an extensive treatment as that carried out by Georg Spitzlberger in the province of Raetia (Spitzlberger, 1968).

Stamped tile and brick has received a great deal of attention as can be seen from the standard works on Roman epigraphy and from such recent works as that by T. Helen who investigated the organization of the Roman brick industry by studying some 9000 stamped tiles from the Rome region (Helen, 1975). Both military and civilian tileries stamped tiles although the extent to which they did varies from one province to another, and within a province. In Britain, for example, the number of civilian tile-stamps probably does not exceed 600, the majority coming from the Cotswold region where some 250 have been recorded in and around Cirencester and another 260 in Gloucester (McWhirr and Viner, 1978). Despite this comparatively small number from Britannia no complete list has yet appeared in print, but no doubt this will be remedied when the second volume of RIB appears. Military tile-stamps on the other hand have been studied more fully and two recent papers by R. P. Wright show what can be done when stamps are carefully investigated (Wright, 1976 and 1978).

This type of study is just one example of a series of projects that are now underway in Britain and elsewhere. There has been a noticeable increase in the research being undertaken in problems relating to the production and distribution of tile and brick and into the techniques which help to resolve these problems, as the contributions to this volume indicate.

I became aware of this increasing interest after giving a paper to the Society of Antiquaries in March 1977 and the discussion which followed prompted the conference at which the papers in this volume were given. The conference was organised by the School of Humanities of Leicester Polytechnic and took place in April 1979 at the Scraptoft Campus of the Polytechnic. It could not have taken place without the help and encouragement of Mr. David Bethel, Director of the Polytechnic and Mr. John Hiorns, Head of the School of Humanities.

All but three of the papers given during the weekend have been included in this volume and the fact that it has appeared within nine months of the conference is a tribute to the hard work and enthusiasm of the contributors and to the publishers. It should be stressed that many of the papers represent work which is still in progress and some of the views put forward may have to be modified later when more research has been carried out. This volume cannot claim to be a definitive account on the subject, but it is a statement of our present knowledge in some fields. In planning such a conference not all areas of the subject can be covered and there were obvious gaps as there are in this volume. Many problems were raised, but only a few answers offered and hopefully the production of this volume will act as a catalyst for further research to be undertaken in this often-neglected subject.

One such problem which concerned practising excavators who attended the conference was what to do with the considerable quantity of tile and brick which is found during the course of excavation. As a result of this concern Dr. C. J. Young was prompted to write an additional paper covering some of these problems utilising his experience gained working with the Romano-British Course Pottery Study Group, in particular in his editing of their guidelines for the Processing and Publication of Roman Pottery, a DOE Occasional Paper.

There has been no attempt by the Editor to standardise the use of the words tile and brick throughout the volume. In some papers the authors have attempted their own definition whereas others have used the words synonymously.

A great deal of help in editing papers and reading proofs was given by Helen McWhirr; Carole Shaw assisted with the typing and some illustrations were redrawn by Pat O'Halloran. Nick Griffiths drew all the plans of the tile-kilns and Jack Jeffs took the photographs of Brixworth and contributed in a number of ways. To all of these people the Editor is most grateful for without their advice and assistance it would not have been possible to produce this volume. Finally I should like to thank all those within Leicester Polytechnic who helped with the conference.

Alan McWhirr
Editor, September 1979

REFERENCES

Goodchild, 1937. R. G. Goodchild, 'The Roman Brickwords at Wykehurst Farm in the Parish of Cranleigh,' Surrey Archaeological Collections, 45 (1937), 74-96.

Grimes, 1930. W. F. Grimes, Holt Denbighshire. The Twentieth Legion at Castle Lyons. Y Cymmrodor 41 (1930).

Helen, 1975. T. Helen, The Organisation of Roman Brick Production in the First and Second Centuries A. D.: an Interpretation of the Roman Brick-Stamps. Helsinki 1975.

Lowther, 1948. A. W. G. Lowther, A Study of the Patterns on Roman Flue-Tiles and their Distribution. Research Papers of the Surrey Archaeological Society No. 1. Farnham 1948.

McWhirr and Viner, 1978. A. McWhirr and D. Viner, 'The Production and Distribution of Tiles in Roman Britain with particular reference to the Cirencester region', Britannia 9 (1978), 359-377.

Spitzlberger, 1968. G. Spitzlberger, 'Die römischen Ziegelstempel im nördlichen Teil der Provinz Raetien.' Saalburg Jahrbuch 25 (1968), 65-

Wright, 1976. R. P. Wright, 'Tile-stamps of the Sixth Legion found in Britain', Britannia 7 (1976), 224-235.

Wright, 1978. R. P. Wright, 'Tile-Stamps of the Ninth Legion found in Britain', Britannia 9 (1978), 379-382.

AN ETHNOARCHAEOLOGICAL APPROACH TO THE STUDY OF ROMAN BRICKS AND TILES

David Peacock

Daniel Wilson (1861) once said that to confine onself to mere antiquities was like trying to read by candle-light at noon-day. His sentiments have been well taken and today ethnographically derived models form an essential part of the prehistorian's tool kit. Classical archaeologists have also used ethnographic analogy on occasions but its impact has been less because the basic framework of thought has been derived in other ways, particularly from written sources. However, documentary information relating to brick making is decidedly scarce and epigraphic data from stamps and graffiti are the exception rather than the rule; it seems that ethnography could play a much greater role than has been hitherto allowed. However, it should be stressed at the outset that modern practices will prove nothing about the past. What they will do is to provide models which can then be tested by critical evaluation of the Roman evidence. This approach may reveal contrasts as well as congruences and both are equally valuable in assessing Roman practice.

In the past ethnographic analogy has been used to fill in details of techniques that may have been used in the manufacture of Roman bricks and tiles. This seems to be a very fruitful line of approach. For example, if we wish to understand how imbrices were made we cannot do better than to look at recent production in, for example, Sicily, well documented by Hampe and Winter (1965). Modern methods seem to explain nearly all the technical details seen in Roman examples—the standardisation, the longitudinal striations and the form of the edges.

Turning to another aspect of technology, ethnography can feed us with ideas about firing practices. Long ago Loeschcke (1931) used his knowledge of modern Turkish kilns to suggest a reconstruction of a Roman brick kiln excavated at Speicher in Germany. In this case the analogy lacks conviction because the Turkish examples are crowned with a clay dome and we do not usually find remains of these in kiln excavations. However, there are a number of other possibilities and in particular it is worth noting that the nineteenth century kiln from Broadmayne in Dorset has a ground plan remarkably similar to the normal square Roman kiln (Young, 1968): perhaps this provides a more realistic reconstruction.

Clamp firing without a kiln has been postulated for the Roman period but few potential clamp sites have been throughly investigated apart from Green's (1970) work at Itchingfield in Sussex (see also Chapter 7). Again ethnology can tell us what to expect for recent clamps are fairly well documented (e.g. Dobson, 1850).

This discussion of technology is brief because the main object of this note is to concentrate on a more important aspect that seems to have been largely ignored until now: ethnography can also help us to understand and to classify the different types of organisation underlying brick production and distribution. It is possible to identify a number of different modes of production, each of which seems to be favoured by a particular set of social and economic circumstances. I would suggest that this is of fundamental importance and if we wish to understand the economics of Roman brick and tile production we must first identify the mode of production. This is no easy task for Roman practice need not correspond exactly with what we can see today, but at least ethnographical models provide us with a yardstick with which we can compare out Roman data. Consideration of the ethnographical, historical and industrial archaeological literature suggests that there are five distinct modes of production for bricks occurring at the present day or in the recent past.

My FIRST CATEGORY can be called household production—where each household makes all the bricks it requires for its own consumption—none are destined for the market. This situation is very rare but it may exist in the southern Yugoslavian province of Kosmet where in 1975 I observed numerous peasants firing small quantities of bricks in their back gardens, presumably for their own use. This activity does not seem to have been described in the literature and clearly it is worthy of further investigation. This mode has not been recognised in the Roman period but this is hardly surprising for nobody has yet considered the possibility.

My SECOND MODE of production is very common indeed. It comprises the small rural brickyard employing about six men and distributing the bricks over a radius of about 5-10 miles, with occasional rare exotic orders travelling 50 or even 100 miles. Most of the bricks and tiles used in Britain during nineteenth and twentieth centuries seem to have been produced in this type of yard and the same pattern can be seen all over the world wherever satisfactory clays are to be found. The reason for this marked division of the industry into small units is simple: both clay and bricks are heavy and thus incur high transport costs. Unlike pottery however, the end product must be cheap in relation to its weight and hence brickworks must be located near to both clay and to markets if it is to remain viable. Small yards with local distribution make economic sense even in Britain today (Collier, 1966; Gleave, 1965). In Northwestern Europe activity is traditionally seasonal because of climatic constraints. The clay is dug and weathered during the winter and brick making begins during March when extra labour will be employed. Frequently activity is linked to farming which provides employment in the slack season (see e.g. Gault, 1952).

It is also worth noting that there is often considerable mobility of labour—much more than in the pottery industry because brick making equipment is much simpler and more portable. Seasonal workers, particularly the more skilled members of the team, may be recruited from a considerable distance. At Civry-la-Forêt in the Ile de France labourers were recruited locally but the moulder often came from as far away as Normandy (Gault, 1952). This migrationary tendency is particularly well exemplified among the Walloon brick makers of Belgium who used to travel annually all over Europe—to France, Spain and even Russia to work in the brickyards (Vandereuse and

Pinon, 1967). Mobility is an important factor and worth bearing in mind when considering die links between stamps used in the Roman period for the probability of movement of personnel rather than products is very high.

Alternatively the brickyard itself can be a mobile entity. In Kenya today bricks can be produced on the building site by teams of itinerant brick-makers, firing in surface clamp kilns (information Mr. S. Wandibba). This seems to be exactly the way in which many of the housing estates of Victorian Britain were constructed and has been documented for London, Southampton and Portsmouth (White, 1971; Woodforde, 1976).

A THIRD MODE of production is the nucleated brickyard complex—where small concerns are grouped together in one particular area, usually a region where there is a plentiful supply of very good clay. It is very common to find potteries clustering in this way but it is comparatively rare in brick-working. However, it can be seen, for example, at Nabeul in Tunisia, but it is particularly well-exemplified in the Pueblo-Tlaxcala basin of Mexico (Seele, 1968). Here there is a very tight concentration of activity and it is surprising to see so many brickworks working in a competitive situation. Some of the produce is consumed locally, but some is transported over the mountains to Mexico City 140 km distant. Nucleated brickworks such as this are rare because clustering is only viable if one of three conditions holds, viz:

1. that adequate clays are only available in one area;
2. that there are good efficient communications enabling the produce to be widely distributed without making the cost prohibitive; or
3. that the works are located in an area of very high population and hence demand.

A FOURTH MODE of production is the estate brickworks in which the produce is essentially not for sale but is destined for use on the estate itself. There is no doubt that estate production was very common in eighteenth and nineteenth century Europe but unfortunately few examples have been recorded in detail apart from the Ashburnham Estate works in Sussex, which had a pair of wood burning kilns very similar in design to Roman examples in use until 1968 (Leslie, 1971; Gordon, 1969). The history of the Ashburnham works is instructive and gives an indication of what might have been general practice. The kiln was constructed in 1840 to provide for estate requirements and for the rebuilding of Ashburnham Place. This was again its main function in the period 1897-1927, but both before and after this commercial sales were important. Much of the produce was consumed locally but much was also used in the Victorian building of Eastbourne 12 miles distant. One load even went to Faversham in Kent, 50 miles away, but this was quite exceptional. The main raison d'être of the kiln was the needs of the estate, but commercial sales kept the activity alive when internal demand was low so that the facility was there when needed by the estate. This is the pattern we might encounter in the Roman period also. It could well be that many of the commercial brickworks I mentioned earlier began life as estate works, later becoming more and more preoccupied with the market. This is certainly the case in nineteenth century Sweden as Bruno (1954) has shown.

One FINAL MODE of production worth mentioning is the Municipal brickworks, controlled and run by the town authorities. An example of this could be seen among the Moravian communities of eighteenth-nineteenth century North Carolina, but this is a special case for the Moravians were, by any standards, an unusual group (Bivins, 1972). However, a further example of a town brickworks has been described from Mölln in Germany and this could have been typical of many other Medieval towns on the Continent (Hermann, 1961). The works continued in operation until 1906 and provided for official buildings, churches and fortifications. They were owned by the council and leased to a brick-worker who operated on their behalf. The total output was always low and never exceeded 40,000 building bricks and 10,000 roofing tiles per annum, the majority of which may have been destined for official projects, although the possibility of private sales cannot be discounted.

Thus in the recent past bricks have been made and distributed in a number of different ways and any or all of these possibilities might apply to the Roman period. If we are to understand the industry in economic terms, we should begin by attempting to understand the mode of production and fortunately each has its own special characteristics of location, technology and product distribution, all of which should be identifiable archaeologically. Unfortunately in in practice it is difficult to do this for several reasons. Firstly the distribution of Roman bricks has not yet been seriously studied although we now have the required petrological and other scientific tools. Secondly excavation has largely concentrated on kiln structures so we are unable to assess properly the nature of many a Roman brickworks.

However, it is possible to make a few tentative suggestions for Britain thanks to the work of McWhirr and Viner (1978). Not least of their valuable contributions is a list of Roman brick making evidence from Britain and the distribution presents a number of curious features. If we put aside those brickworks with obvious military associations, the majority of civilian kiln sites are in the country away from the main centres of population and there is a marked dearth of evidence from around the towns. For example, nearly 60% of all kilns are more than 20 km from a civitas capital. This is surprising for we might anticipate a concentration of evidence around the large urban centres not only because demand would be greatest in Roman times but also because today archaeological activity is most intensive around the major Roman towns. If we accept the lessons of ethnology, it would seem probable that most of the isolated rural kilns were producing for the locality and it is very unlikely that they distributed much of their produce very far. Either they are commercial rural brickworks or they are essentially estate kilns. I favour the latter view for in a few cases it is possible to show connections with villas: this has been suggsted for Eccles and Ashstead in Britain and has been demonstrated at sites such as Hoheneck or Mürlenbach in Germany (Detsicas, 1967; Goodchild, 1937; Paret, 1911; Reusch, 1958). A further point worth noting is that Romano-British kilns often occur as single examples. Today most commercial brickworks have at least two kilns so that one can be loaded as the other cools, permitting more or less continuous production during the season. If the majority of Roman kilns are really as isolated as they appear to be, this could be an indication that the brickworks were not designed for continuous production but are best explained as devices to meet intermittent estate requirements. To what extent this British estate production had been commercialised as in nineteenth century Sweden is unclear. However in Italy, the well known evidence of Roman brick-stamps appears to reveal an extreme development in this direction (Helen, 1975).

In Britain, Colchester, London and Verulamium have produced small clusters of kilns around them, and some could have been on estates contributing to the urban market. However, most of the major towns seem to have produced very little evidence from their immediate environs, which I find surprising. This could be a reflection of the need for more intensive fieldwork outside the town walls but equally it could be that urban supplies were met in a different way. Perhaps rural brickworkers migrated seasonally to the towns and fired bricks on site in clamp kilns as in nineteenth century London and Hampshire or perhaps brickworking was concentrated in municipal works analogous to the Mölln town brickyard. This certainly seems to be at least a partial answer in Gloucester where the RPG stamps are generally read as Rei Publicae Glevensium. The distribution, which may be restricted to the Colonia and its territorium is exactly what we might ancitipate. The PR BR LON stamps from London are also clear evidence of some form of official administration and imperial ownership has been suggested for the Pamber tilery near Silchester.

If one of our aims is to understand how the brick industry was organised in Roman Britain, we are clearly far from our goal for at present we do not have the evidence to make judgements except at a few specific points. What is needed is more analysis of fabrics, more fieldwork, more excavation, particularly of the areas around brick kilns and generally more attention to brickwork—the unstamped as well as stamped varieties. This paper has attempted nothing more than to suggest some of the things we could be looking for when we study bricks and tiles. Let us not forget that in economic terms brick was always much more important than pottery although today it holds less attraction for the typologist.

Bibliography

Bivins, J., Jr., 1972. The Moravian potters in North Carolina, Chapel Hill, Univ. N. Carolina Press.

Bruno, W., 1954. 'Tegel industrien i Mälarprovinserna 1815-1950,' Geog. skrifter Uppsala Univ., 28.

Collier, J. L., 1966. Development and location of the clay-brickmaking industry in the south-east Midlands of England since 1800, Ph. D. thesis London University.

Detsicas, A. P., 1967. 'Excavations at Eccles 1966'. Arch. Cant., 82, 162-78.

Dobson, E., 1850. A rudimentary treatise on the manufacture of bricks and tiles, London.

Gault, A., 1952. 'L'ancienne tuilerie de Civry-la-Forêt (Seine et Oise)', Bull. folkl. d'Ile de France, 14, 428-9.

Gleave, M. B., 1965. 'Some contrasts in the English brick-making industry', Tijdschrift voor Economische en Sociale Geographie , 56, 54-62.

Goodchild, R. G., 1937. 'The Roman Brickworks at Wykehurst Farm in the Parish of Cranleigh. With a note on a Roman tile-kiln at Horton', Epsom'. Surrey Arch. Coll., 45, 74-96.

Gordon, H. A., 1969. 'The brick kilns at Ashburnham', Country Life, 24th July, 240-4.

Green, T. K., 1970. 'Roman tileworks at Itchingfield', Sussex Arch. Coll., 108, 23-38.

Hampe, R. and Winter, A., 1965. Bei Töpfern und Zieglern in Süditalien, Sizilien und Griechenland, Mainz.

Helen, T., 1975. Organisation of Roman brick production in the first and second centuries A.D. An Interpretation of Roman Brick Stamps, Helsinki.

Hermann, S., 1961. 'Von der Möllner Stadtzeigelei', Lauenburgische Heimat, NF 35, 13-19.

Leslie, K. C., 1971. 'The Ashburnham Estate Brickworks 1840-1968, ' Sussex Ind. Hist., 1, 2-22.

Loeschcke, S., 1931. 'Die römischen Ziegelöfen im Gemeindewald von Speicher,' Trierer Zeit., 6, 1-7.

McWhirr, A. and Viner, D., 1978. 'Production and distribution of tiles in Roman Britain with particular reference to the Cirencester region'. Britannia, 9, 359-77.

Paret, O., 1911. 'Ein römischer Gutshof mit Ziegelei bei Hoheneck OA Ludwigsburg', Fundb. aus Schwaben, 19, 90-118.

Reusch, W., 1958. 'Jahresbericht 1945 bis 1958, Kreis Prüm', Trierer Zeitschrift, 24/6, 561-3.

Seele, E., 1968. 'Die Ziegelherstellung im Becken von Puebla Tlaxcala', Die Ziegelindustrie,17 (18), 380-6.

Vandereuse, J. and Pinon, R., 1961. 'Contributions à l'etude de la briqueterie en Wallonie', Ann. Comm. royale belge Folklore,12, 322-81.

White, W. C. F., 1971. 'A gazetteer of brick and tile works in Hampshire', Proc. Hants, F.C. & Arch. Soc., 28, 81.

Wilson, D., 1861. Prehistoric man: researches into the orign of civilisation in the Old and New Worlds. Edinburgh.

Woodforde, J., 1976. Bricks to build a house. London.

Young, D., 1968. 'Brickmaking at Broadmayne', Proc. Dorset Nat. Hist. & Arch. Soc., 89, 318-24.

BRICK AND TILES IN ROMAN SICILY

R. J. A. Wilson

In two well-known passages of the De Architectura, Vitruvius, writing during the reign of Augustus, displays a remarkable ignorance of the properties of kiln-fired brick and a total lack of intuition about the important role that it was to play in the architectural revolution that was soon to sweep Rome.[1] Vitruvius' experience was quite clearly limited to the buildings of central and northern Italy. He can hardly have travelled to south Italy or Sicily where kiln-fired brick had for some time been in regular use. A long tradition had existed there, of course, in the manufacture of terracotta roof-tiles and other decorated roof ornament, going back to the very beginning of the sixth century B. C. ;[2] and it was already in the fourth century that the Greeks had begun to experiment with terracotta for other parts of their buildings, as for example the floor paving of the public baths at Gela, erected at the end of the fourth century B. C.[3] Brick did not, however, become widely used until the third and second centuries B. C., both before and after Rome's annexation of Sicily as the first provincia in 241 B. C. It was during this period that it came into common use for constructing peristyle columns in domestic town houses and public stoas. At Morgantina, for example, several houses have such columns composed entirely of individual circular bricks, such as those in the House of Ganymede of 260/50 B. C. which are between 38 and 40 cm in diameter, and 8 cm thick[4] (Plate 2. Ia). The purpose of the hole in the centre of each brick is obscure: it has been suggested that each column had a central timber upright,[5] but the hole may only be present for ease of manufacture and drying out before firing. An alternative method for building columns involved the use of segments of brick, as in a second-century-B. C. stoa at Halaesa[6] (Plate 2. Ib). The columns there are rather larger, about 63 cm in diameter, and the bricks are 7. 5 cm thick. Both methods are employed in a house at Tindari, where the columns of the lower storey of the peristyle are made up of segmental bricks 60 cm in diameter overall, but the upper storey consisted of whole circular bricks (but without the central hole) 35 cm in diameter.[7]

The second major use of bricks in Hellenistic and Roman republican Sicily was for paving, either public areas such as piazzas, stoas and roads, or else the central courts of peristyled buildings. The paving of the stoa at Halaesa is composed of rectangular bricks 33 cm by 50 cm and about 6 cm thick; the adjacent agora has bricks 33 cm square[8] (Plate 2. Ib). At Solunto, the agora was likewise paved in brick, probably in the second century B. C. (as at Halaesa), and the main street leading to it was given the same treatment.[9] Here most of the bricks, also 33 cm square, were laid diagonally to the axis; but the pattern was varied by portions with the bricks laid parallel to the axis, both in continuous strips along each side and at intervals in the central part of the road. There was even some attempt at colour variation (the bricks along the sides were fired grey), and at one point bricks were laid to form an

Plate 2.I a. Morgantina, house of Ganymede, circular bricks in peristyle, third century B.C.

b. Halaesa, stoa, column- and paving-bricks, second century B.C.

elegant star design. In private houses there are examples of brick-paved courtyards of Republican date at Solunto[10] and Morgantina, the latter firmly dated to the second century B. C.[11]

The third way in which bricks were used during Republican times in Sicily was in walling; not of course brick-faced concrete walling which did not appear until the Empire, but walls built of brick laid as though they were blocks of masonry. In south Italy, this method was used for a variety of structures, such as the 'ritual' containers on the acropolis at Velia adjacent to the sanctuary there,[12] but in Sicily fired-brick walling of this type has only been found (as far as I am aware) in tombs, especially in the cemeteries of Centuripe[13] and Messena. One tomb reported at the latter in 1945 was built of robust bricks 10 cm thick and 40 to 45 cm long. A date in the second century B. C. was indicated by the finds, and similar tombs recorded in 1929 were likewise of the third or second century B. C.[14] Three remarkable tombs of the same date, found just across the Straits of Messina at Reggio Calabria in 1909, were also constructed of bricks 35 cm by 38 cm by 9 cm, and roofed by the earliest known examples of brick barrel vaults in mainland Italy.[15]

It is clear from the examples mentioned that kiln-fired brick, specially manufactured for the purpose, was used in Sicily, as well as in Magna Graecia, for columns, paving, and in some cases walling, from at least the first half of the third century B. C. onwards. There are few clues to the centres of manufacture. Brick stamps do occur throughout the Roman Republic, but all too often the corpora do not differentiate between tile and brick and, anyway, the names recur too infrequently to enable many inferences to be drawn about the distribution of products.[16] A kiln found at the end of the last century at Spadafora on the north coast between Milazzo and Messina produced bricks measuring 29 cm by 50 cm by 10 cm, but no illustrations accompany the published report and the date suggested, the third century B. C., was put forward on the evidence of a single coin.[17] More important is the kiln excavated at Morgantina in 1955, so far published only in a brief preliminary note[18] (Plate 2. IIa). It is built on a rock-cut ledge and consists of a central firing chamber and stoking areas at both ends, the whole being 13 m long overall. No charcoal was found in the débris, only great masses of white or grey ash, and the excavator suggested that straw may have been the principal fuel, a material still used today in Sicilian kilns. The slightly-bowed central flue, up to one metre wide, is brick-built (the bricks are 33 cm by 16 cm by 8 cm) and spanned by seven brick arches, of which the two furnace arches are intact. These in turn supported the floor of the elliptical firing chamber above, but this has disappeared. The kiln was producing circular column bricks and rectangular bricks as well as roof tiles and water pipes, and could be dated to the early third century B. C. It is very similar to a roughly contemporary, better preserved brick kiln at Velia in south Italy.[19]

A third Sicilian kiln of Roman Republican date, which may also have produced bricks and tiles, was found in 1961 one kilometre from the walls of the Greek city of Camarina on the south coast[20] (Plate 2. IIb). The floor of the firing chamber here was almost intact, supported on three brick arches and pierced by four rows of vents. The kiln was dated to the second or first century B. C., but the evidence for brick and tile production here is less certain than it is from the Spadafora and Morgantina kilns.[21]

Plate 2.II a. Morgantina, brick kiln from north, third century B.C.
b. Camarina, kiln, second-first century B.C.

This comparatively widespread use of brick in hellenistic and republican Sicily has prompted the suggestion[22] that central and north Italy actually derived its knowledge of brick-making from the south and so had a fundamental influence on the subsequent explosion of its use in brick-faced concrete walls all over the Roman Empire. Be that as it may, the new concrete architecture which emerged made surprisingly little impact on the staunchly conservative building traditions of Sicily, which remained largely unaffected by the new concrete techniques to the end of antiquity. Only Taormina, Catania, Centuripe and Termini Imerese seem to have embraced the new style at all widely, and of these only Taormina and Centuripe used brick to face their concrete. Taormina has, in fact, a wealth of standing brick-faced concrete monuments which make it the most Romanized of the Roman cities of Sicily. Pride of place must surely go to the theatre with its spectacular backdrop of Mount Etna and the sea (Plate 2.IIIa). A building originally of the third century B.C., it was entirely rehandled during the Empire, first probably under Augustus, and then more radically under Hadrian.[23] It is probably to that date that the brick scaenae frons belongs, and also the double corridor running round the back of the cavea. Bricks used at ground level in the outer portico here are identical to those used in the Odeon at Taormina (Plate 2.IIIb)—they measure 40 cm by 33 cm by 5.5 or 6 cm and are scored on the diagonals for cutting into triangular bricks—and this suggests that the Odeon is also Hadrianic rather than Augustan, the date proposed for it by Lugli.[24] Another substantial brick-faced monument, probably of the second half of the second century A.D., is the so-called Naumachia, the apse- and recess-relieved façade of a gigantic reservoir, 122 m long[25] (Plate 2.IVa). Other brick-faced monuments[26] in Taormina include further reservoirs, an aqueduct, baths and several house-tombs of the type familiar from the Isola Sacra near Ostia and underneath St. Peter's in the Vatican.[27]

However superficially similar these monuments might appear to those of any moderately prosperous Italian city of the middle Empire, it is quite clear that the Taormina brick industry followed rules of its own that bear no relation to the brick-faced concrete of the Italian monuments. Firstly, Sicilian bricks are nearly always considerably thicker than their Italian counterparts. In Italy, bricks of the second century, for example, are not normally thicker than 3.5 cm,[28] but in Sicily at any period they are rarely less than 5 cm thick, and 7 to 10 cm is quite normal. It is possible that this represents the lingering on in kiln-fired bricks of the same dimensions as had been used for their sun-dried predecessors.[29] Secondly, Italian bricks are, by and large, relatively homogeneous, and there is a general rule-of-thumb guide that bricks get thinner the later on in the Empire they were produced, decreasing from an average thickness of 4 cm in the early first century A.D. to less than 2 cm in the early third.[30] In Sicily such rules are not applicable. In the casa Margherita reservoir at Taormina, for example, there are four different thicknesses of brick used in the twelve surviving courses: 4.25 cm, 6.25 cm, 7.5 cm and 8.5 cm[31] (Plate 2.IVb). Such a mixture in Italy would be taken as proof that the wall contained re-used material of different periods of manufacture, but there is no reason in Sicily to think that the bricks were not made contemporaneously. Thirdly, Sicilian bricks continue to be made rectangular as well as square, in contrast to normal practice in central Italy, and the dimension of 33 cm, which cropped up frequently in Republican bricks, enjoyed continuing popularity

Plate 2.III a. Taormina, Roman theatre, scaenae frons, second century A.D.

b. Taormina, Roman Odeon, second century A.D.

Plate 2.IV a. Taormina, 'Naumachia' reservoir facade, second-third century B.C.
b. Taormina, Casa Margherita, interior of reservoir

during the Empire. The standard sizes of central Italy, such as those of the *sesquipedales* and the *bessales*, are comparatively rare in the island.

All these factors make it highly probable that the bricks used in Sicilian cities came from their own, local brickyards. At Taormina, for example, there are extensive clay deposits below the rock on which the city stands, in the area of the former Greek colony of Naxos. Here in 1975 five kilns were uncovered, two oval-shaped ones being destroyed when three rectangular kilns were built to replace them. The date of all five apparently falls within the fifth century A. D., but few details are available and the site has not yet been fully published.[32] The kilns were certainly producing coarse pottery, but roof tiles as well, of the late Roman type discussed below; whether they were also turning out bricks is less certain. The Roman brickyards of earlier imperial date are, however, likely to have been situated in the same clay beds. A further indication of a local production for the Taormina bricks is the presence of Greek stamps on many of them. Ἡρακλειοι, Μοσια and Διονυσιου are among those recorded from the Taormina theatre,[33] and other Greek stamps of Roman imperial date are attested elsewhere in the island,[34] a reminder that much of the province remained Greek speaking, or at best bilingual, until the end of antiquity. One brick stamp at Marsala even switches languages mid-stream: B. Abietis| Ν Σεκονδου .[35] A few stamps do document imports from Italy, but they never occur in any significant quantity and never *in situ* in a standing structure. There are stray examples from the imperial brickworks at Syracuse and Catania, and the products of the factory of Q. Oppius Iustus, which had a widespread distribution in central and north Italy, are recorded at Catania, and from a countryside location near Licata on the south coast.[36] By and large, however, the bricks used in Sicily are likely to have been locally made, even though we are mostly ignorant at present of the whereabouts of the brickyards and the range and distribution of their products. The stamps on roof tiles are, by contrast, more informative (below, p. 23).

Brickwork on a smaller scale is also, of course, found in Sicily performing those same tasks to which it was commonly put all over the Roman Empire. These include the use of bricks as arch voussoirs and as bonding courses in masonry work;[37] and as paving in courtyards and walkways. Paving might be carried out either in square or rectangular bricks, or else in herringbone brickwork (*opus spicatum*).[38] This type of flooring (Plate 2. Va), which is of course found very frequently throughout the Roman world, made some of its earliest appearances in Sicily, for it is attested at Morgantina by the middle of the third century B. C.[39] Finally, it is interesting to note the use of whole circular bricks, as well as segmental bricks, in the large latrine of the early-fourth-century-A. D. Roman villa at Piazza Armerina[40] (Plate 2. Vb). Columns were being constructed here in exactly the same way as they had been at Morgantina (which lies 30 km to the north-east) nearly seven centuries earlier.

So much, then, for bricks. Let us now turn to tiles, a term which, for Sicilian artifacts, I propose to reserve exclusively for roof-tiles. They are relatively easy to distinguish from bricks, as they are generally much thinner,

Plate 2.V a. Agrigento, 'House of the Abstract Master', opus spicatum in court
b. Piazza Armerina, Roman villa, circular column-bricks in the large latrine, fourth century A.D.

rarely more than 3.5 cm thick and usually between 2 cm and 3 cm. Such a distinction is clearly not applicable to other areas of the Roman world where bricks much thinner than the Sicilian norm were in standard circulation. Roman roof-tiles in Sicily have not been the subject of much attention, and the account that follows, which is based mainly on observations made during field-work and museum study in central southern Sicily, may need considerable qualification when further work in other parts of the island has been carried out. Three different groups of roof-tile can be distinguished. The first (fragmentary profiles are shown in section A of Fig. 2.1) belongs to the period of the Roman Republic, but it had a long lineage in the island from the beginning of the sixth century B.C. or a little earlier.[41] The tiles are flat, either square or rectangular, and have a flanged edge on two parallel sides. The flange is normally of rounded profile, sometimes rather more angular, and part of it at one end is cut away to allow the roof-tile to interlock firmly with the one it overlaps. The joins between flanges were concealed by cover-tiles of semi-circular profile, not shown on Fig. 2.1. These are the κεραμίδες and καλυπτῆρες of Greek nomenclature,[42] the *tegulae* and *imbrices* of the Roman. As well as for roofing, the flat tiles were used in upright position to form containers for inhumation burials, or else were set in pairs gable-wise to cover the burial (the so-called 'tomba a cappuccina'). Both methods occur widely in the cemeteries of Republican and early imperial date at Lipari, Messina, Catania and elsewhere.[43] In contrast to the situation at Metaponto (p. 45) and elsewhere in south Italy,[44] it is only very rarely that we find in Sicily at this period examples of an alternative method of roofing to the flat *tegulae* and curved *imbrices*, the so-called 'Laconian' system. This method of roofing employed imbrices only, each alternate one being inverted.[45] A re-roofing of the stage building of the theatre at Monte Iato, carried out in the second or first century B.C. (below, p. 23), seems to have used this type of roof, as did also the east stoa at Morgantina; the latter is, however, pre-Roman (*c.* 300 B.C.).[46] This is the preferred method of roofing in Sicily even today: as no doubt in antiquity, the weight of the tiles themselves, together with the generally shallow pitch of the rafters, keeps the roof intact without the need for nailing the tiles.

At a later stage, at least in some parts of the island, the Laconian system seems to have taken over as the principle method of roofing. The tiles used, however, were not semi-cylindrical but have a much shallower curving profile, and form the second main group of Roman roof-tiles found in Sicily (Fig. 2.1, group B, latitudinal sections on right). These tiles taper slightly, as usual, to enable them to lock into one another, but a more distinctive feature is the crudely fashioned rudimentary flange at the narrower end (Fig. 2.1, group B, longitudinal sections on left; also Plate 2.VIa). The date at which the change-over occurred is at present far from clear, but the type is certainly absent on sites which were abandoned at the end of the first century B.C. It seems to have been introduced, however, not much later, probably sometime during the first century A.D., and then is common throughout the whole of central southern Sicily where the almost complete absence of flat *tegulae* on sites of imperial date indicates that the Laconian system must have prevailed.[47] A group of the same type of roof-tiles (which it is proposed to call the Sicilian 'imperial' type) has also been published from the Partinico area west of Palermo.[48] But the picture in the towns is less clear, and flat *tegulae* were

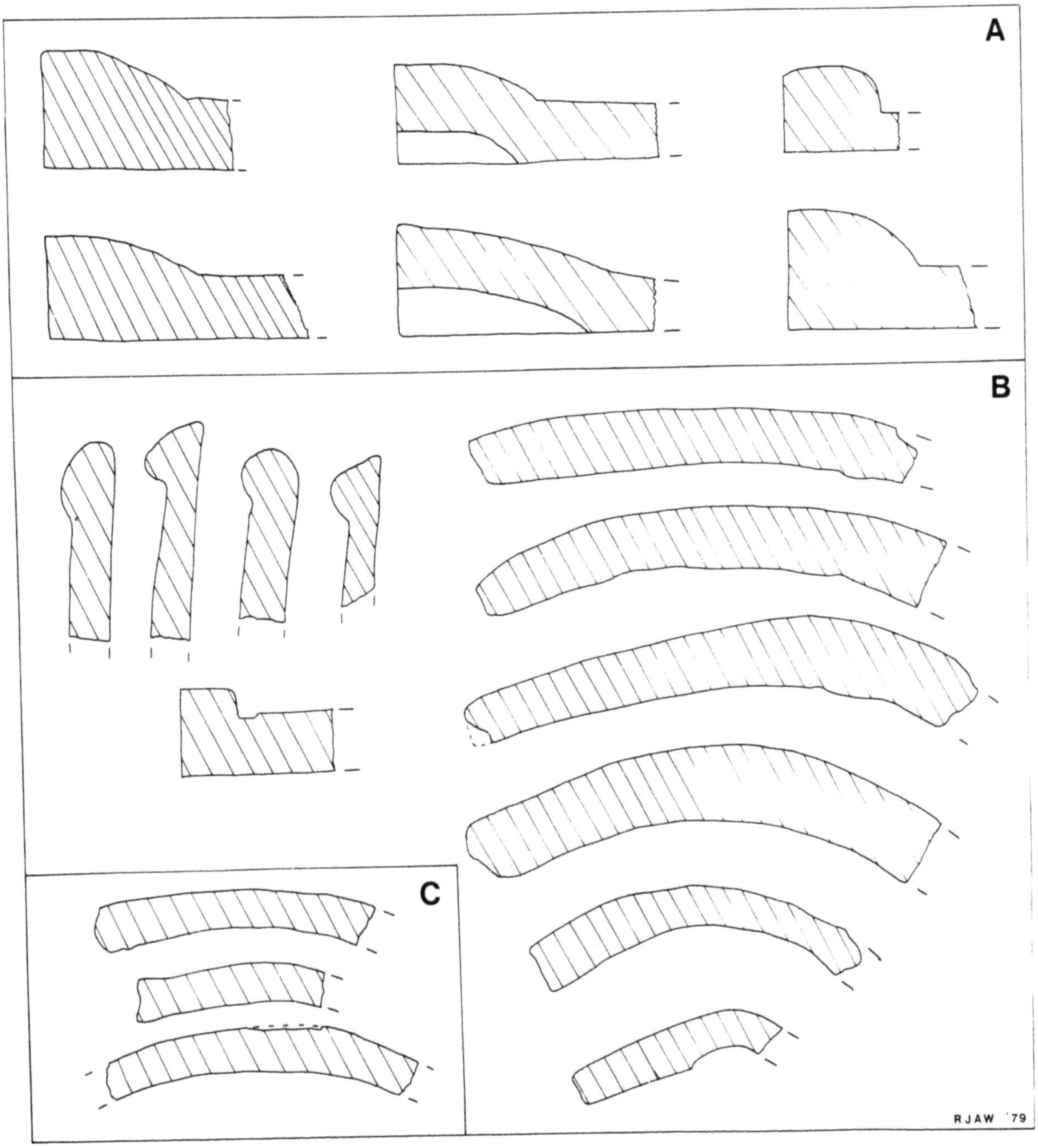

Fig. 2.1 Profiles of Roman roof-tiles from central southern Sicily

Plate 2.VI a. Gela Museum, Roman roof-tiles stamped PAE
b. Heraclea Minoa, late Roman villa, fragments of 'Byzantine'-type roof-tiles

certainly still being used for burials in the second and third centuries A. D. in the north-eastern corner of the island.[49]

The third group of Sicilian roof-tile is a variant of the curved 'imperial' type, but it lacks the rudimentary flange and is usually decorated with combed lines (Fig. 2. 1, group c and Plate 2. VIb). This was introduced in the late Roman period, at least by the early fifth century A. D., and survived in use into the Byzantine period.[50] The combed 'Byzantine' tiles are certainly found elsewhere in the central part of the Mediterranean,[51] but the earlier 'imperial' type seems to be more distinctive, and may be peculiar to Sicily.

The practice of stamping marks or names on tiles goes back in Sicily at least to the fifth century B. C. and becomes common in the third and second centuries B. C. Many examples of Republican date have been recorded in Sicily, most commonly those of proper names in the genitive case, but most of the names are recorded only once or twice and little can be said about distribution, or the size of their respective industries.[52] Two groups of tile-stamps, however, one of Republican date, the other imperial, call for special comment. The first is the large collection of stamped imbrices which has come from the Swiss excavations at the hill-town of Monte Iato in western Sicily, where the first four campaigns (1971-4) yielded 482 stamped fragments which have been the subject of a full study; many more have been found in more recent excavations.[53] The majority of these were 'official' inscriptions, presumably stamped in the municipal tilery or, less likely, in a private tilery which was supplying tiles for 'official' buildings. Some indicate the name of the city, Ἰαιτας, others the building for which the tiles were destined, such as the θεατρον. in both cases, the possessive genitive is used (Plate 2, VII a-b). Also found were many stamped ἱεραι, 'holy' (here the feminine plural noun, κεραμίδες 'tiles' has to be supplied), presumably manufactured for a sanctuary building (Plate 2. VIII a); and indeed a temple of Aphrodite has been identified by the excavators. Also discovered were seven stamp-types giving names of eponymous magistrates sometimes in conjunction with the stamps already mentioned.[54] Nearly all these tiles seem to have been made in the third, second or first centuries B. C.: Monte Iato, in common with other city communities in Sicily continued to be self-administering during the Roman Republic, retaining the same Greek institutions and customs they had used before Roman involvement in the island. Messina has yielded tiles or bricks stamped Μαμερτινων, produced during Mamertine control of the city in the first half of the third century B. C., as well as others stamped Ἀπολλωνος presumably originally destined for a shrine of Apollo.[55] Tiles for sanctuaries (ἱεραι or ἱερων) have made appearances elsewhere, too, at Messina, Halaesa, Acrae and Calacte.[56] But none of these places has produced clear and detailed evidence of a well-organized municipal tileworks to match that from Monte Iato.[57]

The second group of tile-stamps comes from various sites around Gela in southern Sicily, and all are found on the second, 'imperial' type of roof tile (Plate 2. VI a). The stamp is usually just below the flange, and parallel to it; it is sometimes at right-angles to it, and it may also occur on the flange itself. Over a hundred examples are known from this area, representing 16 different manufacturers. A full study of this group will be published elsewhere,[58]

Plate 2. VII a. Monte Iato, roof-tile stamped IAITOY
b. Monte Iato, roof-tile stamped ΘEATPOY

Plate 2. VIII

a. Monte Iato, roof-tile stamped IEPAI

b. Agrigento Museum, 'sulphur-tiles'

and only a few salient points need be made here. Of those shown on Fig. 2.2 and 3, those stamped GALB or GALBA, CAL or CALVI, and SAB and SIRF are of particular interest. Tile-stamps which record just a single name are most likely to indicate the tile maker, or else the proprietor of the tile works, if he was a different man from the manufacturer. GALBA is not a servile or freedman name and this stamp is therefore likely to record the estate owner. In fact the cognomen Galba, never common, is unattested after the first century A.D. for an individual, and it is possible that the four places where these stamps were found—two of them four kilometres apart, a third fifty kilometres further west, and a fourth seventeen kilometres north-east of the third (see Fig. 2.4)—were all part of imperial estates. At any rate the praedia Galbana on an Italian inscription certainly were.[59] Another type, the CAL or CALVI variety, clearly stands for CALVISIANA, i.e. the latifundia or praedia Calvisiana which makes an appearance in the Antonine Itinerary (95.7) as a road-station on the south coastal road. Little is known about these settlements, as so few have been excavated either in Sicily or elsewhere in the Mediterranean, but it is well known that a single landowner, especially in the late Empire, might own vast areas which included whole villages within his property, and it is best to regard such settlements as lying largely in the hands of private enterprise and private ownership. Even when such places were chosen as official mansiones of the cursus publicus we need not assume that anything more than the buildings erected in connection with this function were maintained by the central government. In other words, these CALVI stamps are not municipal products like those of Monte Iato (p. 23) or of Gloucester: they are the products of a privately run tilery supplying the buildings of the Calvisiana estate. If this is accepted, it is tempting to look at their distribution—unfortunately only at three sites, one 3 km east of the first, the third 8 km north of the first (Fig. 2.4)—and to assume that all three lay within a single tract of the latifundia Calvisiana, an estate which should therefore be over 20 square kilometres in size. But it would be rash to jump to such conclusions. Firstly it is well-known that a latifundia might comprise not one single tract but several separated pieces of land in the same area, all belonging to the same landowner and bearing the same name. Secondly, there is always the possibility that one landowner might sell his surplus tile stock to an adjacent estate owner who did not have the resources to manufacture his own roof-tiles.[60] The problem is further complicated by the presence of two quite different tile-stamps at all three sites, marked SIRF and SAB respectively. Here we have no check against the ancient sources. They might represent either estate names or their owners, or else tile manufacturers who might have been quite independent of the estate domini. The SIRF stamp, for example, might well stand for Sir(inius?) f(ecit) or something similar, and belong to the latter category. SAB, on the other hand, is more difficult. Does this refer to a manufacturer called Subucius or Sabinius or Sabinianus? Or does it refer to a latifundia Sabuciana? And what is the relationship of all these stamps to the latifundia Calvisiana? Did a hypothetical latifundia Sabuciana change hands and become known later as the latifundia Calvisiana? Or were there praedia Sabuciana which were subdivisions of a much larger latifundia Calvisiana, yet praedia which were themselves extensive enough to make tile-making in their own right a viable concern? All these questions will remain unanswered until many more tile-stamps have been found in this region, preferably in closely dated contexts.[61]

Fig. 2.2 Tile-stamps from central southern Sicily

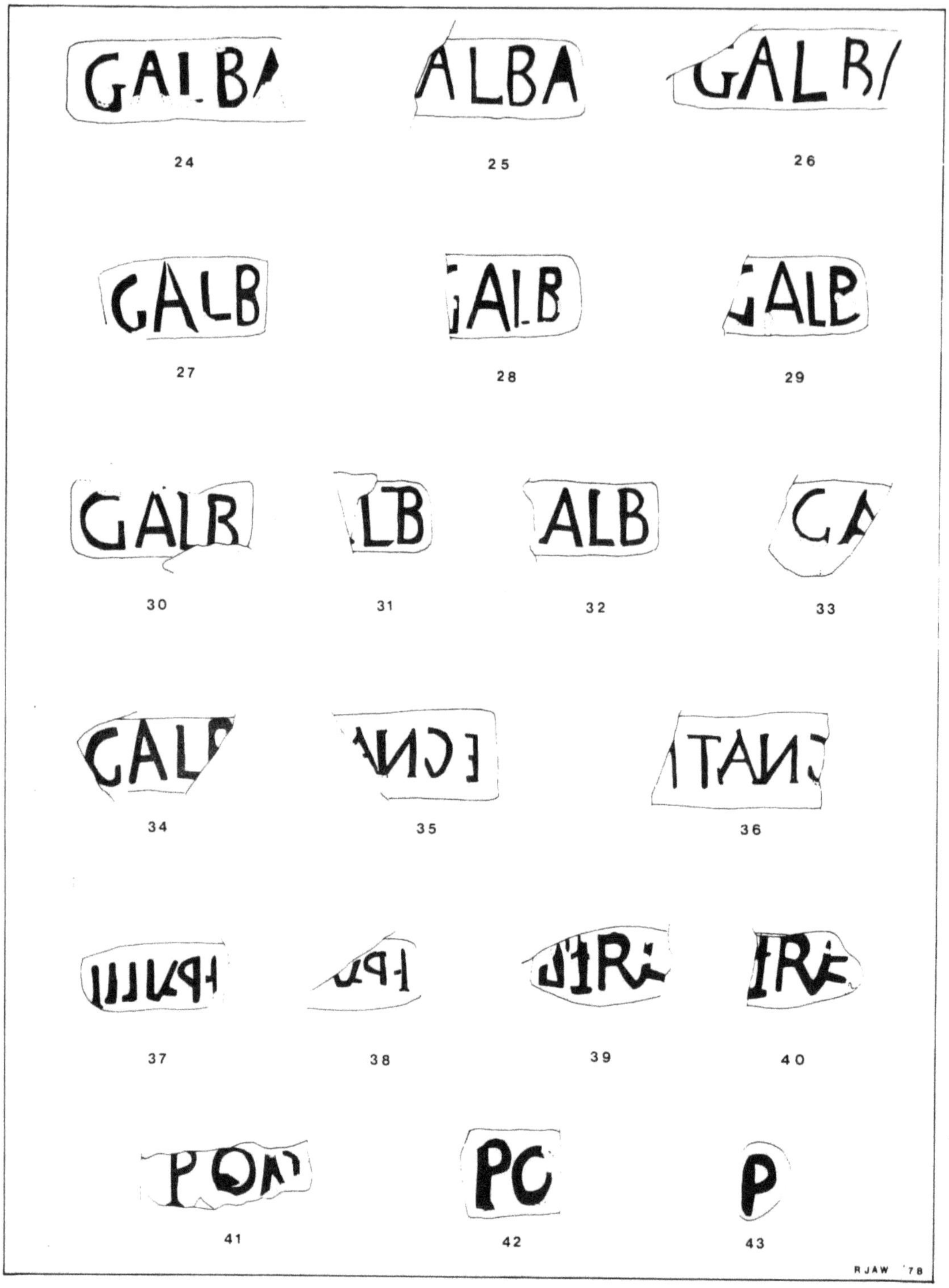

Fig. 2.3 Tile-stamps from central southern Sicily

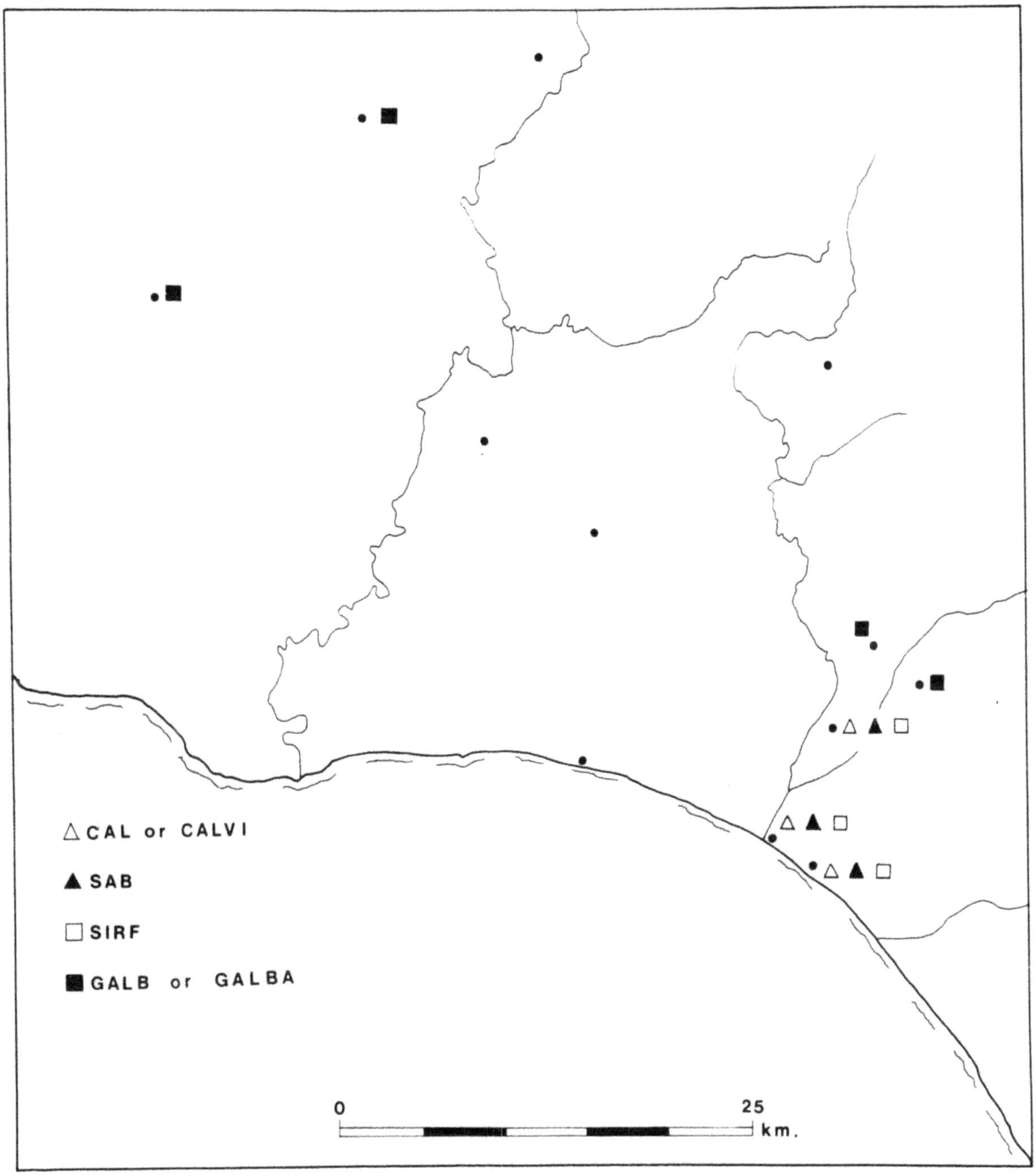

Fig. 2.4 Distribution of tiles stamped CAL(V), SAB, SIRF and GALB(A)

Plate 2.IX a. Syracuse, 'Sessanta Letti', terracotta vaulting-tubes
b. Piazza Armerina, Roman villa, terracotta vaulting-tubes in furnace of baths, fourth century A.D.

Plate 2.X Morgantina, terra-cotta vaulting-tubes in bath-building, third century B.C.

Finally, to conclude, two rather more specialized products. The first are the so-called *tegulae sulfuris*, 'sulphur tiles', from Agrigento and its immediate hinterland. Until sulphur was discovered in the United States at the end of the last century, the Agrigento region was the world's sole supplier of this material. The Roman mines have probably been entirely destroyed by later workings, and these stamps are therefore our only guide to the organization of the industry. The letters appear in relief and in retrograde on terracotta slabs usually about 35 cm square. The lowest one in Plate 2. VIIIb, which measures 38 cm by 43 cm, is rather larger than average. They were apparently used to stamp each consignment of plastic sulphur before shipment. A full discussion of these stamps must be reserved for elsewhere: they often refer to the reigning emperor (IMP in line 1 of top tile of Plate 2. VIIIb) and/or to an *officina* (Cassiana, Porciana and Gellius Pelorus appear in the three illustrated), and/or to a *conductor* who leased rights from an *officina* (in the last line of the top two tiles). The last of our terracotta products are the curved interlocking tubular tiles used in vaulting. This method occurs very widely in Roman north Africa from the beginning of the second century A. D. onwards, but it is found also in Sicily in several locations on or close to the south coast, which suggests direct influence from north Africa. The tiles are documented at Syracuse (Plate 2. IXa), Marsala, Motya, Vito Soldano, Catania and also at Piazza Armerina where they are used to roof over the furnace areas of the baths[63] (Plate 2. IXb). The discovery by Dr. Parker of vaulting tubes in a wreck with other African products off the south coast of Sicily suggests that some at least of the Sicilian examples were not locally made but imported direct from Africa. As regards the origins of this form of vaulting, some important fresh light has been shed by the recent discovery of tubes of this sort, rather larger in size, in a circular building 6 m in diameter at Morgantina[64] (Plate 2. X). The date of this structure falls quite clearly within the third century B. C., and the bath-building to which it belongs was not occupied after the end of that century. A similar building of the same date with round rooms and more terracotta vaulting tubes was excavated in Syracuse in the 1930s, but its importance was not recognised then for what it was.[65] Once more we see Sicily, possibly here inspired by hellenistic Egypt, in the forefront of experimentation in the use of terracotta materials which were later adopted and developed on a much wider scale in other areas of the Roman world.

NOTES

1. II. 3. 1-4 and II. 8. 19.
2. Van Buren 1923, *passim*.
3. Orlandini 1960. The bricks are each 50 cm square (*ibid.*, 187). For an early treatment of Sicilian bricks and tiles, Avolio 1829, 23-62.
4. Sjoqvist 1960, 131. Other buildings at Morgantina using brick columns include the propylon at the west corner of the agora (Stilwell and Sjoqvist 1957, 154), the House of the Tuscan Atrium (Sjoqvist 1958, 160), the House of the Arched Cistern (Sjoqvist 1962, 139) and the House of the Official (Stilwell 1963, 167-8).

5. Sjoqvist 1960, 131, cf. semicircular holes in bricks at Reggio Calabria, probably third century B. C. (Spinazzola 1907, 705). Circular bricks with central holes also occur in hypocaust pilae of later date, for example at Pompeii, St. Bertrand de Comminges, and Wroxeter: information from Dr. G. A. Webster and Prof. A. L. F. Rivet.

6. Unpublished in detail. For a brief note Bernabò Brea 1972-3, 176, and for a plan Scibona 1971a, pl. I. For circular bricks at Himera, Lugli 1957, 548.

7. Bernabò Brea and Cavalier 1965a, house C, probably early first century B. C.

8. These measurements of my own are slightly different from those given in Carettoni 1961, 288 and 292, which describes the results of trial trenching in the area before the large-scale excavations of 1970-1 (see note 6).

9. Lugli 1957, 538-9.

10. In the first house on the left (west) of via Cavallari, with bricks 35 cm square (unpublished). The bricks in the courtyard of the House of the Criptoporticus at Agrigento (most 28 cm square; some 58 cm square) may belong to a repair carried out during the Empire (for the quarter in general, De Miro 1957).

11. E.g. in the House of the Official, post c. 211 B. C. (Stilwell 1963, 168); House of the Doric Capital, with paved court 5. 30 m by 6. 70 m, second century B. C. (Stilwell and Sjoqvist 1957, 156-7). If the paving of the building identified as the Prytaneion is original and not part of a refurbishing, then its bricks (35 cm square) belong to the first half of the third century B. C. (Sjoqvist 1958, 161; Stilwell 1959, 167-8; Miller 1978, 115-7).

12. Napoli 1973, 31 ('third or second century B. C.'). For a general account of early brick-making in Magna Graecia, Blake 1947, 284-6 with earlier bibliography.

13. Libertini 1942, 276-7: tomb constructed of bricks 30 cm by 28 cm by 9 cm, second-first century B. C. Most of the brick tombs in the necropolis at Lipari are built of mud-brick, only one being built of kiln-fired bricks, and those probably are cut roof-tiles (Bernabò Brea and Cavalier 1965b, 48, c. 335/280 B. C.).

14. Griffo 1942, 73-7; Orsi 1929, 50-1.

15. Orsi 1909. Orsi's suggested date was 'fourth-third century, more probably third', but a slightly later date has also been proposed (Delbrueck 1912, 95).

16. The best examples on what are certainly bricks come from Messina and Halaesa: from Messina there are 'government' stamps from the Mamertine rulers (283-264 B. C.) and hints of a power struggle on those stamped Ῥηγινων ὀσθων ('true Reggians'); both occur at Reggio as well, as do bricks for the city walls (τειχεων). On these, Salinas 1886; Orsi 1916, 194; Guarducci 1969, 494-6; Mingazzini 1970, 408. For brick stamps from Halaesa, Carettoni 1959, 321 (Διων retrograde), 324

(ἱερων, Πα, Ἱππαδ..), 333-4; Carettoni 1961, 277 (Πα), 293 (Νυμφη, Διων retrograde, Δαμα, Με retrograde). See also notes 52 and 55-7.

17. Salinas 1898.

18. Stilwell and Sjoqvist 1957, 158.

19. Mingazzini 1954, 22-5, 32-9. One stoking area was well preserved, and the presence of a second flue arch on the other side of the kiln strongly suggests that, as at Morgantina, there was a second stoking area there too, even though Mingazzini (ibid., 23) ruled out that possibility.

20. Pelagatti 1966, 96 with fig. 22, largely repeated in van Buren 1966, 357, with pl. 85, fig. 8.

21. Cuomo di Caprio 1971-2, 457-61 lists 46 examples of kilns in Sicily of all dates from prehistoric times to late Roman, but none of these (except for Spadafora, Morgantina and Camarina) seems to have been used exclusively or predominantly for brick or tile making: the vast majority were primarily pottery kilns.

22. By van Deman, reported by Blake 1947, 291.

23. The building has never received the full architectural study it deserves. The most useful discussions are Libertini 1930a, Santangelo 1950, 35-57, Lugli 1956, 89-92, Lugli 1957, 628-30, and Blake and Bishop 1973, 263.

24. Lugli 1956, 93-5; Lugli 1957, 630, where he notes that while the first period of triangular bricks in Rome is Neronian-Flavian, the theatre 'could go back' to Augustus. On the odeon, see too Santangelo 1950, 57-60.

25. On this monument see Rizzo 1928, 355-70; Pace 1930; Santangelo 1950, 76-9; Lugli 1956, 95-8; and Lugli 1957, 607 and 630.

26. No complete survey is available; for a brief account, Wilson, forthcoming a. For brick-faced concrete at Centuripe, Libertini 1926 and Rizza 1949.

27, Toynbee 1971, 82-91, 132-43.

28. But they can be as much as 4.3 cm. For a general summary of brick thicknesses as a guide to dating, Meiggs 1973, 542-3 and Lugli 1957, 600-14, cf. 628-30; for brickwork of the first and second centuries in detail, Blake 1947, 292-8; Blake 1959, 162-3; Blake and Bishop 1973, 302-3.

29. As suggested by Blake 1947, 284, n. 21.

30. See note 28. Cf. in Taormina, the first-century baths (Pelagatti 1964), bricks 7.5 cm thick; the Naumachia (see note 25), late second century (?), bricks 7.5 cm-10 cm; Centuripe, second-century baths, bricks 7-8 cm thick (Rizza 1949).

31. Personal observation; the reservoir is unpublished except for its dimensions, recorded by Rizzo 1928, 351. The precise sequence of brick courses, from present earth level upwards, is: one row of 8.5 cm bricks, three rows of 7.5 cm; one of 6.25 cm; one of 4.25 cm; one of 8.5 cm; two of 4.25 cm; two of 8.5 cm; and one of 6.25 cm, which is the spring course for the concrete vault.

32. For a preliminary account, Fallico 1976-7, 632-3.

33. *IG* XIV, 2396. 2-3; Lugli 1957, 629.

34. E.g. *IG* XIV, 2395. 1 (Haluntium), 2399. 2 (Alcamo: Εὐτυχης, very common late Roman name), Pucci 1971, 255-6 (Piazza Armerina).

35. Bivona 1972-3, 194-5 with pl. XLII, fig. 3. A small kiln of imperial date at Marsala, rectangular in shape (very unusual in Sicily), may conceivably have been used for making bricks: for an inadequate account, Bisi 1966, 321, with plan on p. 314.

36. *CIL* X, 8045. 1 (Syracuse); *ibid*. 15 (Licata = *CIL* XV, 1344. 4); cf. OPP. . . from Catania, *Notizie degli Scavi* 1891, 361. For unprovenanced imperial brick stamps in Catania, Libertini 1930b, 319-20.

37. The former occurs too commonly to list examples here; the latter is much rarer in Sicily: examples at Taormina (tombs), Catania (Odeon, Rotunda baths) and Centuripe ('La Dogana', Augustales building, walls near the church of the Crucifix). The use of brick bonding courses comes into fashion in the East in the course of the second century (Boethius and Ward Perkins 1970, 389), and is no earlier than that in Italy, where it was never common (cf. Lugli 1957, 645), or in Sicily.

38. Examples at Marsala (Capo Boeo, 1930s excavations, unpublished except for Marconi Bovio 1940 and Di Stefano 1976) and Agrigento (examples in 'House of Abstract Master' (Plate 2. Va), and house with atrium and peristyle in south-west portion of the excavated quarter; also apparently for paving a *decumanus* on the north side of this *insula*. For a general account of the excavations, De Miro 1957).

39. In the House of Ganymede, 260/50 B. C. (Sjoqvist 1960, 131), the House of Palmento and the House of Pappalardo (both mid-third century B. C.; Stilwell 1967, 247).

40. Unpublished in detail; in brief, see Gentili 1966, 19. It is, I suppose, possible that the bricks are re-used material taken from older buildings rather than freshly manufactured (cf. note 55; and for re-used marble and capitals at the villa, Ampolo *et al*. 1971, 207-224).

41. van Buren 1923, xviii-xix; for profiles of tiles from the Syracuse Olympeion, mid-sixth century B. C. if from the original roof, Lissi 1958, 211, fig. 18. For a profile of a third-century B. C. example, also from Syracuse, Cultrera 1938, 289, fig. 16, and on the longevity of this so-called 'archaic' type of roof-tile in Italy, Kahane *et al*. 1968, 8.

42. The so-called 'Sicilian' type, as distinct from the 'Corinthian', the method generally employed in mainland Greece, which had gable-shaped cover tiles: Dinsmoor 1950, 43-4; Lawrence 1973, 108. For an extraordinary ridge-pole roof tile from a farm of Republican date (to cover the junction between separate roofs meeting at an angle?), Curchio and Pelagatti 1970, 452 and fig. 19.

43. For the types, Toynbee 1971, 101-2. Lipari: Bernabò Brea and Cavalier 1965b, esp. 261-8. Messina: Orsi 1916, esp. 127-8, 130, 142, 149, 160,

165; Orsi 1929, 50-1; Griffo 1942, 66-77. Catania: Libertini 1937 and 1956. Agrigento: Salinas 1901, tomb XI.

44. Cf. discussion by E. M. Wightman in Small 1977, 208-9, with bibliography of other south Italian sites.

45. Dinsmoor 1950, 43-4; Lawrence 1973, 108.

46. Monte Iato: Müller 1976, 49-50 with bibliography. Morgantina: Sjoqvist 1964 1964, 137-8.

47. Detailed dating evidence has not so far been forthcoming, but the general picture is clear from examination of the tiles found in the plain of Gela during Adamesteanu's fieldwork (Adamesteanu 1955) and of those found during fieldwork in the hinterland of Heraclea Minoa in 1977-8 (for a first report, Wilson and Leonard, in press).

48. Giustolisi 1976, 37-8.

49. Especially Messina, for bibliography of which see note 43. I have not yet had an opportunity to examine roof-tiles from urban structures; but the fallen débris in a house at Tindari, which collapsed early in the fourth century A. D. (Bernabò Brea and Cavalier 1965a), has curved roof tiles, but without the rudimentary flanges.

50. It is used exclusively on a Roman villa-site, founded *c.* 425 A. D.: Wilson and Leonard, in press.

51. They are found in Basilicata, and also (in small quantities) in Carthage: information from Prof. D. Adamesteanu and Dr. D. P. S. Peacock.

52. *IG* XIV, 2394-9; *CIL* X, 8045, and discussion of these by Scramuzza 1937, 356-7, but it is not clear how many are brick stamps and how many stamps on roof-tiles. Some of these may be of Italian manufacture, and a tile stamped VINICI from a Palermo burial of the first century A. D. may be Campanian (Tusa 1954, 146, cf. *CIL* X, 8042. 18 and *CIL* XV, 1527-9). Agathocles of Reggio Calabria, who made tiles in the third or second century B. C., exported to Lipari and Tindari; contemporary products of Aristocrates and Aischrion have also been found at both Lipari and Tindari (Bernabò Brea and Cavalier 1965b, 331-2); and the stamp Nikon has been found at both Lipari and Halaesa (Carettoni 1959, 316). See also notes 16 and 61.

53. Müller 1976; for the more recent finds, see especially Isler 1978, 7, 9 and 11.

54. Ἐπι Συμμαχου, ἐπι Ταμμαρου, ἐπι Διονυσιου, ἐπι Νικια Δειανια, ἐπι Λακωνος, ἐπι Φιλωνος and ἐπι Τιτου; Müller 1976, 57-62.

55. See note 16: some of the stamps appear to be on bricks, others on tiles, or on both. For a re-used brick stamped Ἀπολλωνος at Piazza Armerina, Pucci 1971, 255.

56. Messina: *IG* XIV, 2394. 1; Salinas 1886, 461; Orsi 1916, 194. Halaesa: Carettoni 1959, 324, fig. 30b, 334. Acrae: Mingazzini 1970, 408. Calacte: *IG* XIV, 2395. 7, of which the precise reading is disputed by

Scibona 1971b, Fiore 1971 and Mingazzini 1972 (this is a brick, not a roof-tile, as the photograph in Scibona 1971b, pl. VII. 1, makes clear).

57. Other 'official' municipal tile or brick stamps are suspected at Taormina (*IG* XIV, 2396. 1, Δαμοσιος), Acireale (there is an unpublished tile stamped Ἀκιος in the Accademia Zelantea collection, cf. Antonine Itinerary 87. 3 and *AE* 1923, 75) and Troina (but the reading is disputed: Manganaro 1964, 251-2, cf. Manni 1976, 606-7).

58. Wilson forthcoming b; for an earlier study, Adamesteanu 1955.

59. *CIL* VI, 30983 = *ILS* 3840.

60. For a modern analogy with the Ashburnham estate, see D. P. S. Peacock, page 7.

61. Of the published group of stamped tiles of 'imperial' type from contrada Raccuglia near Partinico (Giustolisi 1976, 37-8, with pl. XXI-XXIV and XXVIII), one stamped Πορτ appears to be the same as that recorded at Monte Iato (Isler 1978, 9), and others stamped Ὀνασου (possessive genitive, probably the name of the maker rather than the owner) are known as well at Palermo (Gabrici 1921, 201) and in the environs of Segesta (examples shown to me on site, 1972). For other fragmentary stamped roof-tiles of imperial date in western Sicily, including slight evidence for manufacture on site, Traselli 1970.

62. *CIL* X, 8044, 1-14; *Notizie degli Scavi* 1900, 659 (= *AE* 1901, 166) and 1925, 434; Griffo 1963, 165-9; and unpublished tiles in Agrigento Museum.

63. Tubular tiles are also used to lighten the weight of the vault of the adjacent *tepidarium*, but there they are mixed indiscriminately with lava pieces as *caementa*. Documentation for the others: Cultrera 1954, 116, Houel 1785, 92 and pl. CLXXXX (=Plate 2. IXa), Frel 1950 (Syracuse); Crema 1959, fig. 755 (Marsala); V. Tusa in Ciasca *et al.* 1965, 38, pl. 26, Ciasca *et al.* 1966, 16, pl. II, Ciasca *et al.* 1970, 10-11, pl. 4. 1 (Motya); La Lomia 1961, 158 (Vito Soldano); Orsi 1906, 235, fig. 21 (Priolo near Syracuse); Orsi 1912, 412 (Catania).

64. Allen 1974, 376-82.

65. Cultrera 1938.

ACKNOWLEDGEMENTS

Plate 2. IIb was kindly supplied by Prof. ssa P. Pelagatti, Superintendent of Antiquities for Eastern Sicily, Pl. 2. VII-2. VIIIa by Prof. H. P. Isler of the University of Zurich, and Plate 2. X by Prof. H. L. Allen and the University of Princeton. Plate 2. IXa is taken from J. Houel, *Voyage Pittoresque dans les Isles de Sicile, de Lipari et de Malte* III, Paris 1785, by permission of the Board of Trinity College, Dublin. All other photographs are by the author, who thanks especially Prof. E. De Miro, Superintendent of Antiquities for Central Southern Sicily, for permission to publish Pl. 2. Va, 2. VIa, 2. VIb and 2. VIIIb. The ranging-rod which appears in some plates is imperial (foot-measures), not metric; the scale in Pl. 2. VIa and b is in centimetres.

BIBLIOGRAPHY

Adamesteanu 1955: D. Adamesteanu, 'Due problemi topografici del retroterra gelese', Rendiconti Accademia dei Lincei 10 (1955), 199-210.

Allen 1974: H. L. Allen, 'Excavations at Morgantina 1970-2. Preliminary Report XI', American Journal of Archaeology 78 (1974), 361-82.

Ampolo et al. 1971: C. Ampolo, A, Carandini, G. Pucci and P. Pensabene, 'La villa del Casale a Piazza Armerina. Problemi, saggi stratigrafici, ed altre ricerche', Mélanges de l'École Française de Rome, Antiquité 83 (1971), 141-281.

Avolio 1829: F. Di Paola Avolio, Delle antiche fatture di argille in Sicilia, Palermo, 1829.

Bernabò Brea and Cavalier 1965a: L. Bernabò Brea and M. Cavalier, 'Tindari, area urbana. L'insula IV e le strade che la circondano', Bolletino d'Arte 50 (1965), 205-9.

Bernabò Brea and Cavalier 1965b: L. Bernabò Brea and M. Cavalier, Meligunis Lipara II, Palermo, 1965.

Bernabò Brea 1972-3: L. Bernabò Brea, 'Attività della Soprintendenza alle antichità per la Sicilia orientale', Kokalos 18-19 (1972-3), 161-179.

Bisi 1966: A. M. Bisi, 'Lilibeo (Marsala). Ricerche archeologiche', Notizie degli Scavi 1966, 310-347.

Bivona 1972-3: L. Bivona, 'Epigrafia latina', Kokalos 18-19 (1972-3), 193-200.

Blake 1947: M. E. Blake, Ancient Roman Construction in Italy from the Prehistoric Period to Augustus, Washington, 1947.

Blake 1959: M. E. Blake, Roman Construction in Italy from Tiberius through the Flavians, Washington, 1959.

Blake and Bishop 1973: M. E. Blake, Roman Construction in Italy from Nerva through the Antonines, edited and completed by D. T. Bishop, Philadelphia, 1973.

Bloesch and Isler 1976: H. Bloesch and H. P. Isler, Studia Ietina I: Die Stützfiguren des grieschischen Theaters; Gestempelte Ziegel; Rezepte vom Monte Iato, Zurich, 1976.

Boethius and Ward-Perkins 1970: A. Boethius and J. B. Ward-Perkins, Etruscan and Roman Architecture, Harmondsworth, 1970.

Carettoni 1959: G. Carettoni, 'Scavi di Halaesa', Notizie degli Scavi 1959, 293-349.

Carettoni 1961: G. Carettoni, 'Scavi di Halaesa, seconda relazione', Notizie degli Scavi 1961, 266-321.

Ciasca et al. 1965: A. Ciasca, M. Forte, G. Garbini, S. Moscati, B. Pugliese and V. Tusa, Mozia I, Rome, 1965.

Ciasca et al. 1966: A. Ciasca, M. Forte, G. Garbini, S. Moscati, B. Pugliese and V. Tusa, Mozia II, Rome, 1966.

Ciasca et al. 1970: A. Ciasca, M. G. Guzzo Amadasi, S, Moscati and V. Tusa, Mozia VI, Rome, 1970.

Crema 1959: L. Crema, L'architettura romana, Turin, 1959.

Cultrera 1938: G. Cultrera, 'Rovine di un antico stabilmento idraulico in contrada Zappalà', Notizie degli Scavi 1938, 261-301.

Cultrera 1954: G. Cultrera, 'Siracusa. Il Bagno Dafne', Notizie degli Scavi 1954, 114-30.

Cuomo di Caprio 1971-2: N. Cuomo di Caprio, 'Proposta di classificazione delle fornaci per ceramica e laterizi nell' area italiana', Sibrium 11 (1971-2), 371-464.

Curchio and Pelagatti 1970: G. Curchio and P. Pelagatti, 'Palazzolo Acreide, Contrada Aguglia 1960-2, La fattoria tardo -ellenistica', Notizie degli Scavi 1970, 447-499.

Delbrueck 1912: R. Delbrueck, Hellenistische Bauten in Latium II, Strasbourg, 1912.

De Miro 1957: E. De Miro, 'Il quartiere ellenistico-romano di Agrigento', Rendiconti Accademia dei Lincei 12 (1957), 135-40.

Dinsmoor 1950: W. B. Dinsmoor, The Architecture of Ancient Greece, 3rd edition, London, 1950.

Di Stefano 1976: C. A. Di Stefano, 'Marsala, ricerche archeologiche al Capo Boeo', Sicilia Archeologica 9, no. 32 (1976), 25-31.

Fallico, 1976-7: A. M. Fallico, 'fornaci tardo-romane 'Naxos', Kokalos 22-23 (1976-7).

Fiore 1971: P. Fiore, 'Acquedotto sacro a Demetra', Sicilia Archeologica 4, no. 14 (1971), 37-9.

Frel 1950: J. Frel, 'Sessanta letti', Archeologické Rozhledy 2 (1950), 108-11.

Gabrici 1921: E. Gabrici, 'Ruderi romani scoperti alla Piazza della Vittoria in Palermo', Monumenti Antichi 27 (1921), 181-204.

Gentili 1966: G. V. Gentili, The Imperial Villa of Piazza Armerina (Itinerari no. 87), 3rd edition, Rome, 1966.

Giustolisi 1976: V. Giustolisi, Parthenicum e le Aquae Segestanae, Palermo, 1976.

Griffo 1942: P. Griffo, 'Messina. Necropoli ellenistico-romana agli "Orti della Maddalena" e nelle zone ad essi adiacente', Notizie degli Scavi 1942, 66-91.

Griffo 1963: P. Griffo, 'Contributi epigrafici agrigentini', Kokalos 9 (1963), 163-184.

Guarducci 1969: M. Guarducci, Epigrafia Greca II, Rome, 1969.

Houel 1785: J. Houel, Voyage Pittoresque dans les Isles de Sicile, di Lipari et de Malte, III, Paris, 1785.

Isler 1978: H. P. Isler, 'Monte Iato, L'ottava campagna di scavo', Sicilia Archeologica 11, no. 38 (1978), 7-29.

Kahane et al. 1968: A. Kahane, L. Murray Threipland and J. B. Ward-Perkins, 'Ager Veientanus, North and East of Rome', Papers of the British School at Rome 23 (1968), 1-218.

La Lomia 1961: M. R. La Lomia, 'Ricerche archeologiche nel territorio di Canicattì: Vito Soldano', Kokalos 7 (1961), 157-65.

Lawrence 1973: A. W. Lawrence, Greek Architecture, 3rd edition, Harmondsworth, 1973.

Libertini 1926: G. Libertini, Centuripe, Catania, 1926.

Libertini 1930a: G. Libertini, 'Il teatro di Taormina', Bolletino del Istituto Nazionale del Dramma Antica 1930, 1-11.

Libertini 1930 b: G. Libertini, Il Museo Biscari, Rome, 1930.

Libertini 1937: G. Libertini, 'Catania. Scoperte varie', Notizie degli Scavi 1937, 75-82.

Libertini 1942: G. Libertini, 'Scavi nella necropoli in contrada Casino', Notizie degli Scavi 1942, 259-311.

Libertini 1956: G. Libertini, 'Catania. Necropoli romana e avanzi bizantini nella via Dottor Consoli', Notizie degli Scavi 1956, 170-189.

Lissi 1958: E. Lissi, 'Siracusa. Scavo presso l'Olympeion (anno 1953)', Notizie degli Scavi 1958, 197-223.

Lugli 1956: G. Lugli, 'L'architettura in Sicilia nell'età ellenistica e romana', Atti del VIII° Congresso Nazionale di Architettura, Palermo, 1956, 89-107.

Lugli 1957: G. Lugli, La Tecnica Edilizia Romana, 2 volumes, Rome, 1957.

Manganaro 1964: G. Manganaro, 'Per la identificazione di Troina con Engyon', Siculorum Gymnasium 17 (1964), 251-2.

Manni 1976: E. Manni, 'Proposte di identificazione di centri antichi della Sicilia', Mélanges offerts à Jacques Heurgon II, Rome, 1976, 605-17.

Marconi Bovio 1940: J. Marconi Bovio, 'Marsala, villa romana', Le Arti 18 (1940), 389-90.

Meiggs 1973: R. Meiggs, Roman Ostia, 2nd edition, Oxford, 1973.

Miller 1978: S. G. Miller, The Prytaneion, its Function and Architectural Form, Berkeley and Los Angeles, 1978.

Mingazzini 1954: P. Mingazzini, 'Velia, scavi 1927: fornace di mattoni ed antichità varie', Atti e Memorie della Società Magna Grecia, new series 1 (1954), 21-60.

Mingazzini 1970: P. Mingazzini, 'Elenco di bolli di mattoni pubblici', Rendiconti Accademia dei Lincei 25 (1970), 403-29.

Mingazzini 1972: P. Mingazzini, 'Un bollo greco impresso su di una conduttura d'acqua rinvenuta presso Caronia', Sicilia Archaeologica 5, no. 17 (1972), 7-8.

Müller 1976: P. Müller, 'Gestempelte Ziegel', in Bloesch and Isler 1976, 49-77.

Napoli 1973: M. Napoli, Guide to the excavations at Velia, Cava dei Tirreni, 1973.

Orlandini 1960: P. Orlandini, 'Impianto greco di bagni pubblici presso l'Ospizio', Notizie degli Scavi 1960, 181-202.

Orsi 1906: P. Orsi, 'Priolo, le catacombe di Riuzzo', Notizie degli Scavi 1906, 218-43.

Orsi 1909: P. Orsi, 'Reggio Calabria. Sepolcri ellenistici in contrada Borace', Notizie degli Scavi 1909, 314-8.

Orsi 1912: P. Orsi, 'Catania, avanzi di edificio termale ai Quattro Canti', Notizie degli Scavi 1912, 412-3.

Orsi 1916: P. Orsi, 'Messana, la necropoli romana di S. Placido', Monumenti Antichi dei Lincei 24 (1916), 121-218.

Orsi 1929: P. Orsi, 'Nuovi indizi di una grande necropoli di Messana', Notizie degli Scavi 1929, 46-51.

Pace 1930: B. Pace, 'La cosidetta naumachia di Taormina', Bolletino d'Arte 9(1930), 376-80.

Pelagatti 1964: P. Pelagatti, 'Scoperta di un edificio termale a Taormina', Cronache di Archeologia e Storia d'Arte 3 (1964), 25-37.

Pelagatti 1966: P. Pelagatti, 'Attività delle Soprintendenze (1961-5): Sicilia', Bolletino d'Arte 51 (1966), 88-116.

Pucci 1971: G. Pucci, 'I materiali epigrafici del magazzino della villa del Casale', in Ampolo et al. 1971, 249-56.

Rizza 1949: G. Rizza, 'Centuripe. Edificio termale in via Adriano', Notizie degle Scavi 1949, 190-2.

Rizzo 1928: P. Rizzo, Tauromenion, Riposto, 1928.

Salinas 1886: A. Salinas, 'Messina', Notizie degli Scavi 1886, 460-2.

Salinas 1898: A. Salinas, 'Spadafora. Fornace antica scoperta presso Spadafora (provincia di Messina)', Notizie degli Scavi 1898, 257-8.

Salinas 1901: A. Salinas, 'Agrigento, necropoli Giambertone a S. Gregorio', Notizie degli Scavi 1901, 29-39.

Santangelo 1950: M. Santangelo, Taormina e Dintorni, Rome, 1950.

Scibona 1971a: G. Scibona, 'Epigraphica Halaesina I (Schede 1970)', Kokalos 17 (1971), 3-20.

Scibona 1971b: G. Scibona, 'Nota a IG XIV 2395.7 (Instrumentum Publicum Calactinum)', Kokalos 17 (1971), 21-25.

Scramuzza 1937: V. M. Scramuzza, 'Roman Sicily' in T. Frank (ed.), Economic Survey of Ancient Rome III, Baltimore, 1937, 225-377.

Sjoqvist 1958: E. Sjoqvist, 'Excavations at Serra Orlando (Morgantina). Preliminary Report II', American Journal of Archaeology 62 (1958), 155-64.

Sjoqvist 1960: E. Sjoqvist, 'Excavations at Morgantina (Serra Orlando) 1959. Preliminary Report IV', American Journal of Archaeology 64 (1960), 125-35.

Sjoqvist 1962: E. Sjoqvist, 'Excavations at Morgantina 1961', American Journal of Archaeology 66 (1962), 135-43.

Sjoqvist 1964: E. Sjoqvist, 'Excavations at Morgantina 1963. Preliminary Report VIII', American Journal of Archaeology 68 (1964), 137-47.

Small 1977: A. Small (ed.), Monte Irsi, Southern Italy, BAR Supplementary Series 20, Oxford 1977.

Spinazzola 1907: V. Spinazzola, 'Reggio Calabria. Di alcuni scavi e trovamenti nelle necropoli Reggine', Notizie degli Scavi 1907, 704-715.

Stilwell 1959: R. Stilwell, 'Excavations at Serra Orlando 1958', American Journal of Archaeology 63 (1959), 167-73.

Stilwell 1963: R. Stilwell, 'Excavations at Morgantina 1962. Preliminary Report VII', American Journal of Archaeology 67 (1963), 163-71.

Stilwell 1967: R. Stilwell, 'Excavations at Morgantina 1966: Preliminary Report IX', American Journal of Archaeology 71 (1967), 245-50.

Stilwell and Sjoqvist 1957: R. Stilwell and E. Sjoqvist, 'Excavations at Serra Orlando, Preliminary Report', American Journal of Archaeology 61 (1957), 151-9.

Toynbee 1971: J. M. C. Toynbee, Death and Burial in the Roman World, London, 1971.

Trasselli 1970: C. Trasselli, 'La fattoria romana di Sirignano', Sicilia Archeologica 3, no. 12 (1970), 19-24.

Tusa 1954: V. Tusa, 'Palermo. Rinvenimento di una tomba in Piazza Indipendenza', Notizie degli Scavi 1954, 146.

van Buren 1923: E. D. van Buren, Archaic Fictile Revetments in Sicily and Magna Graecia, London, 1923.

van Buren 1966: A. W. van Buren, 'News Letter from Rome', American Journal of Archaeology 70 (1966), 349-361.

Wilson, forthcoming a: R. J. A. Wilson, 'Towns of Sicily during the Roman Empire', Aufstieg und Niedergang der römischen Welt, II. 11, Berlin and New York, forthcoming.

Wilson, forthcoming b: R. J. A. Wilson, 'Tile-stamps and villa estates in Sicily under the Roman Empire', forthcoming.

Wilson and Leonard, in press: R. J. A. Wilson and A. Leonard Jr., 'Field survey at Heraclea Minoa (Agrigento): first results', Journal of Field Archaeology, in press.

RURAL ARCHITECTURE AND CERAMIC INDUSTRY
AT METAPONTO, ITALY, 350-50 B.C.

Joseph Carter

It is not my intention to give a general introduction to Greek and Roman tile manufacture in Italy but rather to introduce the subject by presenting as a non-specialist, some specific instances of the uses of tiles in rural architecture in a single locale and one example of an unusually well-preserved tile kiln, against the background of the excavation programme which brought them to light.

The Metaponto project

Organized a dozen years ago by Professor Dinu Adamesteanu, the exploration of the chora (territory) of Metaponto is a comprehensive and systematic attempt to describe the Greek rural population (Adamesteanu, 1974). Despite important contributions like the publication of the Dema and Vari houses in Attica (Jones, Sackett, Graham, 1962, 1973) and extensive work on the chora of Chersonesos in the Crimea (Pečírka, 1973), life outside the walls of the polis, has been virtually ignored by classical archaeologists (Humphreys, 1967, Vallet, 1967).

At Metaponto, where the evidence is abundant, still—though perhaps not for long—aerial photographic studies and surface surveys have made possible increasingly accurate plans of the land and the broad patterns of settlement (Schmiedt, Chevallier, 1959, Adamesteanu, 1973, Adamesteanu, Vatin, 1976). Excavation of selected sites within this framework has begun to reveal the complexity of rural life, as well as its development in time.

The Architecture of Rural Settlements

The history of settlement in the territory of chora of the Greek colony, Metapontion (Roman Metapontum, today's Metaponto) embraces the whole period of the use of tiles in Italy down to the present. Sun-dried bricks and fired tiles were the principal materials used by the country-dweller for his house, farm buildings, industrial establishments, the more important sanctuaries, and often for his tomb. This is easy to understand. They were not only resistant to heat, wet and cold and were very durable, but the basic ingredients were also close at hand and relatively economical. The requirements for the manufacture of tiles, a supply of good clay, abundant water, nearby fuel and labour, could and can all be found in different parts of the countryside around Metaponto.

1. Incoronata—The pre-tile phase, eighth and seventh centuries B.C.

At Incoronata (overlooking the Basento River from the West), a site of great historical importance as it is perhaps the first Greek settlement in this part of Italy, no tiles have been found in the early phase. According to its excavator, Professor Piero Orlandini, the mixed settlement of Greeks and indigenous peoples, dating from 750 to 650 B. C. had houses with mud-brick walls footed on stone foundations (Orlandini, 1976). The roofs were of thatch or some other perishable material. The Greeks of Incoronata, whom Orlandini believes to have originated in Asia Minor, disappeared without a trace, about the time of the arrival at the site of Metapontion, of colonists from Achaea (the northern Peloponnesos). This happened early in the second half of the seventh century B. C.

2. Incorontata and San Biagio, the early sixth century sanctuaries

The first known use of tiles in the chora of Metaponto is contemporaneous with the earliest known buildings of the colonists, who were concerned above all with establishing homes and farms in a rich agricultural land, which was not easily surrendered by its native inhabitants.

These first buildings were, however, not farmhouses but rather sanctuaries, like the shrine which sprang up in the ruins of recently abandoned Incoronata in the period 600-550 B. C. Its revetments and antefixes are among the earliest known in this part of Italy (Carter, 1977). Across the valley was an even earlier and vastly more important sanctuary beside the spring at San Biagio. The earliest votive offerings at San Biagio go back to the second half of the seventh century B. C., i. e. almost to the founding of the colony, but the earliest architectural remains, handsome terracotta revetments with heroic scenes in relief, are like those of Incoronata, of the early sixth century B. C. (Adamesteanu, 1974).

Systems of roofing: The colonial sanctuaries and farmhouses of the archaic period employed both major systems of roofing then in use in the Greek motherland; the heavier, more costly and more impressive Corinthian system, and the Laconian, which was far more popular. The tiles of this period are often of greater size and finer workmanship than those of later periods. The hybrid type, known as Sicilian, does not appear to have been commonly used at Metaponto.

Settlement in the chora in the sixth century B. C. was widespread. The houses, sanctuaries, and tombs attest the wealth of the rural population at this time. For reasons not yet clear there is a hiatus in the history of the chora. Farmhouses of the Classical period are virtually unknown.

3. Saldone—Tile tombs of the fifth century B. C.

A family burying ground, discovered in 1976 by the University of Texas team, at Saldone, 8 km from the walls of Metaponto is so far the major evidence for life in the chora in the fifth century B. C. (Carter, forthcoming a). Several of the tombs contain vases by the early masters of South Italian vase painting, the Pisticci Painter and his more original colleague, the Amykos Painter. At Saldone, significantly, the use of tiles for tombs was introduced a century earlier than its first appearance in the great urban necropolis, just outside the city walls.

Building in the chora in the Early Hellenistic Period

In the later phases of the chora the evidence for the use and the production of roof tiles is particularly rich, and it is to the early Hellenistic and Roman periods that the rest of this paper will be devoted.

Excavation clearly demonstrates that there was a sudden renewal of vitality and intensive development in the countryside, in the early Hellenistic period, concomitant with the arrival of Alexander the Molossian in South Italy and Timoleon in Sicily. It was over by the second quarter of the third century, i. e. by the time of Pyrrhus.

The principal evidence for this shortlived revival is the architecture, which is unusual both for its variety and originality. Besides farmhouses, could be found sanctuaries, industrial establishments, waterworks and necropoleis. Of primary importance were the parallel divisions imposed on the entire landscape. The geometry of this grid (actually several systems of grid) dating back as early as the first quarter of the sixth century B. C., was the major factor determining the orientation of most buildings in the chora. The rigid symmetry of this landscape reflected a high degree of regimentation, of submission to the single overriding purpose of life here, agricultural production.

The Division of the chora: Discovered by aerial photography, the existence of the division lines has been confirmed by observation on the ground and by excavation. In the latter case, however, the evidence is indirect, for example, the rural burying ground at Saldone on the Bradano side of the chora. The tombs, with only one exception, are perfectly aligned with the grid of division lines,

It has been argued very plausibly that roads must have followed at least some of the division lines, and that a project of 'bonifica', i. e. of irrigation and drainage, of the entire chora, dating from the early fifth century B. C., placed its system of canals over the grid (Uggeri, 1969). A project of excavation, geomorphological and chemical analysis of the soil could resolve this major question—hopefully, in the near future.

The existence of a road system is borne out by further indirect evidence, namely the placement and orientation of the farmhouses.

Over 400 farmhouses have been located in surveys of the chora of Metaponto. The surface indication of a farm is sherds, but above all roof tiles. Freshly-ploughed ground over a well preserved ancient farm building leaves a carpet of fragmentary roof tiles. The ability to recognize tiles of different periods has helped to assign provisional dates to the sites. This is a project to which new techniques of sampling could be applied with profit.

1. Fattoria Stefan—An early Hellenistic farmhouse

Of these 400 sites, to date 14 have been excavated, the most completely preserved example, in 1969, by Alexandru Simion Stefan, hereafter referred to as Fattoria Stefan (Carter, 1979). It is just a stone's throw to the west of the principal division line /which bisects the territory between Bradano and Basento Rivers), at a distance of eight kilometers from the city wall. It was sited on a slight rise with a commanding view of the whole chora to the sea, its external walls in near perfect alignment with the division lines.

Though the relation of structure and division is clearest in Fattoria Stefan, it is generally true of the others as well. The farmhouse is oriented parallel to the grid and positioned near the line but not too close, which would make very good sense if the division line were also a road. This is exactly what happens in the chora of Chersonesos, where the division lines are in fact roads.

Turning to a consideration of the problem of the farmhouse structure itself, it is easy to say what it was not. It was not an urban dwelling modified for country living as were the basically pastas-type houses at the Dema Wall or Vari (Jones Sackett, Graham 1962, 1973). Like the farmhouses of the Chersonesos most of its space was apparently designed for farming activity.

The plan of the final phase which dates to the end of the fourth century B. C. fills a nearly square rectangle, whose short eastern side paralleling the central division line is exactly 50 ft of 0. 295 metres (Fig. 3. 1). The ten rooms are organized tightly around an off-centre courtyard. The courtyard occupies a square whose northern side measures exactly 30 ft of 0. 295 metres.

Unfortunately, as with most excavated houses and farmhouses, the function of the various rooms is largely a matter of speculation. No room is without some problem of interpretation.

Was the courtyard, which is at least quickly identifiable, open to the sky as the drain, apparently for rainwater, would indicate? Or was it partially roofed, as the thick tile suggests?

A feature in the room which opens off the courtyard to the northwest gives us one other securely identifiable space. The hearth, constructed of tiles on edge—a system which would have retained heat well—was undoubtedly the kitchen and the heart of the living quarter of the fattoria. The finds of pottery, much storage ware, but also considerable quantities of cooking, table and fine wares, make it clear that the building was a productive unit, which was also intensively lived in. Numerous fragments of pithoi in the room adjoining the kitchen would indicate that this space was used for storage. If the low base against the east wall supported a press, as seems plausible, then it could have been used also for the production of oil or wine. The low L-shaped wall, a feature found also in another room which may also have had a press, perhaps formed a vat or basin.

Unfortunately, the excavation yielded no direct evidence of what Fattoria Stefan produced. The initial results, however, of the first palaeobotanical research in the Metapontine chora, at Pizzica, indicate that barley and the grape were the principal crops of that area at the end of the fourth century B. C.

A very large collection of loom weights from the room just to the south of the room with the pithoi suggests that this was a workroom for women. The large space enclosed, apparently, by only three walls, may have been utilized as a stable. Again we have, unfortunately, no direct evidence of the livestock raised on the farm.

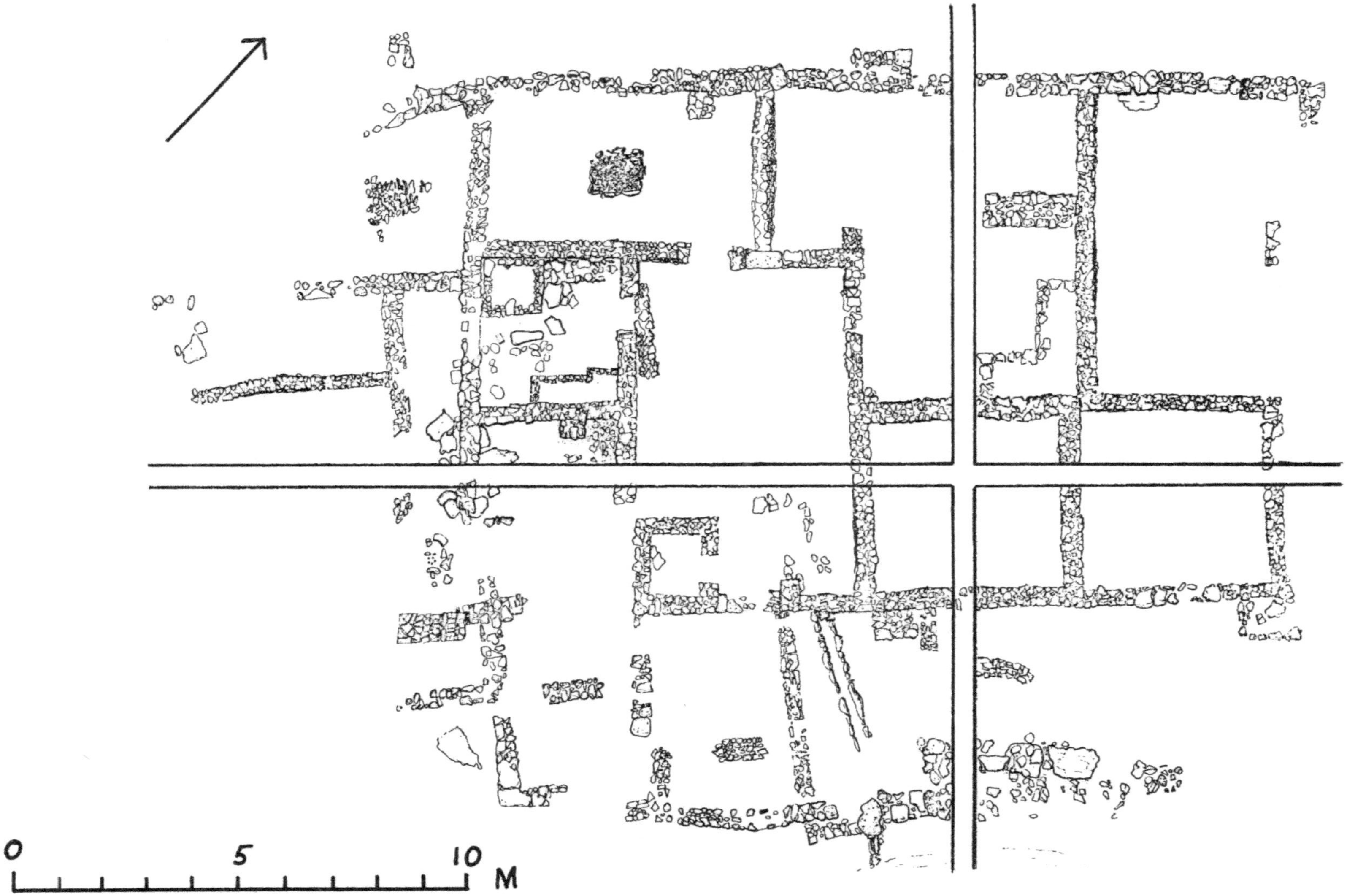

Fig. 3.1 Plan, Fattoria Stefan in the late fourth century B. C.
(Soprintendenza della Basilicata)

Though it is not possible to identify the spaces and their functions with certainty it is clear that the variety in their size and shape was a basic feature of the plan. In sharp contrast to the modified urban design of the Attic country houses where the rooms are all more or less the same size, here the spaces were differentiated in size and shape, apparently according to the uses to which they were to be put—or in modern architectural terminology, the spaces were 'zoned.' '

Successive modification, made by further subdivision of the original rectangle, could and probably did respond to changes in the agricultural activity of the unit. The plan of the farmhouse, like the arrangements in the landscape, reflects a dominant concern with efficient organization of production. What is remarkable is the inventive way the builders seem to have modified and arranged basic units within a framework to meet specific needs.

The materials employed for the construction of Fattoria Stefan were those used for every other farm structure in the territory. On a foundation of field stone, levelled with several courses of fragmentary roof tiles rose a wall of mud brick. This was protected by a roof of well-made Laconian tiles, bearing the stamp of a certain APISTIΩN, while the floors were uniformly of beaten earth, in other words Xenophon's ideal house. It was identical in its choice of materials to the majority of urban dwellings. Foundations and walls were fashioned on the spot from materials close at hand. Even the tiles were probably locally produced. The use of this material not only for roof, but in the foundation and elsewhere in the structure is another indication of the economy and flexibility of the design.

2. A Spring-reservoir Complex at Pizzica

The essential condition for life in the country is an abundant, steady supply of water. The presumed absence of this resource is the theoretical argument most often cited to deny the existence of the isolated farm in the Greek world.

My second example of early Hellenistic rural architecture is a spring-reservoir complex discovered in 1977, by the University of Texas excavation team working in close collaboration with the Soperintendenza alle Antichità della Basilicata, at Pizzica (Carter, 1974, 1977, 1979). It is located at the foot of a low spur the same distance from the city as the Temple of Hera (Tavole Palatine) on the Bradano. Just to the south would have flowed the channel of the Basento, which Professor Adamesteanu identified on aerial photos, as supplying the moat of the city.

The excavation is still in progress. The northern third of the reservoir is as yet only partially excavated owing to flooding. The waters of the spring which today flows where it did in 600 B. C., had, until the introduction of a "well point" pumping system in 1977, effectively blocked the discovery of its ancient course. The main feature of the area and its development, however, are clear (Fig. 3. 2).

The spring mouth, channels and two collecting basins were lined with stones about 600 B. C. This is the date of the earliest votives (and this is about the date of the earliest evidence of Metaponto in the <u>chora</u>).

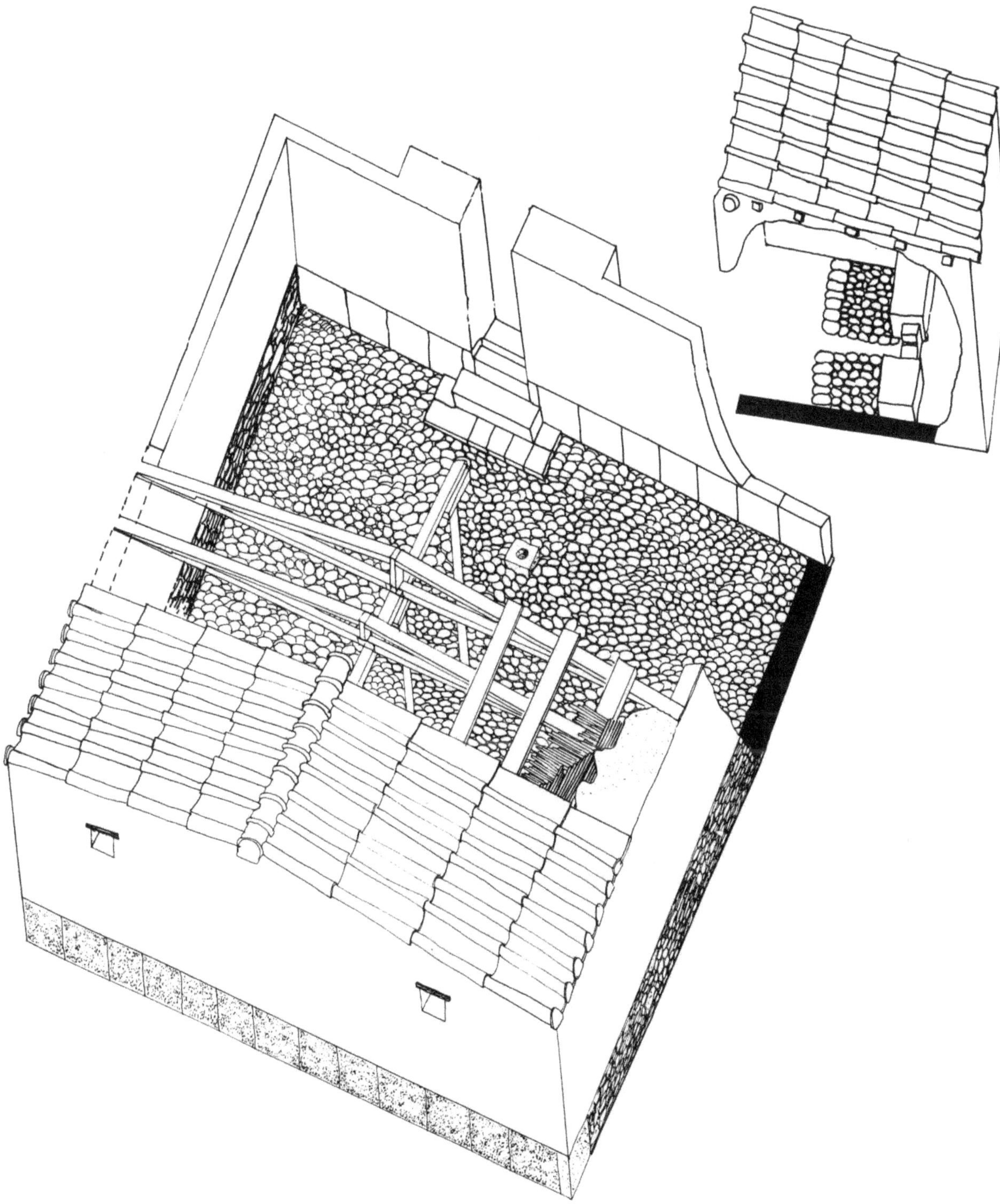

Fig. 3.2 Reconstruction of the spring-reservoir complex at Pizzica
(Michael Guarino)

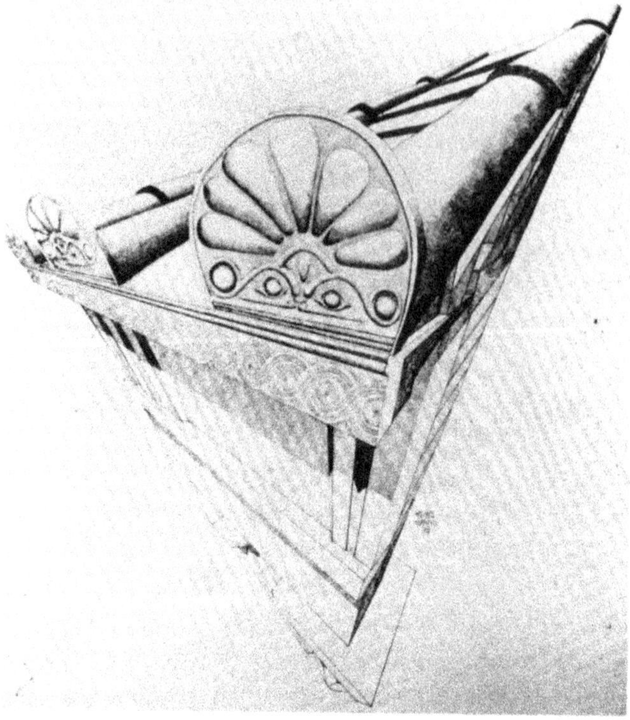

Plate 3. I a. Reconstruction of the mid-sixth century B. C. Sanctuary at Incoronata (Michael Guarino)

b. Roof-tile with stamp APISTIΩN from Fattoria Stefan

At the end of the sixth century a building of unusual plan with a fieldstone foundation course and pavement rose on an extensive fieldstone terrace many times its size. The finds indicate both religious and industrial activity (iron and ceramic manufacture) around this building.

The structures of the second half of the fourth century are a spring house and a reservoir. The spring house was built directly on top of the spring of 600 B. C. Votive figurines of the fourth century mingle with examples in the Daedalic style, while the walls of the reservoir were aligned with those of the late sixth-century building.

The reservoir consists of a fieldstone pavement measuring approximately 11 x 14 metres. It is well below the spring and below the contemporary ground level. It was approached from the west by a narrow stairway constructed of a large squared blocks of local stone, and set into a parapet wall of similar material. A corresponding wall was very probably to be found on the east. Only the negatives of three large blocks remain, their dimensions roughly the same as those of the western wall.

The southern extent of the fieldstone pavement was indicated by a line of debris including re-used squared blocks, one with an archaic mason's mark. The pavement itself was covered with numerous fragments of large vessels for collecting water, of very high quality red-figured ware, perhaps from votive offerings, a column base, and roof tiles. It seems certain that the reservoir was at least partially roofed.

Are there parallels for this kind of structure? It is unlike the typical urban fountain house with its parallel settling tanks, draw basin and colonnaded façade. The well-reservoir, Minoe, on Delos is perhaps the closest, both in form and function. Basically a utilitarian structure, it, like the spring at Pizzica, had its religious aspect, as a shrine of the nymphs.

Yet there is every reason to believe that the spring-reservoir was not an isolated phenomenon in the chora of Metaponto, and that similar structures existed outside any Greek city which depended heavily on agriculture. I suspect that this and similar reservoirs may have been a feature of the "bonifica" of the chora.

The experience of the traveller, recounted in an epigram of Leonidas, who stops at a spring to offer a drinking cup to the nymphs, must have been a common one, in the countryside.

In the chora of Metaponto ancient habitations tend to cluster along the valleys of secondary streams feeding the two great rivers, where springs would naturally have been found. The spring at San Biago, associated with the rich and architecturally impressive sanctuary has been called the only all-weather spring in the chora. It was until last year, simply the only excavated one.

Continuity in the chora in Roman times—The Tile Factory at Pizzica

The spring at Pizzica continued as a focus of rural life into Roman times and even later. The revival of Metapontum as a military camp in the second century B. C. was accompanied by a limited revival in the countryside. In

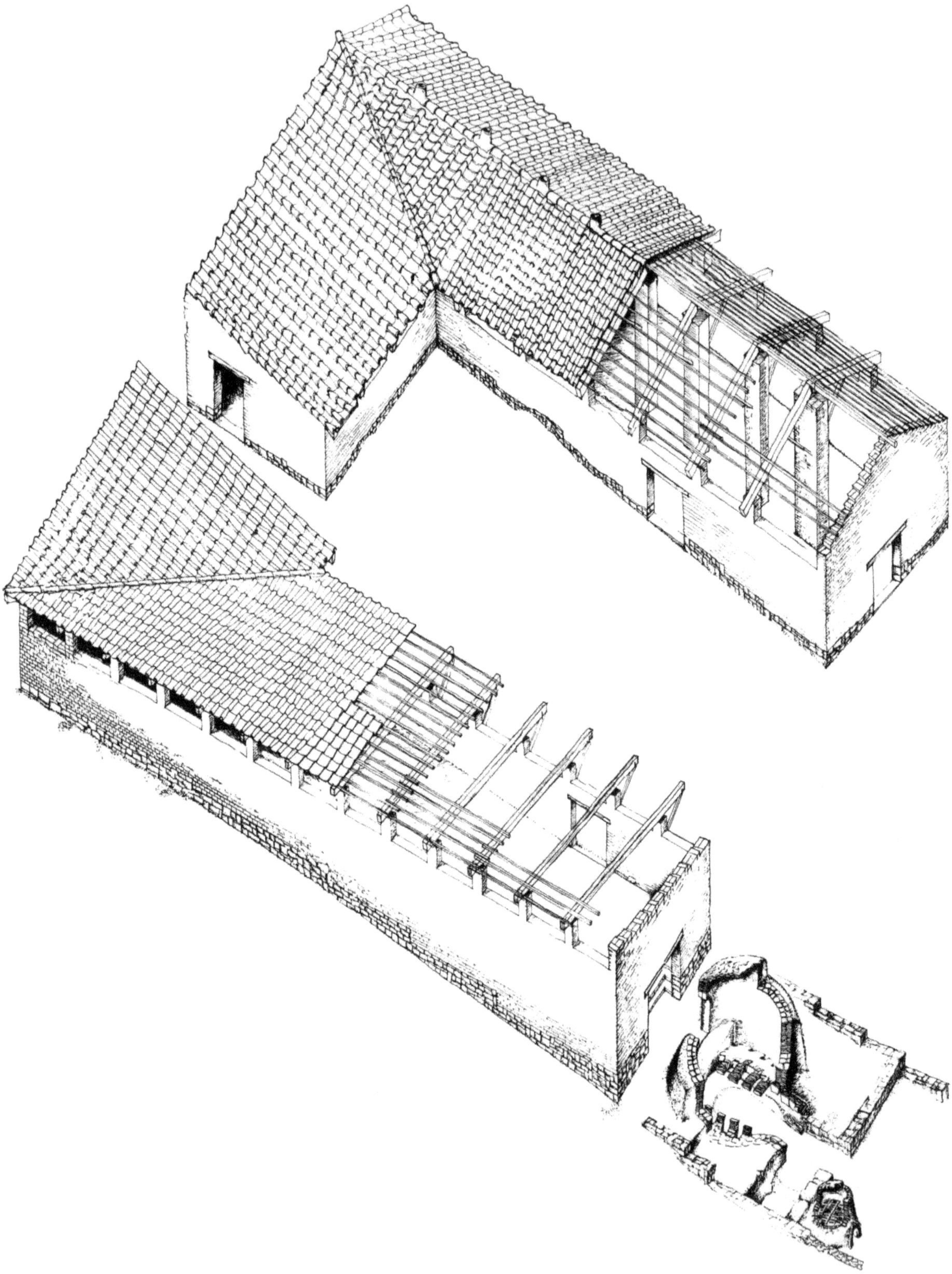

Fig. 3.3 Reconstruction of the tile factory (Bruno Perini)

contrast to the numerous scattered farm units of the fourth century B. C. , the settlements of the late Republican period were fewer and on a vaster scale, reflecting the economic changes which accompanied the Roman domination of the South.

The best preserved establishment of this period yet to have been found at Metaponto was excavated at Pizzica between 1974 and 1976 by the University of Texas team (Carter, 1977, forthcoming b). It is a tile factory with attached factory building, clay quarry, a rubbish deposit, and an abundant supply of water. The sacred spring, which had gone underground by this time, was tapped by a shallow well, with a very neat tile liner, especially made for the purpose (Fig. 3.3).

The factory building was discovered first. It is an extensive structure, consisting of three wings around a courtyard which is open to the south, and entered by an off-centre passage through the north wing. The wings had no partition walls, in fact, no features at all except for the row of pilaster bases in the eastern wing. Partially overlapping the western wing was a small rectangular building of later date, and to the east the negative of the foundations of an earlier building whose stones were re-used in the factory.

Each of the wings measures 33 metres in length. This is slightly more than 100 Doric feet (of 32.68 cm, a plethron) or slightly more than 120 Campanian feet (of 27.30 cm, or an actus). The Ionian-Roman foot of 29.40 cm employed for Fattoria Stefan would have seemed a more natural choice in this area of Italy.

The date of the building was determined by cutting soundings to bedrock across the entire site. The stratigraphical situation is complex. The site was apparently levelled and filled at several different times. The lowest stratum contained material of the fourth century B. C. This is certainly related to the fourth century phase of the spring-reservoir complex on the slope just below. Fortunately, a thick fall of tiles associated with the destruction of the factory sealed some areas of the site. Sections of charred beams and iron nails had been preserved just below the tiles.

The latest material from below the fall dated to the mid-first century B. C. It consists of some fine unguentaria and large quantities of a local Metapontine grey-ware, similar in its forms to the widely diffused Campana C.

An occupation of the site in the early first century A. D. is also well-documented by finds of good quality terra sigillata and coins. The superimposed rectangular building is in all likelihood of this period. Unfortunately, the plough had thoroughly destroyed the stratigraphic evidence at this point.

The factory building has been provisionally restored on paper. A mud-brick wall, some 50 cm thick rose over the stone foundations. These were thicker near the southern end of the building. The large re-used blocks of limestone, from the building to the east, were found in the south end of the eastern wing of the factory, but especially around the kilns in the western wings. Here the walls would have been thicker and higher so that it would not have been necessary for the roof to slope very much with the slope of the hill from north to south. The rooms which run the whole length of a wing would, of course, have been much higher on the south. There was no indication,

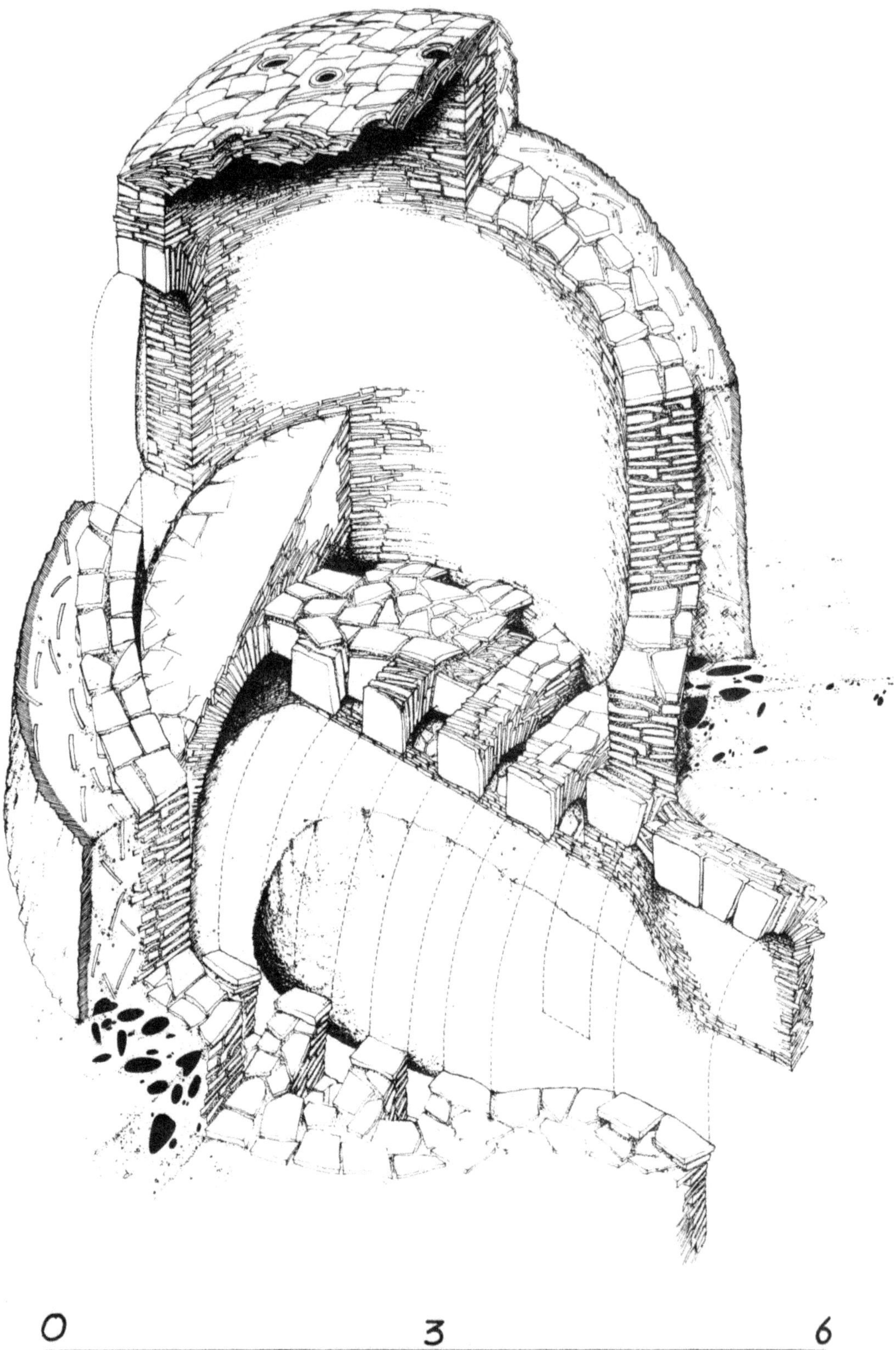

Fig. 3.4 Reconstruction of the tile kiln (Bruno Perini)

of a second storey, though in the east wing the pilasters could have helped to support one.

The orientation and dimensions of tiles in the fall has suggested the roof design, seen in the reconstruction; a single slope for the west wing, and north wing, and a pitched roof for the east. The Laconian system was employed throughout.

The lack of interior divisions or of objects relating to domestic life, in sharp contrast to the situation at Fattoria Stefan leads to the conclusion that this was a utilitarian structure for the production of tiles, being a warehouse or workrooms. If it was lived in, too, then the habitation must have been very austere. Its distance from the city, about three kilometers, would have made it possible for the work force to commute daily. Of course, it is always possible that the living quarters were nearby and have either been thoroughly destroyed or not as yet discovered.

The materials employed in the factory, stone foundations, mud-brick walls, and a roof over a wooden frame, are those of Fattoria Stefan.

The construction of the working wings of the old masserie (or plantations) in this part of Italy is not far different. Tufa is sometimes employed for the walls. Further modern analogies from Hampe's handsomely illustrated work (Hampe, and Winter, 1965) on the ceramic industry of southern Italy, Sicily and Greece can suggest the uses to which a large covered space near the kiln would have been put. It could have both protected the moulded tiles or bricks before firing, and served as a warehouse for the finished product.

2. The Kilns

A doorway in the centre of the south end of the factory building opened into the area of the kiln, or rather of the kilns. Factory building and kilns were clearly planned as a unit (Fig. 3.4).

The interior of the kiln at ground level (which corresponds to the upper part of the combustion chamber) was nearly four metres. A crude kiln in a poor state of preservation was found built into the northern end of the combustion chamber on top of several metres of fill. The fill included a stamped tile with the name HPAKAS. A similar tile has been found inland near Matera. This later kiln may have been associated with the rectangular building which succeeded the tile factory on the site.

Because the combustion chamber of the big tile kiln was cut deeply into the slope of the hill, it was in a remarkably good state of preservation. The building material, broken tiles joined with a very hard mortar also contributed to the good condition. Kilns in this area of Italy are often less carefully made (e.g. the kilns at Apani near Brindisi). The ash pit was undamaged, and the praefurnium (or stoke hole) walls plus an extension on the east side were intact, though the vault, unfortunately had been destroyed.

The arches which supported the perforated floor of the firing chamber numbered five. Enough of their springing exists to estimate their original height. The sections also illustrate the two phases of the kiln. In a second phase the floor of the combustion chamber was raised to create a high-walled ash pit. A former tile maker from the nearby village of Bernalda believes

Plate 3. II a. Excavation of the reservoir at Pizzica, late fourth century B. C.

b. General view of the tile factory at Pizzica 150-50 B. C.

Plate 3. III a. Detail of the fallen roof of the tile factory with Laconian pan and cover tiles in evidence

b. View of kilns from the south

Plate 3. IV a. Stamped tile with name HPAKAΣ

Plate 3. IV b. Detail of the tile kiln showing the earlier phase

Plate 3. IV c. Phallic relief from the entrance to the praefurnium that this modification would have decreased the efficiency of the kiln.

The reconstruction takes into account all existing remains of the kiln, but of course, the actual pavement of the perforated floor, as well as the whole structure of the firing chamber may, in reality, have been somewhat different from the solution proposed here. It is quite possible that the firing chamber was a temporary structure, taken down and built up again after each firing.

The solution adopted for the reconstruction of the perforated floor was the one actually employed in the smaller kiln found, later on, immediately to the south of the tile kiln. Here the arches were intact and a grill-like floor for the firing chamber was created by placing tiles between and perpendicular to the arches. Mortar was not used here.

The small kiln is an integral part of the whole complex. It was clearly for pottery and not for tiles, and it is possible that the distinctive local grey ware, found all over the site, may have been produced here.

An enormous stratified deposit of this material was excavated about 50 metres to the south-east of the kilns. Among the fragmentary vessels were a number of obvious "wasters". The deposit lay immediately over an ancient clay quarry. The earliest material in the deposit antedates the construction of the factory. The implication is that the site was used for ceramic manufacture even before the erection of the factory. It is worth noting that the rural sanctuary by the spring at San Biagio had its own kilns as early as the sixth century B. C.

Few ancient kilns have the firing chamber intact. The best preserved tile kiln at Locri (Cuomo, 1972, 1974), well-published by Cuomo, apparently did not have a permanent cupola to contain the material to be fired (page 83).

Several examples of relatively modern kilns provide an analogy for the reconstruction with permanent firing chamber. One from Buonabiticolo in Calabria is illustrated by Hampe. The other is a recently abandoned kiln at Bernalda. The structure of the combustion chamber and firing chamber is a single cylinder of bricks divided by a perforated floor, which differs from the Pizzica kiln's floor in the fact that no separate arches support it. It was constructed as a dome, much like a modern example from nearby Pisticci, photographed by Hampe.

Hampe's book, sad to say, documents a dying art. Mass-produced sheet-roofing, which is more economical, is replacing terracotta tiles in the dwellings being constructed today in this region of Italy. The aesthetic impoverishment is all too obvious.

The sturdy construction and careful modification of the kiln at Pizzica would indicate that it was used for a long period of time. The ceramic evidence from the factory suggests a span of about a century from 150 B. C. to 50 B. C. Kiln and factory are precious evidence not only of the ceramic industry of the late Republican period in Italy but also for a period in the history of Magna Graecia and its rural population about which all too little is known.

At the close of the 1976 campaign, steps were taken immediately to protect these unusually well-preserved kilns, and it is hoped that they will become part of a large archaeological park devoted to the rural population of the Metapontine chora. The architect's original project called for a tile roof, but considerations of economy prevailed.

The description of the tile kiln is not quite complete. Its original builders had also taken steps to protect it. Vase paintings document the use of phallic herms beside the kiln (Noble 1965), to ward off the daimons who were supposed to wreck the kiln and its contents in various ways, but there was no archaeological confirmation of this practice, previously (Plate 3. IVc).

The final discovery made after the sheet-roof was in place and a torrential rain had deluged the area has something of the miraculous about it, and makes the tile kiln at Pizzica, as far as I know, unique. The rain washed clean the seemingly ordinary upright slab at the entrance to the praefurnium, revealing the phallic relief with its vivid cloring still intact.

With these observations on the uses of ceramic tiles and their production, in the territory of Metaponto, I hope to have suggested, also, some of the complex problems of rural life in general, in this part of ancient Italy. It is a field, I believe, which more than the city, presents the classical archaeologist with an urgent challenge to discover, before the fragile evidence disappears altogether.

D. Adamesteanu 1973, 'Le suddivisioni di terra nel Metapontino', in Problèmes de la terre en Grèce ancienne (Paris 1973).

D. Adamesteanu 1972, Basilicata Antica (Cava dei Tirreni 1974).

D. Adamesteanu, C. Vatin 1976, 'L'arrière-pays di Metaponte', CRAI 1976 110-125.

J. C. Carter 1974, 'Saggi di Scavo a Pizzica (Metaponto)', in Atti del XIV Convegno di studi sulla Magna Grecia (Naples 1975) 277-283.

J. C. Carter 1977, Ancient Crossroads: The Rural Population of Classical Italy (Austin, Texas 1977).

J. C. Carter 1979, 'Architettura rurale nella chora di Metaponto nel primo periodo ellenistico', Magna Grecia, 14, 1-2, January-February 1979, 9-11.

J. C. Carter (forthcoming a), 'Burying Ground in the Territory of Metaponto', Studi in onore di Dinu Adamesteanu (1979).

J. C. Carter (forthcoming b), Pizzica (Metaponto), Notizie degli Scavi.

N. Cuomo di Caprio 1972, 'Proposta di classificazione delle fornaci per ceramica e laterizi nell'area italiana', Sibrium 11 (1971-1972) 371-464.

N. Cuomo di Caprio 1974, 'Fornaci per ceramica a Locri', Klearchos 61-62 (1974) 43ff.

R. Hampe and A. Winter 1965, Bei Töpfern und Zieglern in Süditalien Sizilien un und Griechenland (Mainz 1965).

S. Humphreys 1967, 'Archaeology and the economic and social history of Classical Greece', Parola del Passato 116 (1967) 374-400.

J. E. Jones, L. H. Sackett, A. J. Graham 1962, 'The Dema House in Attica' ABSA 57 (1962) 75-114.

J. E. Jones, L. H. Sackett, A. J. Graham 1973, 'An Attic Country House below the Cave of Pan at Vari', ABSA 68 (1973) 355-452.

J. V. Noble 1965, The Techniques of Painted Attic Pottery (New York 1965).

P. Orlandini 1976, 'Scavi archeologici in località Incoronata presso Metaponto', Acme 29, 1 (January-April 1976) 29-37.

J. Pečírka 1973, 'Homestead Farms in Classical and Hellenistic Hellas', in Problèmes de la terre en Grèce ancienne (Paris 1973) 113-147.

G, Schmiedt, R. Chevallier 1959, Metaponto e Caulonia (Florence 1959).

G. Uggeri 1969, 'ΚΛΗΡΟΙ arcaici e bonifica classica nella ΧΩΡΑ di Metaponto', Parola del Passato 118 (1969), 51-71.

G. Vallet 1967, 'La cité e son territoire dans les colonies grêques d'occident', in La città e il suo territorio, Atti del VII Convegno di studi sulla Magna Grecia (Naples 1968), 67-158.

IMPORTATION AND MANUFACTURE OF STAMPED BRICKS AND TILES IN THE ROMAN PROVINCE OF DALMATIA

J. J. Wilkes

1. Introduction

The stamped bricks and tiles of Roman Dalmatia are among the most interesting and varied to be found in any province of the Empire. Here I can offer no more than a superficial and patently incomplete review of the evidence. Much of what is said in this paper is based on the work of others, and this is especially true for the tentative conclusions. So far the study of the material has been confined to those items which bear identifiable stamps. Little or nothing has been done to describe the many varieties of form, design and function, of the bricks and tiles which abound in the debris of ancient settlements along the entire coastline of Dalmatia.

2. Military factories

The stamps of four legions and one auxiliary cohort are known in Dalmatia. All except one or two strays can be dated securely to the middle and later first century A. D. When the province of Dalmatia was first organized after the Great Rebellion of A. D. 6-9 two of the five legions that made up the army of Illyricum were based in the province: leg. VII appears to have been established at Tilurium (Gardun) for some years before that date but leg. XI came to occupy the base at Burnum following the transfer of leg. XX to Germany in the aftermath of the Varus disaster in A. D. 9. After the failure of the rebellion by the Dalmation governor L. Arruntius Camillus Scribonianus in 41 both legions were awarded the titles Claudia pia fidelis by a grateful Claudius, and these were retained by both legions throughout their subsequent history. Since all the known stamps of legg. VII and XI bear these honorofic titles (C P F) it may be inferred that the practice of stamping bricks and tiles, though not necessarily their manufacture, by the legions did not begin in Dalmatia before the reign of Claudius.

The only known stamps of the seventh legion in Dalmatia (LEG VII C P F) were recorded by the writer in the archaeological collection of the Franciscans at Sinj (at least six separate stamps, complete and fragmentary, on tegulae). Enquiries revealed that some of the stamps came from the nearby Claudian colony at Aequum (Čitluk) but it cannot be ruled out that they came from the legion's base at Tilurium, not much farther away. The discovery of these stamps may have some bearing upon the date when the seventh left Dalmatia for Moesia. The later date that had been suggested, the winter of 56/7 when leg. IV Scythia left Moesia for the east, seems now to be more likely than a departure around 45 (which most seem to have favoured), which would leave only the few years between the conferring of the Claudian titles in 41 or 42 and a departure in 45.

All the known stamps of the eleventh legion bear the Claudian titles (LEG XI C P F) and their manufacture can be dated to the reigns of Claudius and Nero (from 41 until the legion left Dalmatia in the Autumn of 69). They are found in quantity at the legionary base Burnum (near Kistanje) on the river Titius (Krka) in southern Dalmatia, with stray finds from near the base of the seventh legion at Tilurium (Dolac) and from a Christian basilica in the provincial capital Salona.

Possibly a short interval elapsed between the departure of the eleventh legion and the arrival of the newly-constituted leg. IIII Flavia firma at Burnum in A. D. 70, which stayed there until c. A. D. 86 when it was moved up to the Danube to strengthen the army of Moesia against the Dacians. The larger number of stamps recorded for this legion (LEG IIII F F) in Dalmatia from a sojourn of around fifteen years suggests that the rate of brick-and-tile production was stepped up under the Flavians. The stamps are more widely dispersed than those of the eleventh. Apart from many examples recorded from the base at Burnum, the finds from Smrdelji a few miles lower down the river serve to identify the location of the legionary tile-factory. Elsewhere stamps of the fourth legion are recorded from the Claudian municipium Asseria (Podgradje near Benkovac), from Promona which was for some time in the first century an auxiliary base, and from Andetrium (Muć), then the base of the cohors VIII Voluntariorum civium Romanorum. There is a stray find from the provincial capital Salona. All these have been recognized as products of the Smrdelji factory: an exception may be the single stamp found at Vitina near the auxiliary fort Bigeste, some distance down the coast. Comparison of the stamp and fabric with other products of the legion suggests that it does not come from the Smrdelji factory.

At some period tiles produced at the Smrdelji factory were stamped by the eighth legion Augusta (LEG VIII AVG). Elsewhere there is a single example from the base at Burnum, many from the Claudian municipium Asseria (which is close to Smrdelji), and a stray (from the Smrdelji factory) from near the auxiliary fort Bigeste in the south (Gradčine near Ljubuški). The eighth legion formed part of the army in Illyricum under Augustus but this seems on general grounds too early for manufacture and stamping of tile. Moreover suggestions that all or part of the legion was temporarily removed in the second century from its base at Argentorate (Strasbourg) in Upper Germany (either under Antoninus Pius or during the Marcomannic wars under Marcus Aurelius) may be discounted. The two epitaphs of its serving soldiers from Dalmatia (from Burnum and Alvona in western Istria) appear to date to the first century A. D. It also seems probable that its production of stamped tiles at Smrdelji took place around the time when those fo the fourth Flavian legion (which they closely resemble) were produced—that is the years 70-c. 86. Early in A. D. 69 the eighth legion marched from its base on the lower Danube in Moesia to Aquileia in north-east Italy, then in the Autumn moved into the Po valley to fight for the Flavians at Cremona. If the epitaphs belong to this period, it seems likely that a vexillation may have been left at Burnum. This had been temporarily vacated by the eleventh legion early in 69 when it marched into Italy and fought for Otho at the battle of Bedriacum, after which it was ordered back to Dalmatia by the victorious Vitellius. Equally a detachment of the eighth legion could have occupied Burnum in the likely interval between the final departure of the eleventh legion at the end of 69 and the arrival of the fourth Flavian sometime in the following year.

In the first century A.D. the presence of many auxiliary units in Dalmatia is recorded by epitaphs of serving soldiers. Only one of these is known to have stamped tiles, the cohors VIII Voluntariorum civium Romanorum (COH VIII VOL). Its products are known at Andetrium (Muć) and at the old base of the seventh legion at Tilurium; and there is a stray from a Christian Basilica in the provincial capital Salona, which is close to both places. Those from Andetrium may be dated to the second half of the first century A.D. when the cohort was in garrison there, after which it moved to Tilurium where it remained throughout the second and third centuries.

It emerges that the manufacture and stamping of tiles by the Roman army in Dalmatia was carried on from around the middle of the first century A.D., by a succession of legions using the tile factory at Smrdelji not far from the fortress at Burnum. Around the same time one auxiliary cohort produced and stamped tiles somewhere in the vicinity of its base at Andetrium. After the departure of the fourth legion around A.D. 86 the garrison of Dalmatia comprised three auxiliary cohorts (I Belgarum, III Alpinorum and VIII Voluntariorum). They may have made tiles and bricks for their own uses but did not stamp them, a practice which may have been imitated by the cohort VIII Voluntariorum only when legions based in the province were doing this.

Military tiles appear to have been used almost entirely for military purposes. In addition to the place of manufacture at Smrdelji, they have been found in the legionary bases Burnum and Tilurium, and in the auxiliary bases at Promona, Andetrium and Bigeste. The Smrdelji products of the eighth and fourth legions found in the nearby Claudian municipium Asseria may represent a special provision for a substantial public-building programme underway in the city at the time. Similarly the products of the eleventh legion and the eighth cohort, some re-used in a fourth-century Christian basilica, are likely to have been obtained from the ruins of nearby public buildings of Salona, for which they had been made available by special provision in the first century A.D. (There is epigraphic evidence for legionary participation in building projects in the larger coloniae of southern Dalmatia during the first century A.D.) It may be affirmed that in the first century military tiles were not made available for private building.

3. Imperial factories

During the first century A.D. large numbers of public and private buildings were constructed partly with bricks, and roofed with *tegulae* and *imbrices*. Many of these were obtained new from the factories, not only in the province but across the Adriatic in north-east Italy. The products of a small number of factories, both imperially and privately owned, were imported in much greater quantities than the rest, from the evidence of their distribution in numerous places along the Dalmatian coast and in the hinterland south of the watershed.

Among imperial factories the most widespread are those of the Pansiana. Once the property of C. Vibius Pansa it appears to have commenced production under Augustus, from somewhere on the coast of Aemilia between Rimini and Pescara. By the reign of Tiberius it had passed into imperial ownership, and from then until the reign of Vespasian its products were stamped with the abbreviated name of the ruling emperor in addition to the marque Pansiana. The products are widespread around all the north coasts of the Adriatic and extend down the Dalmatian coast as far as the Gulf of Kotor in Montenegro. None is recorded from a military site. Many years ago it was observed that the moulds for some of the letters in early Pansiana stamps appear identical to those of two other large factories (Solonas and Cinniana), which were working at the same time and whose products have a similar distribution to Pansiana, albeit in smaller quantities. Use of the same wood or metal letter moulds may indicate that these factories were also in imperial ownership. In the case of Solonas there is also clear evidence for an association with Pansiana, in the form of a few stamps which appear to combine the names of both (PANSIANASO(L)O(NAS)), even signifying an amalgamation.

Like Pansiana, both Solonas and Cinniana began production under private ownership. The former was at one time the joint property of C. Camarius Priscus and L. Plautius (C CAMARI PRIS L PLAVTI SOL), then after a period of sole ownership by the latter it was acquired by M. Ful(cinius) Clemens (M FVL CLEMENTIS SOLONAS DE (figlinis) PLAVTI). The private ownership of Cinniana, by an individual with the cognomen Cinna, is indicated by the stamp CINNAE. Then followed ownership by C. Iulius Pri.... (CINNIANA IVL PRI, CINNIANA C IVL P) after which it probably became imperial property (CINNIANA). Both factories lay probably not far from Pansiana, somewhere on the coast of Aemilia. (The name Solonates is listed by the Elder Pliny among the civitates of the VIIIth region, from which the name of the Solonas factory probably derives.)

A few examples of Cinniana occur in Istria and down the Dalmatian coast, but they are not found in the Neretva valley around Narona nor away from the coast in the hinterland. Solonas is more widespread (common on the coast of Aemilia, especially around Ravenna, Ariminum and Pisaurum), with many examples in Dalmatia at Salona, a few in the Neretva valley around Narona (Tasovčići and Stolac) and further inland at Delminium (Duvno). Along the Liburnian coast of northern Dalmatia the stamp is found at Iader, Asseria, Sidrona and Argyruntum, and was used, along with the products of several Italian factories, in the construction of the Jupiter temple at Vegium. Away from the coast in the north there is a stray at Monetium among the Iapodes beyond the Velebit mountains.

It emerges that during the first century A.D. substantial quantities of tiles from imperial factories in Italy were conveyed across the Adriatic and down the Dalmatian coast to be distributed among many coastal settlements, small and large. A tiny proportion may have been conveyed by road for short distances inland but never across the watershed into northern Dalmatia. None is recorded from a military site. Finally it appears than an imperial origin did not necessarily determine their use in special building projects, or for public rather than private building during the first century. In this respect their distribution cannot be easily distinguished from that of the products of the many privately owned Italian factories which found a market in Dalmatia.

3. Private factories

Among the hundred or more stamps on bricks and tiles of ostensibly private origin recorded in Dalmatia, more than one third were produced by factories located in north-east Italy, notably in the general area of Aquileia, the major port of the northern Adriatic. The products of two factories are notably more widespread than any others, Q. Clodius Ambrosius and C. Titius Hermeros. From the general style and letter forms both factories appear to have been producing around the same period as Pansiana, that is during the Julio-Claudian period. Their stamped products (Q CLODIVS AMBROSIVS and C TITI HERMEROTIS) have a roughly similar distribution, in Italy down the Adriatic coast as far as Firmum, south of Ancona, and in Dalmatia down the entire coast as far as Montenegro. Both marques are especially common around the Gulf of Trieste, and it seems likely that the factories of both are to be sought in the area of Aquileia. Ambrosius in Dalmatia is found as far south as the Flavian municipium at Doclea in Montenegro. Further north in the Neretva valley they appear to have been conveyed from a harbour near Neum to the colony Narona, and then farther inland to the areas of Ljubuški and Stolac. Those of Hermeros are generally fewer in number but in the Neretva valley around Narona have a similar distribution to Ambrosius tiles. Elsewhere Ambrosius is common at Salona, and also reached the cities of southern Liburnia, including Iader, Argyruntum, Asseria and Clambetae. The number of Hermeros stamps at Salona is smaller, and in Liburnia they are recorded only at Iader and inland at Sidrona. In the north of Liburnia Hermeros were used in the Jupiter temple at Vegium and are found on the island Apsorus off the east coast of Istria.

In some cases it may be suspected that the bulk of production from some of these Italian tile-factories was shipped across the Adriatic and down the Dalmatian coast. For example those stamped by the factory of Q. Granius Priscus, working apparently somewhere near Aquileia, are found elsewhere only across the Adriatic, in Liburnia at Iader, Aenona and Clambetae, on the island Corcyra Nigra (Korčula) and in the territory of Narona in a luxurious residence at Stolac. Among the lesser Italian factories whose products are known mainly in Dalmatia one may note the following: M. Albius Rufus, C. Julius Africanus, Euaristus, L. Minicius Pudens and Faesonius.

From my own observations it seems that most of the stamps of the Italian factories in Dalmatia are on tegulae. Among the few exceptions one may note the stamp of Servilia on a massive hypocaust floor tile weighing 53.4 kg, from Stolac in the Narona territory. However both Stolac and Dretelj, where similar tiles have been found, are not so far removed from water transport to require carriage by wagon for any distance.

Apart from a general presumption, based on the style and lettering of the stamps, that the import of Italian bricks and tiles took place in the first century A.D. there is really very little evidence for when the Italian trade was at its most flourishing or for when it declined. It seems probable that gradually, before the end of the first century A.D., local factories began to capture some of the Dalmatian market. A large proportion of bricks and tiles were produced without stamps, and this appears to be especially the case in the late Roman periods: a number of buildings are known in the province of this time which used much brick in their construction and had tile roofs but from whose ruins not a single stamp has been found (for example the large fortified villa of the early fourth century at Mogorjelo near Narona). Among the few recgonizably late stamps found in the area of Narona (notably at Potoci near Mostar), some are badly executed and consists merely of abbreviations (CA[. . .], IVNIVS BA[. . .]ENA, NE).

It may be interesting to note in passing that the movement of bricks and tiles from Italian factories across the Adriatic to the Dalmatian coast, the Hercegovina and Western Bosnia, is not peculiar to the Roman era. There is a record of it in the fourteenth century and there was a similar pattern of brick imports during the years before the First World War. From the latter time one can still see tiles in southern Bosnia which originate from the 'Società Veneta in Pardenone prov. di Udine Italia' which had been conveyed in small ships from the harbour Portogruaro down the coast to Split, whence they were moved overland to Bosnia. In the Hercegovina Italian tiles came through the port of Metković during the years following the Austrian occupation, gradually replacing the stone roofs of the Turkish era with those of red tiles, recalling the red-roofed aspect of the country that must have prevailed in Roman times, through the products of Pansiana and Ambrosius.

Changes of ownership can be detected through the different forms of stamp used by the same factory, although the order of the changes remains uncertain. The Epidiana factory, which was at or near Aquileia, is recorded with what was its first owner or owners C. and M. Epidius (EPIDIORVM C M) and a certain L. Epidius Theodorus (L EPIDI THEODORI). For some time it was the property of a lady named Valeria Magna (VALERIAE MAGNAE EPIDIAN) and another stamp records the name of a leasee C. Petronius Aper (C PETRONI APRI 7 (conductoris) EPIDIAN). The work in the factory, which also produced amphorae, was done by slaves. An amphora made during the ownership of C. Epidius has the name of a slave Eros stamped on the rim (EROS C EPIDI). In Dalmatia amphorae of this form are not recorded, and of its tiles only those with the name of the leasee C. Petronius Aper are recorded.

As the importation of tiles may have declined around the end of the first century A. D. so more factories may have developed locally to serve the needs of the cities in southern Dalmatia. Some of these may have worked on a very small scale indeed, perhaps even for a particular building project. In northern Liburnia the estate of Sex. Metillius Maximus lay probably on the island Curicta (Krk) or perhaps nearby on the mainland not far from the port of Senia (Senj), where the products of the estate tile-works have been found (DE SALT EX SEX M(E)TILLI MAXI), and in south Liburnia a similar origin may be inferred from the stamps of Tettius at Iader, Aenona and Clambetae (L TETTIVS DESEDES). Neither is recorded elsewhere. The same is also true for a number of other stamps which have a limited distribution in the towns of south Liburnia, Iader, Asseria, Clambetae, Aenona and Argyruntum (TRAGVLAE, MVTTIENI). This part of the province of Dalmatia was wholly urbanized and enfranchised during the Julio-Claudian period, and it is to be expected that the local Liburnian proprietors will have developed their own production of bricks and tiles from their resources, probably in the first century A. D. South of Liburnia along the rest of the Dalmatian coast, as far as the border with Macedonia, a clear distinction persisted between the coastal cities, dominated mainly by families of Italian settler origin, and the native peoples of the hinterland and northern interior late into the first century. Here citizenship and municipal organization begins only under the Flavians. Among factories working in the major coastal cities one may note the stamps M IPPOLITVS and M LVTASIVS. The tiles of L. Maltinius Abascantus found at Salona (L MALTINI ABASCANTI), Iader, in the debris from the large residence in the Narona territory at Proboj-Dretelj near Ljubuški, are not known elsewhere. Because it is a rare gentilicium tombstones of the Maltinii family in the territory of Salona may plausibly be connected with the owners of the tile factory. One of the epitaphs must date not earlier than the reign of Hadrian (it records a P. Aelius). The origins of the family may lie across the Adriatic, from the evidence of two epitaphs at Nursia in the fourth region. Thus we have a local Dalmatian family, of Italian settler origin, which flourished in the first and second centuries A. D. (one of the daughters married an eques), with estates near Salona, from which tiles were produced and stamped, and having some connection with a substantial and well-appointed residence of the first century A. D. in the territory of Narona.

Families established in other Dalmatian coastal cities. possibly also of Italian settler origin, stamped their own building materials. The Lurii appear among the leading families of Salona, Risinium (P. Lurius Moderatus Risinitan(us), a witness on the Salona diploma of A. D. 70) and Acruvium. A tile-factory belonging to a member of the family (P LVRIVS FIRMVS) may have been located somewhere near the two latter cities which lay in the recesses of th the Gulf of Kotor on the coast of Montenegro. In the territory of Narona a similar local landowning family is attested by stamps of the Junii (IVNI).

5. Later stamps

Both the imported and locally produced bricks and tiles of the first century A. D. were of such durable quality that they could be used again and again for building, especially in cities. This is the likely explanation for early military tiles being found in the baptistery of a Christian basilica at Salona (noted above). There can be no doubt that the production of bricks and tiles did continue into the late Roman period but it seems clear that very few of the factories either needed or bothered to stamp their products. The stamp GALAX[IDOROS] from the baptistery of the basilica urbana at Salona may have been made when the church was being built in the fourth century. In the case of the villa built for the retirement of Diocletian in 305, on the coast a few miles south of Salona, most of the tegulae used to roof the Mausoleum bear the stamp DALMATIA, which is otherwise unrecorded, and may indicate the product of a state factory that very possibly was established specially for that project.

As has already been observed, Roman institutions and, for that matter, Roman material culture did not spread beyond the Adriatic-Danube watershed before the Flavian period, when a small number of municipia began to be organized in the Bosnia valleys, followed by others during the second century A. D. In the late second and early third centuries larger settlements with substantial buildings did develop in some areas, notably in the north-east, where there was the silver-mining district based on Domavia in the Drina valley. Imported bricks are altogether unknown in the north and the few stamps that have been recorded in this area belong either to municipal tile-factories or to private owners whose names are abbreviated to the point of unintelligibility.

Among late municipal stamps one of Siscia (SISC) recorded in the iron-working district of north-west Dalmatia (Japra Blagaj in the Sana valley) comes from the city of that name in Pannonia, where a colony had been established by Vespasian. A municipal stamp from within Dalmatia (BISTVE) found in the upper valley of the Vrbas in central Bosnia derives from the Flavian municipium Bistue Vetus, which existed somewhwere in the area. Stamps of private factories in northern Dalmatia include P A (ELIVS) S(. . . . .) from the Antonine city at Pljevlja (which occurs in several examples) and also from the mineral spa at Ilidža near Sarajevo, and finally the stamp SATVR[NINVS] from the mining centre Domavia.

6. Conclusions

A number of conclusions may be drawn from this superficial and incomplete review of stamped bricks and tiles found in Roman Dalmatia. In the reigns of Claudius, Nero and the Flavians, the Roman army, both legions and auxilia, manufactured and stamped their own tiles. They were used exclusively for military purposes except for two instances of urban public-building, in the provincial capital Salona and in the Claudian municipium Asseria.

From the reign of Augustus or at least that of Tiberius until Vespasian and possibly later, large quantities of stamped bricks and tiles produced in imperial factories in north-east Italy on the coast of Aemilia, and in the vicinity of Aquileia, were shipped to Dalmatia for use in many coastal cities and settlements of the immediate hinterland. A similar traffic occurred with the products of private factories working in the same area of Italy, especially around Aquileia. How long the traffic in the latter was carried on is uncertain but it seems that the bulk of the imports reached Dalmatia during the first century. When local factories began to develop, their products had a more restricted distribution, with only a few instances of the same local stamp reaching a plurality of cities.

Neither imported products nor the local ones which replaced the former along the coast and immediate hinterland passed across the watershed into northern Dalmatia. Bricks were certainly used widely but from the evidence of the stamps their production did not commence before the beginning of the second century, with some municipal factories and a few private establishments producing for strictly local needs. Long distance transport of bricks and tiles was, and remains, very difficult in Bosnia, except along the lower stretches of the larger rivers.

Editorial Note

Professor Wilkes was unable to attend the conference and the above paper was read by Mark Hassall. For references to many of the sites mentioned see:

Wilkes, J. J., Dalmatia, London, 1969.

POTTERY- AND TILE-KILNS IN SOUTH ITALY AND SICILY

Ninina Cuomo di Caprio

This paper concerns updraught kilns excavated in South Italy and Sicily (Fig. 5.1), both being areas rich in archaeological discoveries, where modern industrial development has not been so intense as in North Italy, thus allowing many sites to remain undisturbed. Most of the kilns to be considered were excavated in the last few decades. In the past, little attention has been paid to kilns; in old publications archaeologists gave scant information on kiln excavations and no technical data at all, thus preventing us today from making a general survey.

For each kiln discussed later the dimensions are stated whenever possible, as well as the class to which each kiln may be assigned. In this attempt to classify kilns, the method proposed is based essentially on structural features. Although restrictive, it offers a series of self-reliable data which are not conditioned by any archaeological and historical controversy, just as a chemical analysis points out properties of a specimen, whatever its origin may be. It is an 'open' classification which may be easily enlarged and adjusted when excavations of new types of kilns make it advisable, and its practical value should not be undervalued as it enables a certain amount of information to be amalgamated under a single figure (Ia, Ib, etc.).

So far, such a classification does not make it possible to outline a chronology, as kilns found in Italy do not seem to develop in any chronological sequence. Some kiln types recur constantly through the centuries, from Roman to modern times. Wherever dating of kilns is possible, it is based on archaeological evidence provided mainly by the pottery found in or near the kiln.

As indicated above, the classification is primarily based on the structural features revealed in kiln excavations, in particular on the form of the substructure built in the combustion chamber (or ovenpit or furnace chamber) to support the raised oven floor. Draught, temperature, efficient passage of heat to the oven and loading capacity could have been directly or indirectly influenced by the size and form of the combustion chamber and the structures built within it. The differing structures built in the combustion chamber and used to support the raised oven floor may be deemed the most significant features and the only ones which can be used as a basis for a kiln classification.

Kilns fall into two major categories according to their geometrical form:

I — round/oval shape
II — square/rectangular shape

Each category is then divided into four classes according to its sub-floor structure (Fig. 5.2).

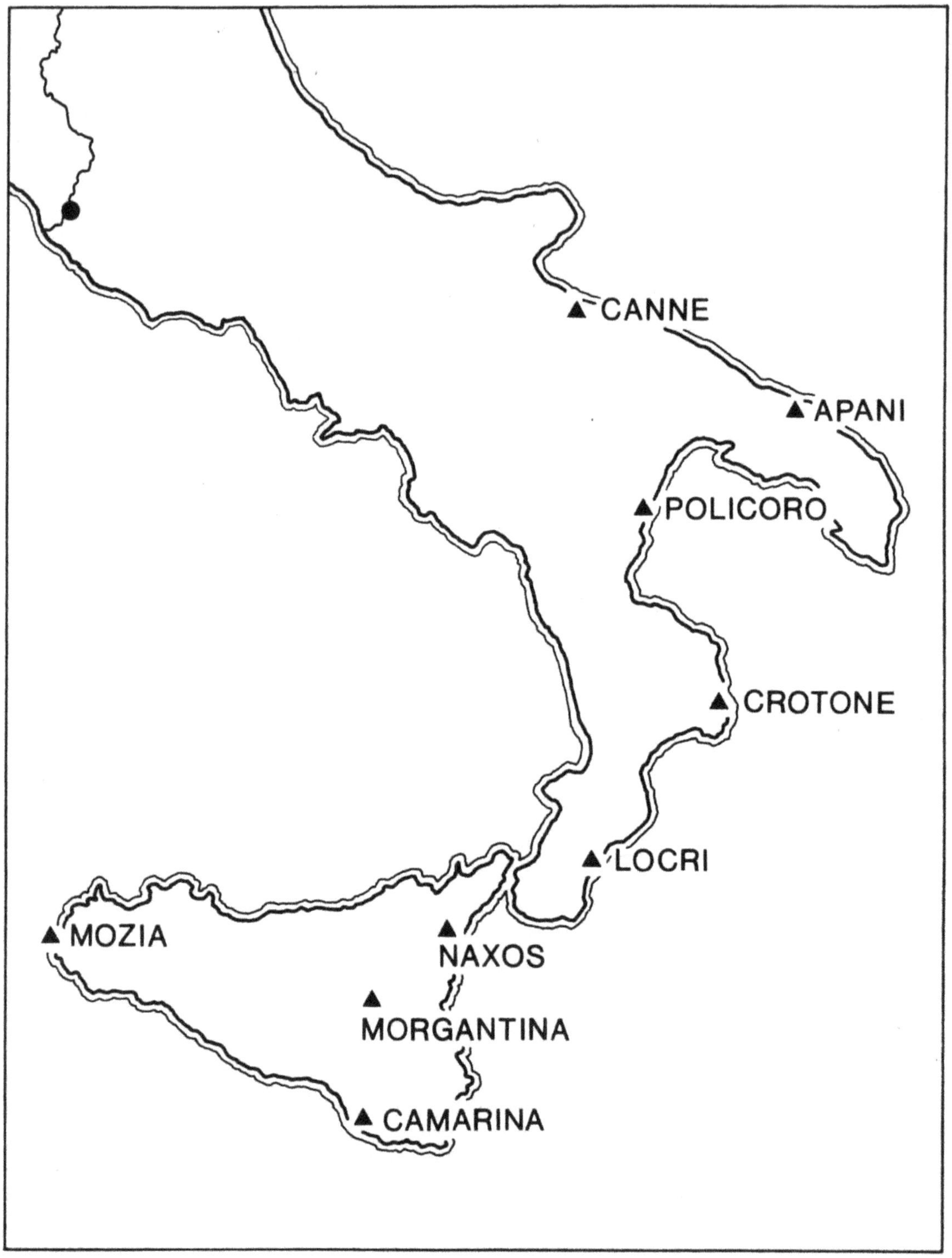

Fig. 5.1 Map of Southern Italy and Sicily

Class Ia

The raised oven floor was supported by a central pedestal (sometimes two or even three small pedestals, closely grouped). Whatever its shape and size, the free-standing support was built in the centre of the combustion chamber to guarantee a strong support to the weakest part of the raised oven floor.

Class Ib

The raised oven floor was supported by radial pilasters, i. e. by thickenings of the furnace chamber wall. In some kilns the strength provided by radial pilasters was increased by a rough tongue which extended from the back wall of the combustion chamber, its axis being aligned with the praefurnium (the flue leading from the stokepit to the combustion chamber).

In some kilns the tongue was the only substructure to support the raised oven floor, acting as its backbone.

Class Ic

The raised oven floor was supported by arches springing from the wall of the combustion chamber. Such arches did not take up much room, thus increasing the space available for bulky fuel.

Class Id

The raised oven floor was supported by a system of cross walls. The subfloor structure consisted of a central corridor (or main under-floor flue) with a number of closely-spaced cross walls carried over by arches. The space between the cross walls (under-floor side flues or cross flues) had flat or sloping floors, and was on the same level or at a higher level than the main under-floor flue. Such a system of under-floor cross walls and cross flues allowed for a heavy loading capacity and an efficient passage of heat to the oven, whatever the size of the kiln.

Class IIa

This corresponds to class Ib of the round kilns. A central wall (or tongue) ran from the back wall of the combustion chamber in line with the praefurnium.

This type does not appear to be common, and so far it is not recorded in South Italy.

Class IIb

This corresponds to class Id of the round kilns. The central corridor (or main under-floor flue), by running the whole length of the combustion chamber, divided it into three compartments, the two outer ones being formed by the under-floor cross flues and cross walls. It thus overcame the difficulty of spanning the width of a large furnace chamber by a series of parallel cross walls.

Class IIc

The sub-floor structure is similar to class IIb in the system of cross walls and cross flues, the difference being two parallel main corridors (or main under-floor flues) instead of one. This double parallel flue kiln had a central wall built in the middle of the combustion chamber in a line with the

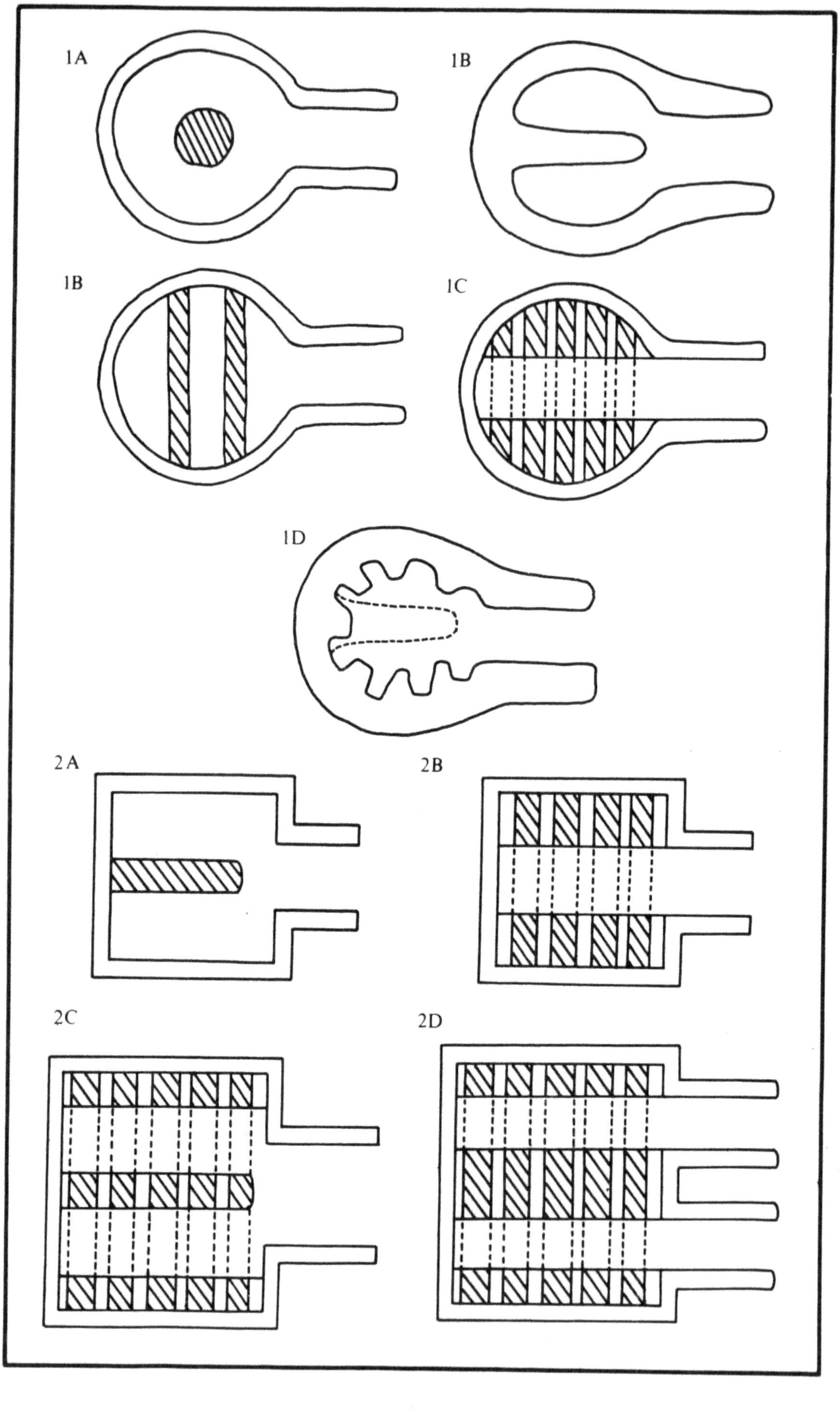

Fig. 5.2 Classification of kilns

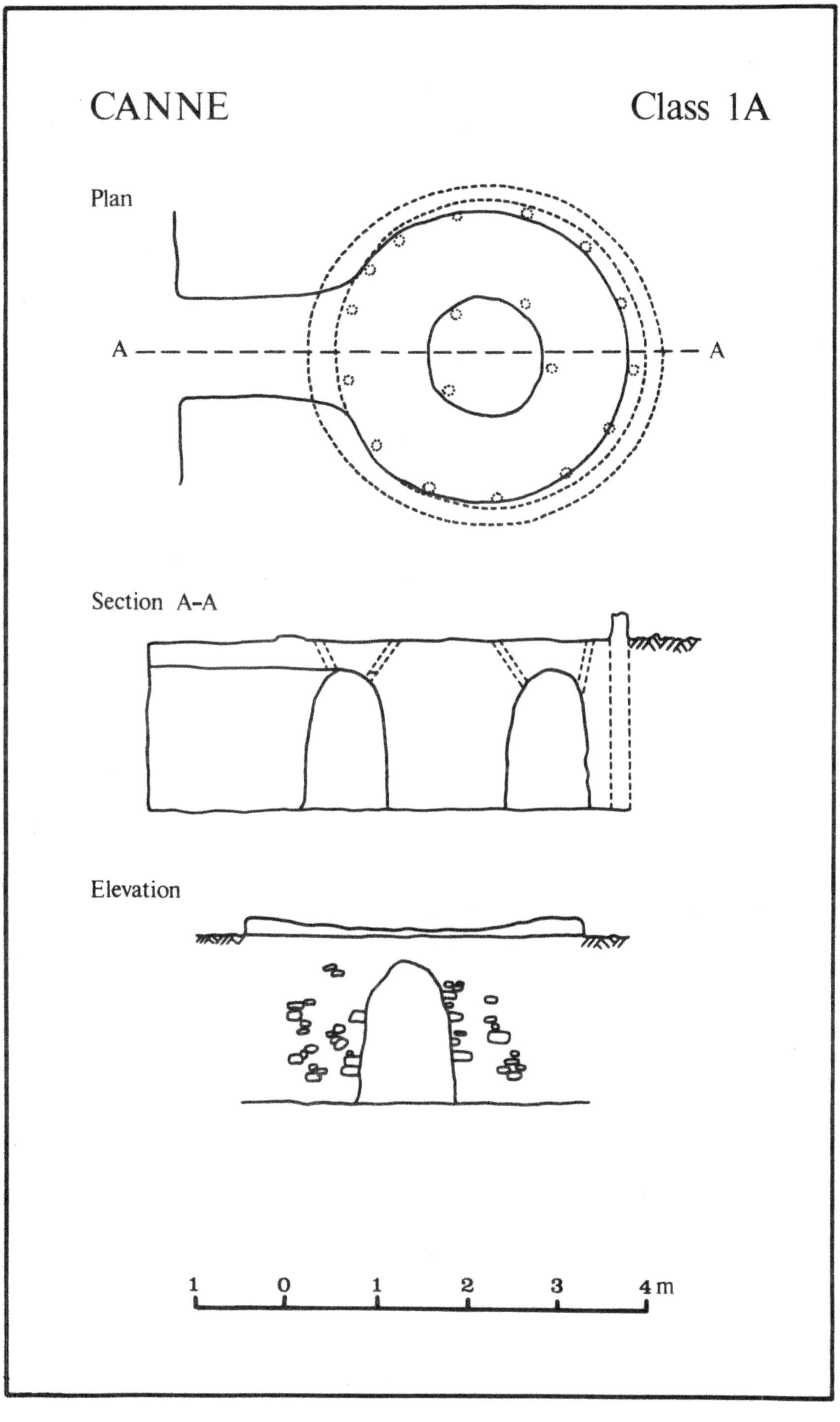

Fig. 5.3

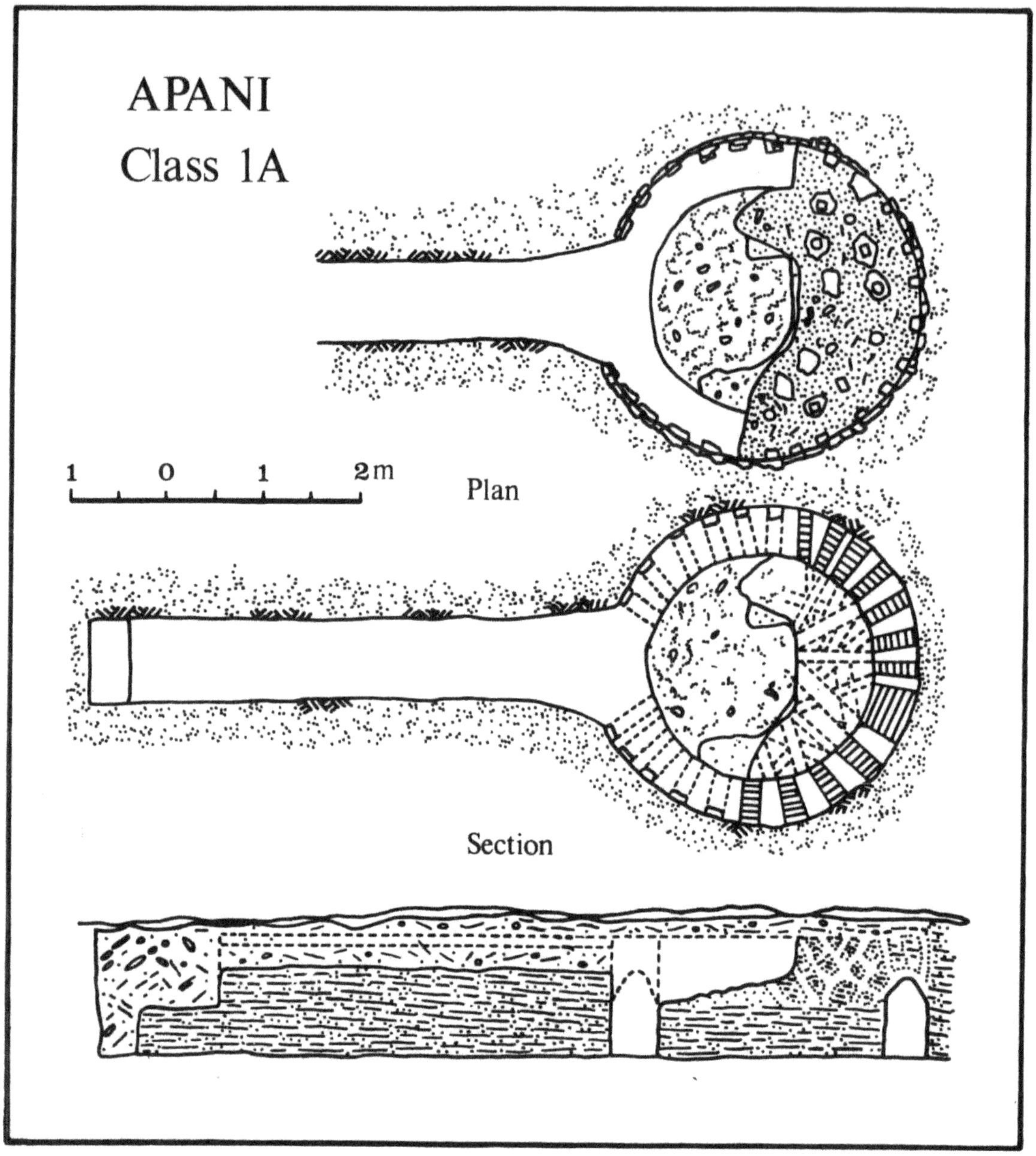

Fig. 5.4

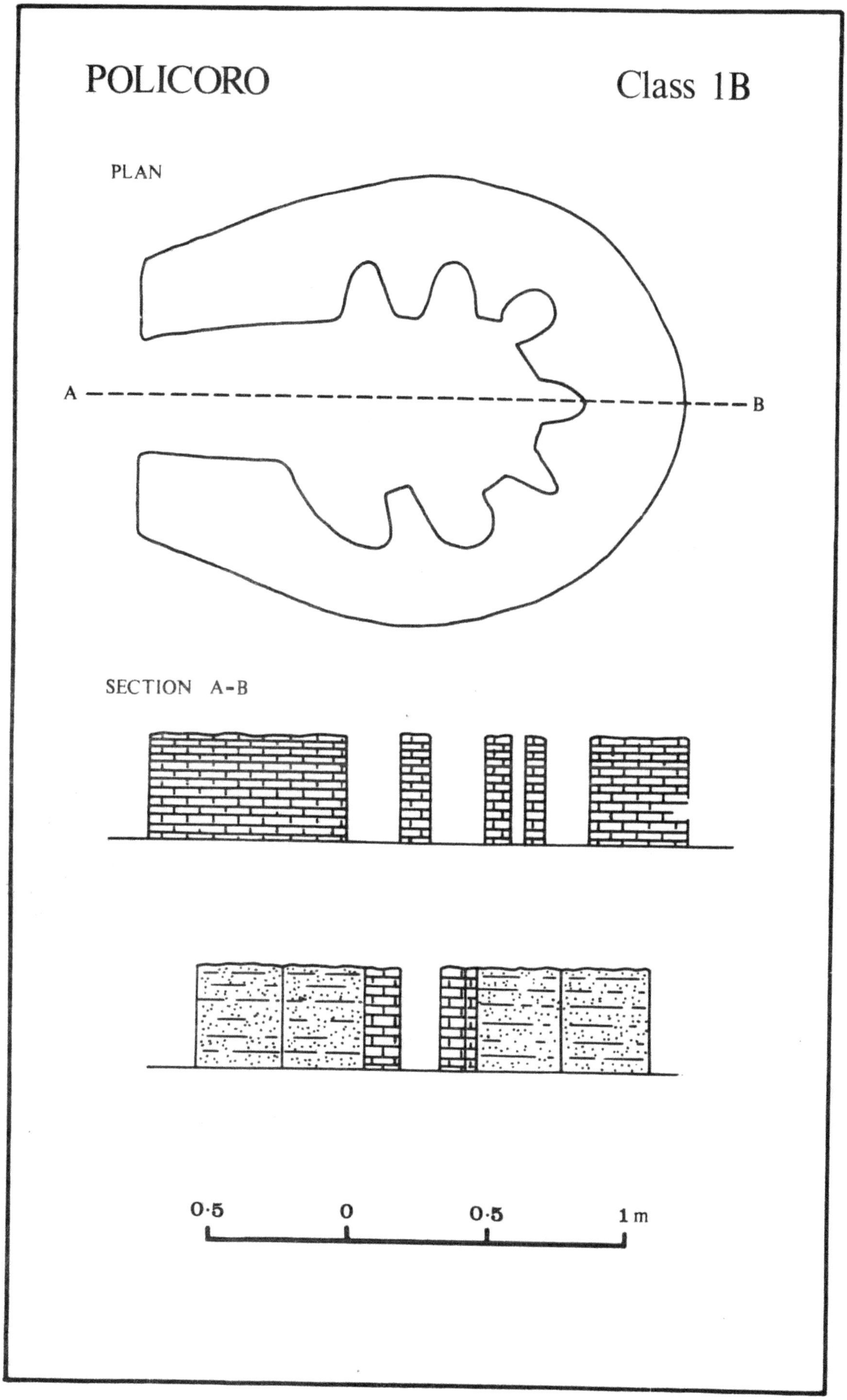

Fig. 5.5

praefurnium, supporting on both sides the arches of the two main under-floor flues.

Class IId

The general layout of this kiln is close to class IIc, the only difference being the two praefurnia, side by side, running from the stokepit to the combustion chamber.

Large-size kilns, heavily loaded and requiring a good draught, may have necessitated such a stout structure which made it possible for fuel to be stoked alternatively in each of the two parallel praefurnia.

SURVEY OF KILNS[1-2]

PUGLIA

Canne, IV-III B.C., class Ia.[3] Fig. 5.3

praefurnium: L—1.70; W—1.10; H—1.55;
combustion chamber: central pedestal ø—1.25; H—1.80
raised oven floor: ø—3.30; vent holes No—18

The kiln was built into a pit, dug below ground level. The huge central pedestal widened out at the top, like a mushroom; four central vent-holes crossed its upper part obliquely, thus allowing an even distribution of heat in the oven.

Apani, Roman Imperial period, class Ia.[4] Fig. 5.4

praefurnium: L—4.90; W—0.80; estimated H—1.10;
combustion chamber: central pedestal ø—2.30; H—1.20;
raised oven floor: ø—3.20; vent holes: estimated number 60

The praefurnium was extremely long and had a pointed vault (now missing). The praefurnium walls as well as the furnace chamber walls and the central pedestal were built entirely of amphorae sherds, carefully and skilfully bonded together with a clay mixture. The central pedestal was squat and very large; it had in its upper part a network of small tunnels through which the heat could reach the centre of the oven from the combustion chamber.

Close to this kiln, a smaller kiln was uncovered, belonging to class Ia. The kilns were used for firing amphorae, many of which were found with Greek and Latin names stamped on the handles.[5] Apani is situated near the seashore, close to the city of Brindisi (Brundisium), with several water-springs and plentiful good clay. In the area the arable fields are literally covered with sherds of amphorae, and recent fieldwork has identified several more possible kilns.

LUCANIA

Policoro, IV-III B.C., class Ib.[6] Fig. 5.5

praefurnium: L—0.70-0.50; W—0.40;
combustion chamber: average internal ø-0.65;
radial pilasters No. 6; overall length 1.93

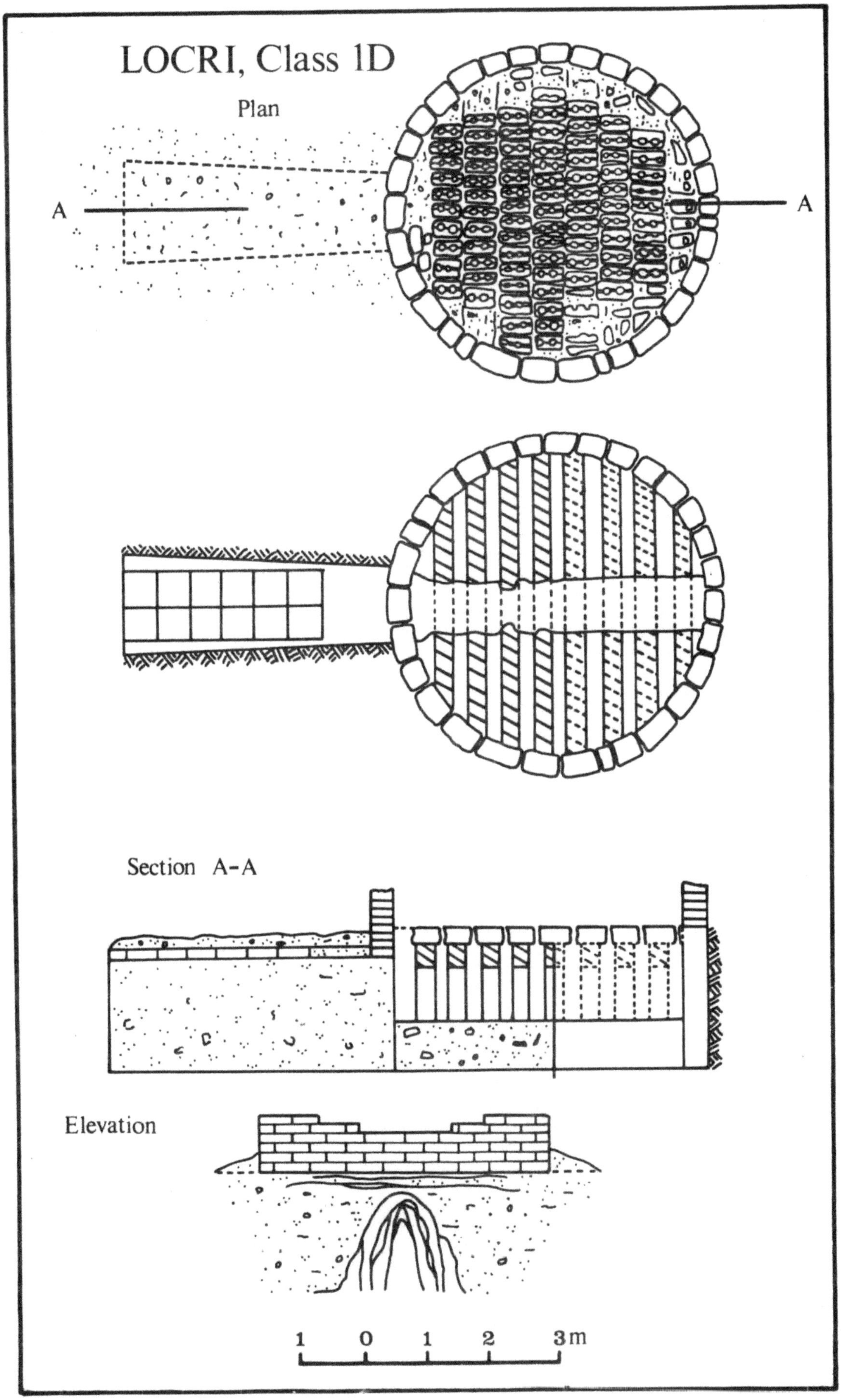

Fig. 5.6

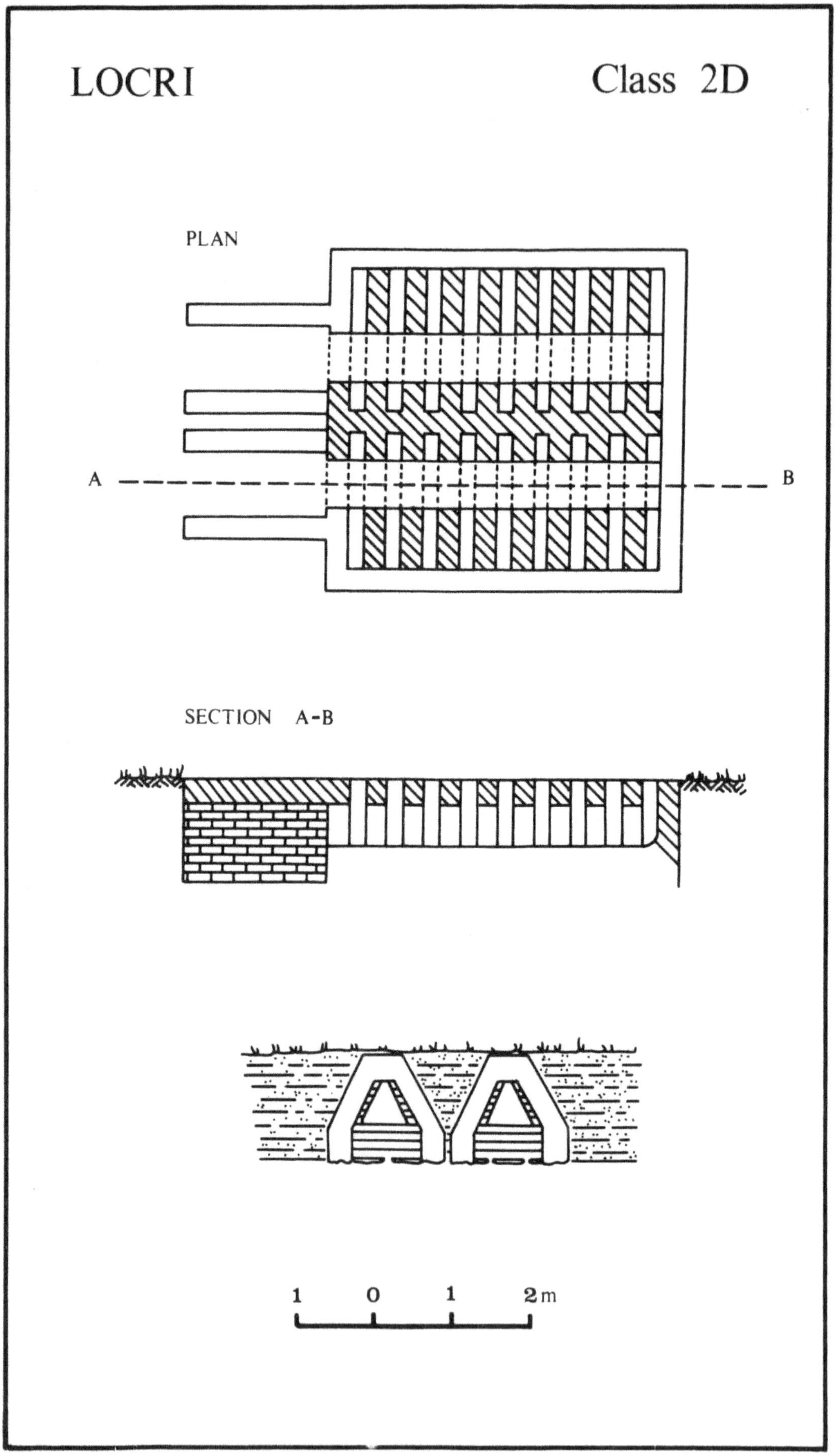

Fig. 5.7

In outline this small kiln was shaped like a horse shoe. The six radial pilasters, uneven in shape and size, were built of tile sherds bonded together with a clay mixture. The upper part of the kiln and the raised oven floor were missing.

In the same area several other kilns have been excavated, belonging to class Ia and class Ib. Also in Metaponto (about 20 km from Policoro-Eraclea) several kilns have been found, with sub-floor structures belonging to class Ia and class Ib.

CALABRIA

Locri, Centocamere, III B. C., class Id.[7] Fig. 5.6

praefurnium: L—3.80; W—1.30; H—1.50;
combustion chamber: main under-floor flue L—3.90; W—0.70; H—1.85;
under-floor cross walls No. 8; under-floor cross flues No. 9;
raised oven floor Ø-3.90
vent-holes: estimated number 190.

This large kiln was built below ground level, with steps leading down to the stoke pit. The raised oven floor was well supported by a system of cross walls carried over the main under-floor flue by arches which were irregular in size and shape. The side flues were at a higher level than the main under-floor flue and had sloping floors. An efficient passage of the heat to the oven was provided by the vent holes, plentiful in number and circular in shape. The walls of the oven were built of large unbaked bricks; they still survived to a height of about 0.80 m. No sign of an entrance was noted, and one must assume that the kiln was loaded by passing the pottery over the kiln walls.

Close to this kiln, a smaller kiln was uncovered, having similar structural features.

Several other kilns, with sub-floor structures belonging to class Id, have been excavated in Locri in the inhabited area called "Centocamere". It is remarkable that the walls of the houses have often been built using many kiln wasters, probably coming from unsuccessful firings in the nearby kilns.

Locri, Area sacra, V B. C., class IId.[8] Fig. 5.7

praefurnium: No. 2; each of them measuring L—1.90, W—0.86; H—1.00;
combustion chamber: two parallel main under-floor flues, each of them
measuring L—4.00; W—0.60; H—0.85;
under-floor cross walls No. 8; under-floor cross flues No. 9;
raised oven floor: L—4.00; W—3.80

This double parallel flue kiln, with two parallel praefurnia was situated in the courtyard of a Greek temple. The two parallel praefurnia, running from the stokepit to the combustion chamber, faced the hills, opposite the seashore; they were built of unbaked bricks like the walls of the oven, very little of which survived. The kiln was dug into the natural soil, and the raised oven floor (now missing) was at ground level.

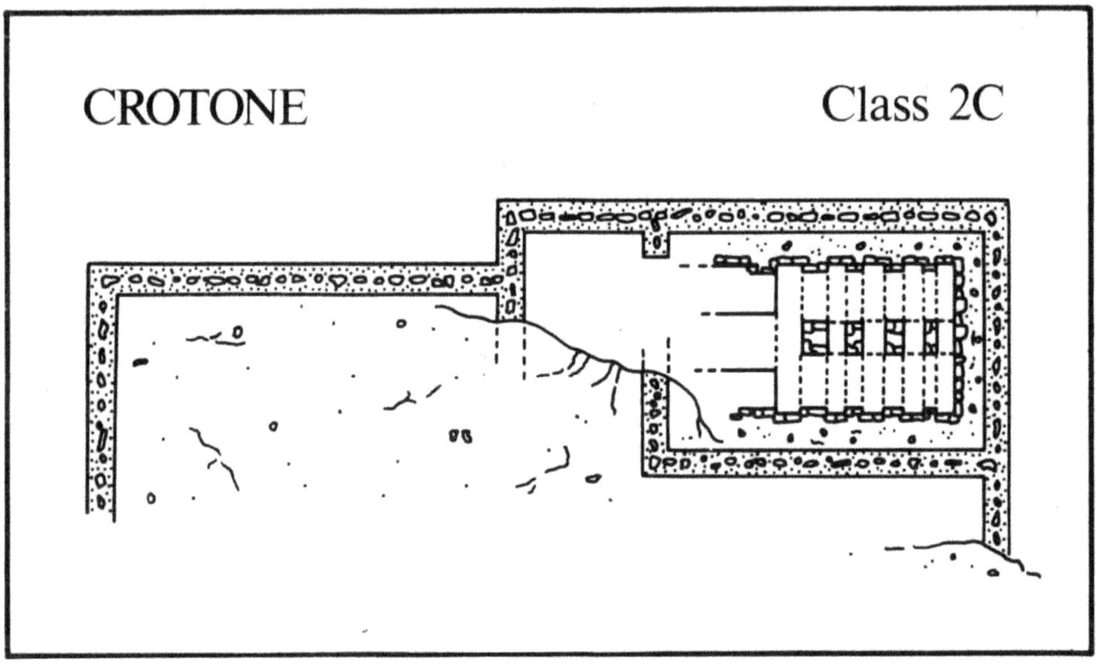

Fig. 5.8

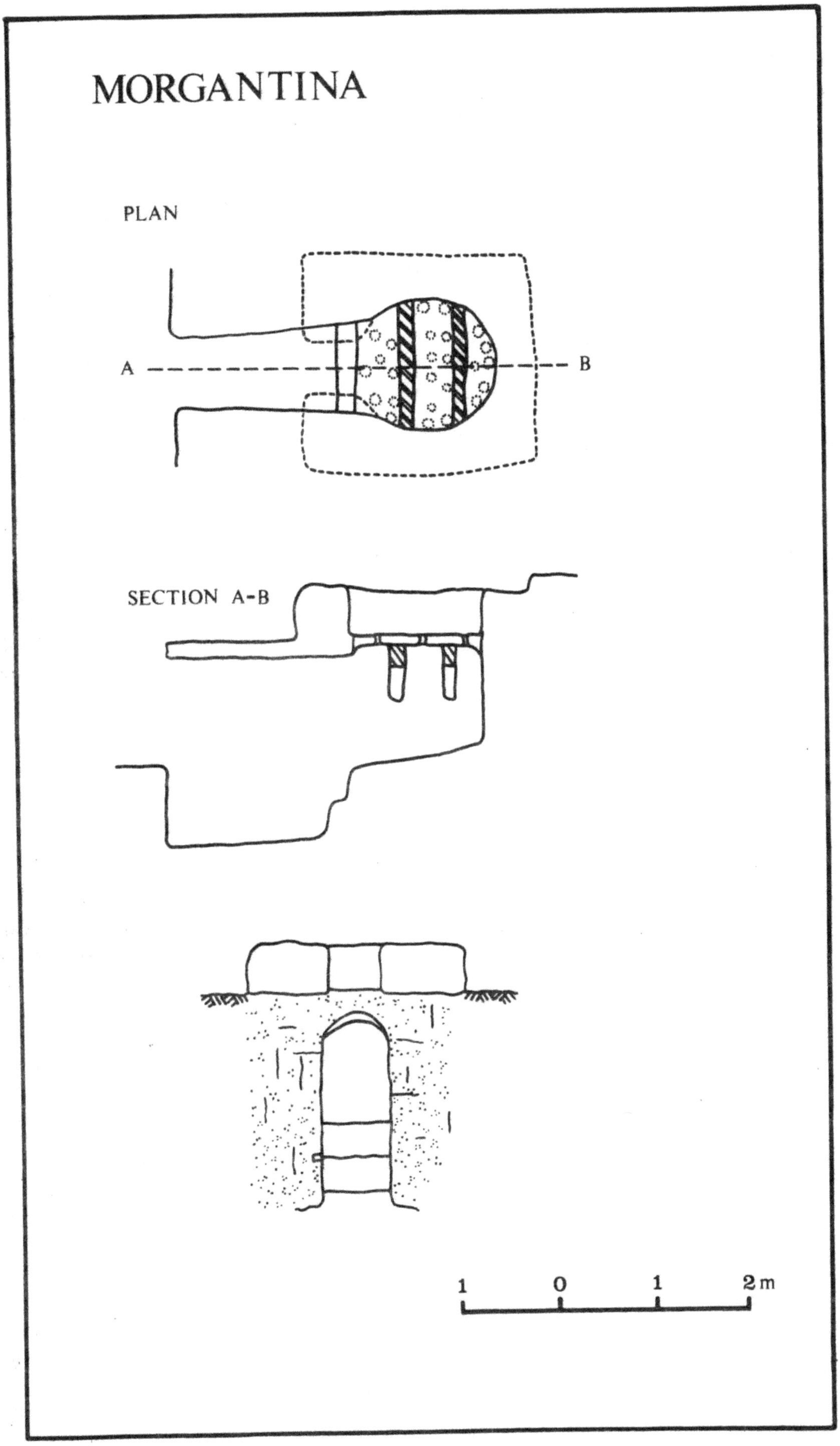

Fig. 5.9

The two main under-floor flues were narrow and low roofed, thus allowing only a minimum amount of space for fuel. There were eight cross walls which ran parallel to the front and rear furnace chamber walls, dividing off 9 under-floor cross flues which were at a higher level than the main under-floor flues.

Crotone, Roman Imperial period, class IIc.[9] Fig. 5.8

praefurnium: L—1.40; W—1.15;
combustion chamber: two parallel main under-floor flues, each of them measuring L—3.45; W—1.25;
under-floor cross walls No. 4; under-floor cross flues No. 5;
raised oven floor (now missing): estimated measures L—3.45; W—3.10

The kiln was situated near the temple of Hera Lacinia, close to a Roman villa, and it may be assumed that it was used for firing the tiles required for building the villa. The kiln was mostly destroyed: the raised oven floor was missing altogether and only two arches in the combustion chamber survived. It was a double parallel flue kiln, with one praefurnium, the kiln being stoked through the praefurnium which divided into two parallel flues beneath the oven floor. A central wall stood in the middle of the combustion chamber, in line with the praefurnium, supporting on both sides the arches of the two main under-floor flues. The arches were built with well-fired voussoirs, bonded together by a thick clay mixture. The space between arches varied in width, becoming narrower the further they got from the praefurnium. The length of under-floor side flues was reduced to a minimum by filling it with tile sherds and bricks, thus preventing wastage of heat.

SICILIA

Naxos, VI-V B.C., class IIb.[10]

In Naxos, in the area called "kerameikos", one rectangular kiln (class IIb) was uncovered together with two round kilns (class Ia). The kilns have been dismantled in order to be reconstructed in a Museum in Sicily, therefore measurements are not available at present. The rectangular kiln was assumed to have been used for firing tiles and architectural terracottas. The sub-floor structure consisted of a main under-floor flue with a number of closely-spaced cross walls carried over the main under-floor flue by arches. The arches were built of broken sherds and clay, none of which were found intact, only the springers remaining in most cases. The raised oven floor was altogether missing.

In an adjacent area two other kilns were excavated, a round kiln (Ø—2.65, class Id) and a large rectangular kiln (5.50 x 4.00, class IId).[11]

Morgantina, II-I B.C., class Ic.[12] Fig. 5.9

praefurnium: L—1.70; W—0.70-0.90; H—1.95
combustion chamber: Ø—1.35; H—1.20;
raised oven floor: Ø—1.35; vent holes No. 25.

The kiln was built well below ground level, access being provided by some rough steps cut into the natural soil. The praefurnium was in a pit, at a lower level than the combustion chamber, two steps leading up from the praefurnium to the combustion chamber.

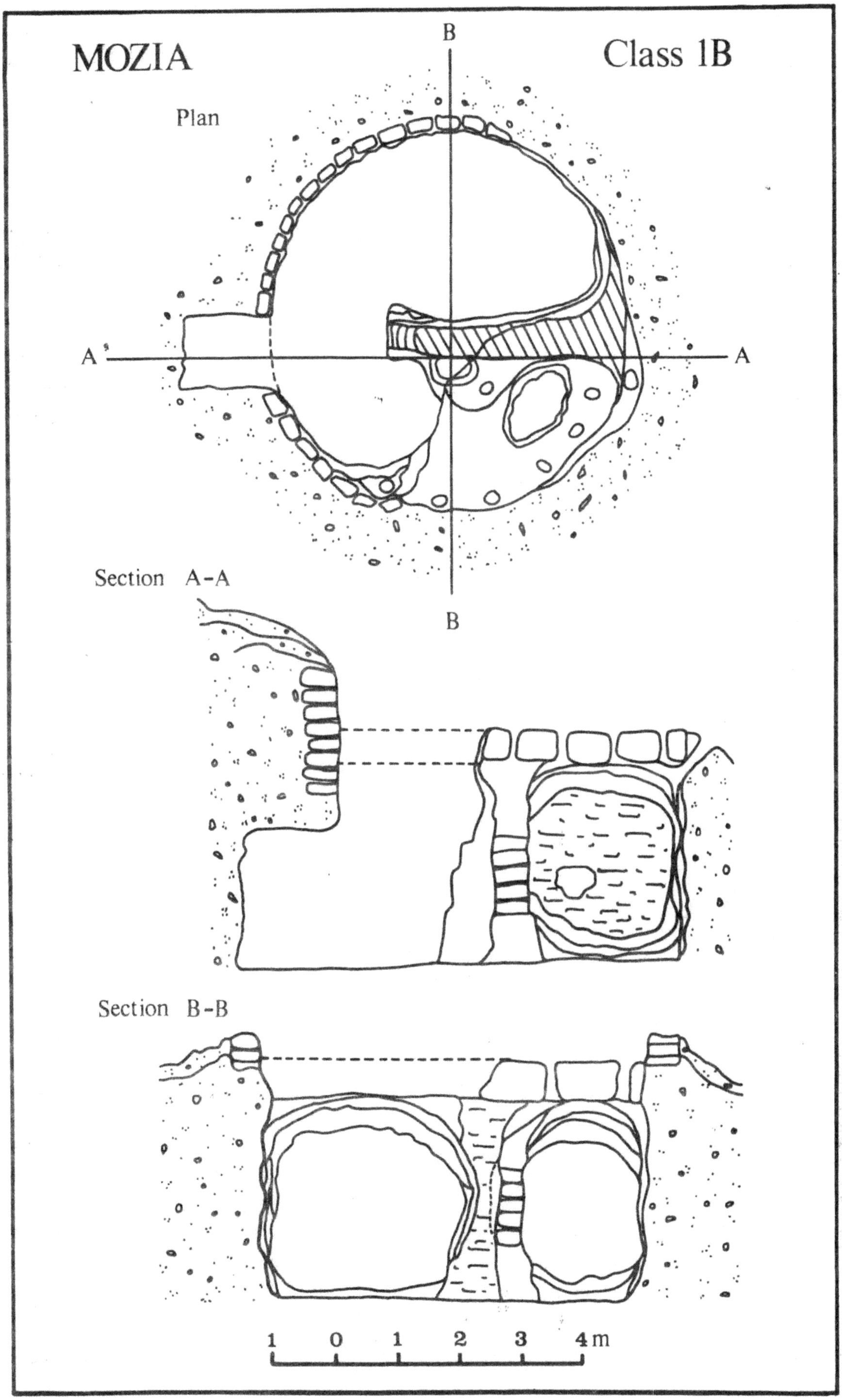

Fig. 5.10

Fuel was thrown down from the stokepit into the praefurnium (such a firing technique was also employed in Medieval kilns in Sicily). The raised oven floor was supported by two arches which sprang from the furnace chamber wall at about 0.60 m above the ground, thus leaving the furnace chamber floor clear. It would have been possible to use bulky fuel, such as straw, for firings, and it may be noted that straw is used nowadays in Sicily for firing pottery and bricks in the few surviving traditional kilns.

In the Morgantina kiln the two arches were parallel and faced the praefurnium; they looked like heavy ropes and were built of sherds bonded together with a clay mixture. The vent holes were arranged in roughly concentric circles in the spaces between the two ribs.

Several other kilns have been excavated in Morgantina area. One is particularly worth mentioning in as much as it is the only kiln so far known in Italy with two praefurnia, one opposite the other, at either end of the kiln.[13]

Camarina, IV-III B. C., class Id.[14]

combustion chamber: main under-floor flue L—2.40
under-floor cross walls No. 3; under-floor cross flues No, 4;
raised oven floor Ø—2.40;
vent holes: estimated No. 35.

The kiln, mostly destroyed, was situated near the coastline, and it is assumed that it was used for firing tiles, although most technical data are missing.

The sub-floor structure consisted of a main under-floor flue, with cross walls carried over it by three brick arches. The raised oven floor was pierced by four series of vent holes.

Mozia, IV B. C., class Ib.[15] Fig. 5.10

praefurnium: W—0.60; H—1.10;
combustion chamber (slightly oval form) Ø—2.90-3.15; H—1.90;
central tongue wall: L—1.90;
raised oven floor: Ø2.90-3.15;
vent holes: estimated number 30.

Owing to its dangerous position, the praefurnium was not fully excavated, therefore its length is unknown. The sub-floor structure consisted of a tongue extending from the back wall of the furnace chamber. The tongue was heavily leaning, owing probably to bad craftsmanship during its construction, and to increase its strength a pedestal was built which could take some of the weight of the raised oven floor. The oven floor was thus supported by both the tongue and the pedestal. The tongue was of a rough, uneven shape and became thinner towards its centre where there was a big gap, perhaps opened intentionally in order to allow a better circulation within the kiln.

Most of the raised oven floor was missing, and only few of the vent holes were found intact; one hole ran through the kiln wall, connecting the combustion chamber directly with outside. One must assume that this vent hole was

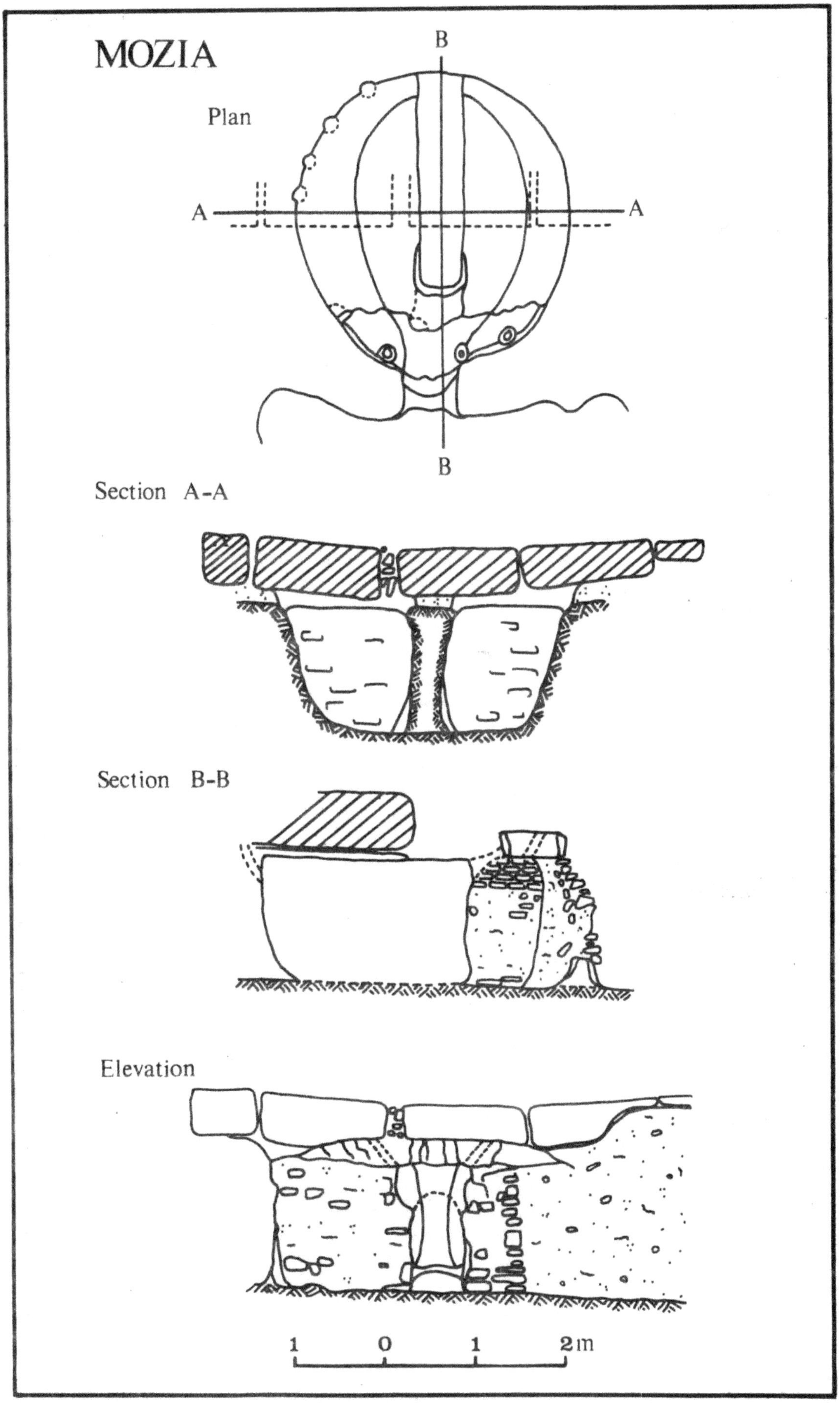

Fig. 5.11

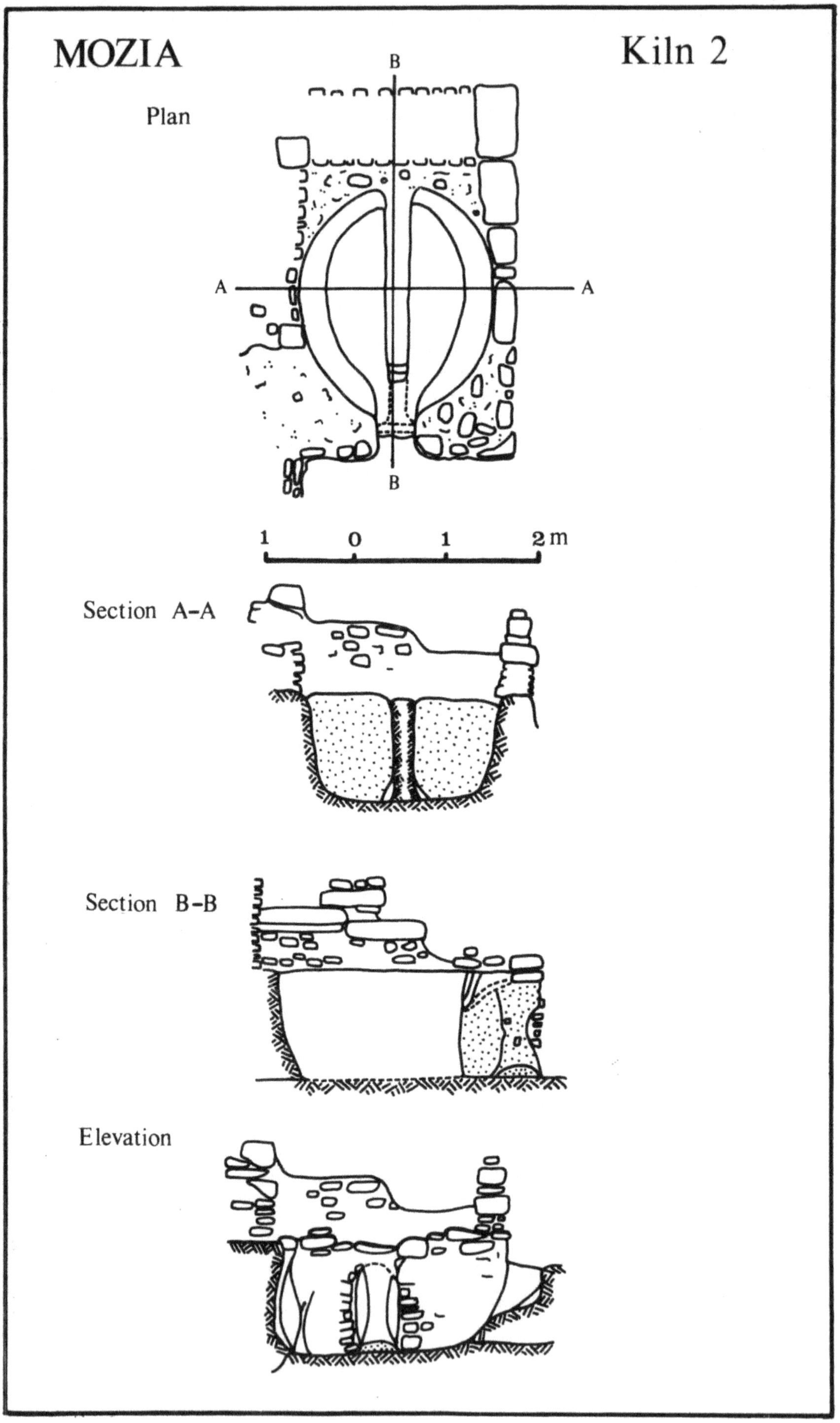
MOZIA
Kiln 2
Plan
B
A
A
B
1 0 1 2 m
Section A–A
Section B–B
Elevation

Fig. 5.12

left open at the beginning of the firing to increase circulation of air, and closed when the fuel was burning well.

The walls of the kiln still survived in a few parts to a height of 0. 90 m, and consisted of unbaked bricks.

In the little island of Mozia two other kilns have been found with sub-floor structure belonging to class Ib[16] (Fig. 5. 11 and 5. 12). In one of them (Ø—3. 10) the remains of several fire-bars were still in situ, spanning the space between the praefurnium and the tongue projecting from the rear wall of the combustion chamber. It is the first kiln excavated in Italy to include such a structural feature. It may be added that in Selinunte as well, some kilns were uncovered with fragments of fire-bars among the debris.[17]

* * * * * *

Following this survey of kilns excavated in South Italy and Sicily a few comments are needed on some kiln features and firing techniques which still survive in old traditional kilns still working in small places in South Italy, although becoming fewer and fewer every year. In fact, traditional working methods, handed down from generation to generation, may help archaeologists to understand structures and techniques which have left no evidence.

The general assumption that bricks and tiles were fired in square and rectangular kilns may require a word of caution. Although from the point of view of stacking the kiln a shape with angles might be deemed most suitable, bricks and tiles could have been fired in round kilns just as efficiently. An example still survives today, and plate 5. I shows a round kiln loaded with bricks at Vibo Valentia (Catanzaro). The brick-maker states that his father as well as his grandfather would rather use a round kiln for heavy pottery as well as for bricks as it allows a better control of the draught.

Very little kiln roofing has ever been found in archaeological excavations, and we assume that the load to be fired was covered by kiln wasters, turf, clay mixture and the like. Again we can learn something from surviving practices: plate 5. II shows a kiln at Gerocarne (Catanzaro) where pots from previous firings are used to encircle and to cover the load of coarse pottery to be fired.

Finally we should consider fuel, which deserves a good deal of attention as so much depended on it. The dimensions of the mouth of the praefurnium as well as the length of praefurnium and of any under-floor flues were influenced by the type of fuel used. Straw was one of the most common fuels in antiquity; it is still used nowadays in pottery kilns where it requires a "calamita", that is a line of box flue tiles set end to end from outside the stokepit to the centre of the combustion chamber. It is a device provided to carry an extra supply of air to the heart of the kiln, so as to allow straw to burn fully. Plate 5. III shows a huge heap of straw at Gr.vina di Puglia (Bari), with a lime-kiln standing in the background.

Plate 5.I a. Modern kiln at Vibo Valentia (Catanzaro) loaded with brick
b. Modern kiln at Gerocarne (Catanzaro)

Plate 5.II Gravina di Puglia (Bari); straw being used as fuel in a lime-kiln (in background)

REFERENCES

1. Most of the kilns to be considered have been described in: N. Cuomo di Caprio, Proposta di classificazione delle fornaci per ceramica e laterizi in area italiana, in Sibrium, XI, 1971-1972, pp. 371-464, with ample references. In the present paper only a restricted bibliography is provided.

2. For each kiln, dimensions of basic features only are given. All measurements are in metres. The following abridgements are used:

   L = Length W = Width
   H = Height Ø = Diameter

   Dimensions are taken internally. Height of the combustion chamber includes the raised oven floor, when found in situ. Height of arches is measured internally.

3. N. Cuomo di Caprio, op. cit., in Sibrium, XI, 1971-1972, p. 414, plate VI.

4. B. Sciarra, Un primo saggio di scavi ad Apani, in Quaderni Museo F. Ribezzo, Brindisi, N. 1, 1964, pp. 39-43, plates 1-5; id., Su un saggio di scavo in contrada Apani, in Studi Salentini, f. XLIII-XLIV, 1973, pp. 127-130; id., Ricerche in contrada Apani, in Recherches sur les amphores romaines, Ecole française de Rome, 1972, pp. 29-34, plate 1; N. Cuomo di Caprio, Una fornace ad Apani, in Notizie Scavi, 1978, pp. 423-428, plates 1-4.

5. B. Sciarra, Alcuni bolli anforari brindisini, in Epigraphica f. 1-4, XXVIII, 1966 (1967), pp. 124-134; id., Bolli anforari brindisini, in Studi Salentini, f. XXXVII-XXXVIII, 1970, pp. 3-15.

6. N. Cuomo di Caprio, op. cit., in Sibrium, XI, 1971-1972, pp. 416-418, plate VII.

7. E. Lissi, Gli scavi della Scuola Nazionale di Archeologia a Locri Epizefiri, anni 1950-1956, in Atti VII Congresso Int. Arch. Classica, Roma, 1961, vol. II, pp. 109-115, N. Cuomo di Caprio, Fornaci per ceramica a Locri, in Klearchos, 61-64, 1974, pp. 43-65, plates 1-9.

8. A. de Franciscis, Scavi e scoperte, in Fasti A., X, 1957, No. 1926; N. Cuomo di Caprio, op. cit., in Sibrium, XI, 1971-1972, pp. 440-442, plate XIV.

9. N. Cuomo di Caprio, Le fornaci di Capo Colonna, in Klearchos, 53-56, 1972, pp. 107-122, plates 1-7.

10. P. Pelagatti, L'attività della Soprintendenza alle Antichità della Sicilia Or. tra il 1965 e il 1968, in Kokalos, XIV-XV, 1968-1969, p. 351, plate LIX.

11. G. V. Gentili, Naxos alla luce dei primi scavi, in *Boll. d'Arte*, 1956, f. 4, pp. 329-331; P. Pelagatti, Naxos, relazione preliminare delle campagne di scavo 1961-1964, in *Boll d'Arte*, 1964, f. 2, p. 153.

12. E. Sjoqvist, Excavations at Morgantina 1959, Rep. IV, in *A. J. A.*, 64, 1960, p. 130, plate 27, fig. 20; N. Cuomo di Caprio, *op. cit.*, in *Sibrium*, XI, 1971-1972, pp. 419-421, plate VIII (the plan of the kiln has been published by kind permission of Princeton University—Illinois University, Urbana).

13. R. Stillwell, E. Sjoqvist, Excavations at Serra Orlando, in *A. J. A.*, 61, 1957, p. 158; E. Sjoqvist, Excavations and discoveries, in *Fasti A.*, X, 1957, No. 2634.

14. P. Pelagatti, Kamarina, in *Sicilia Archeologica*, 1970, p. 14, plate 12.

15. N. Cuomo di Caprio, Una fornace a Mozia, in *Sicilia Archeologica*, X, 34, 1977, pp. 7-14, plates 1-6.

16. V. Tusa, Il Luogo di Arsione, in *Mozia IX*, 1978, pp. 65-90; N. Cuomo di Caprio, Commento tecnico sulle fornaci del cosiddetto "Luogo di Arsione" di Mozia, in *Mozia IX*, 1978, pp. 111-117, plates 10-15.

17. Excavations conducted at Selinunte by the Soprintendenza Alle Antichità della Sicilia Occidentale, publication forthcoming.

ROMAN TILE-KILNS IN BRITAIN

Alan McWhirr

Professor Grimes' masterly account of the work's depot of the Twentieth Legion at Holt, published in 1930, included a general survey of kilns in a section on the Romano-British pottery industry (Grimes, 1930). He classified kilns according to their shape and the arrangement of the flues beneath the oven floor. The only other study of kilns was undertaken by Philip Corder and published in 1957 (Corder, 1957). Corder restricted his study to pottery-kilns, and so there has been no overall study of tile-kilns since the account published by Grimes in 1930. The survey which follows stems from an interest in the production of tile and brick and not from an interest in kilns for their own sake, and so it is limited in most cases to structures where tile and brick can be shown to have been fired. This means that a number of rectangular or square kilns, thought to have been used solely for pottery, have been omitted. Some of the kilns in the Gazetteer are now thought to have been used for pottery, but are included because in the original publication there is reference to the possibility that tile was fired in them. To the kiln specialist, this division between square or rectangular kilns which fired tile and those which did not will be considered heretical, as will the decision not to look at all kilns. However, a survey of pottery kilns undertaken by Vivien Swan is near completion and this study should be seen as complimentary to hers. Once these two surveys have been completed, an overall classification of kilns could be devised if thought desirable and necessary. Similarly, once such surveys have been undertaken in other provinces, there will be a corpus of basic material upon which a major study on kilns from the Roman world could be based.

Since 1930, the number of kilns recorded from Britain has greatly increased. Grimes wrote, 'comparatively few kilns of the rectangular type have been recorded in this country', but now it is possible to list 51 kilns (excluding Holt) used, although not exclusively for firing tile and brick. In addition, there are an increasing number of rectangular or square pottery-kilns such as those from Brampton, Norfolk (Knowles, 1977).

Every attempt has been made in this survey to examine all sites where there is evidence for a kiln-like structure, and these have been listed in the Gazetteer at the end. Even so, some may have been missed and the author would like to know of any such omissions.

No distinction has been made between military and civilian tileries as the techniques of making and firing tile and brick appear to be similar whether they were made for the military or civilian market. The major differences are concerned with distribution. Clearly the legionary tilery at Holt reflects the hand of the organised military unit, but when one looks at the auxiliary tilery at Brampton, then it appears from superficial evidence to be very similar to civilian tileries such as, for example, the one at Minety, Wilts.

The overall shape of a kiln is not a reliable guide as to the material fired in it. In Britain the kilns used for firing tile and brick are in all cases square or rectangular. A circular kiln from Philips Norton, Somerset, found in 1879, had an unusual solid and rectangular sub-floor structure, but there is nothing to show that it was used for firing tile or brick (Corder, 1957, 12). A square or rectangular firing chamber is a sensible design for stacking objects which themselves are of that shape. However, modern examples from Sicily and Italy, as well as examples from elsewhere in the Roman world, show that circular kilns were used for brick (see page 91) and the possibility should not be overlooked in Britain.

The basic features of a typical tile-kiln consists of the FIRING CHAMBER or OVEN in which the material was fired. The oven floor was supported on a series of closely-spaced CROSS WALLS in the COMBUSTION CHAMBER, which were carried across the MAIN CENTRAL FLUE by arches. In between each cross wall was a CROSS FLUE which allowed circulation throughout the combustion chamber and forced the gases into the oven through holes left in the oven floor which coincided with the sub-floor flues. The central flue often projected beyond the combustion chamber to link with the STOKEHOLE and this projection, usually referred to as the FLUE, FIRE TUNNEL or PRAEFURNIUM, was arched or corbelled over. The stokehole was an area outside the kiln from which it could be serviced.

There is a considerable range in size of tile kilns as illustrated by the figures given in Table I. The largest civilian tile-kiln is from Eccles, Kent, which measured 4.9 by 4.7 m internally and the smallest is probably from Lapworth, Warks, which was only 1.8 m square. The same sort of range can be seen with military kilns. The only legionary tilery so far identified as at Holt which stands out from the rest in its size and quality of design and also in having the remarkable 'kiln plant'. Auxiliary tileries can be found at Brampton, Muncaster, South Shields, Quernmore, Grimescar, Mumrills, Gellygaer and possibly Caernarvon. Of these, Brampton has the biggest kiln measuring 3.6 m square internally, and Muncaster the smallest being only 1.8 m square. It is interesting to note that it is not necessarily the largest kilns which were provided with double parallel flues. There are several kilns which are bigger than the double flue kiln at Horton, and it is more likely that the fuel was the factor which necessitated the use of two flues to bring the kiln up to the required temperature.

The sub-floor structure within the combustion chamber was constructed in a variety of ways, as first recognised by Grimes who formulated a simple classification which depends upon the behaviour of the cross flues, i.e. the space left between the cross walls, although a fourth type has been added to cope with one or two examples which do not fit his original scheme.

Type 1	Main central flue and cross flues on the same level.
Type 2	Cross flues at a higher level than main flue with horizontal floors.
Type 3	Cross flues at a higher level than flue but with sloping floor.
Type 4	A sub-floor structure not covered by types 1-3.

There are only a few examples of Types 1 and 2. The kilns at Brampton (1), Muncaster (A), Messingham (1) and Mumnills are of Type 1, and those at Colchester (7), Wiston and Arbury (1) of Type 2. The majority of kilns, where there is evidence to show the sub-floor arrangement are of type 3. There are only two kilns which can be classified as type 4 and they are at Arbury (2), where there is a central pedestal rather than cross walls and flues, and possibly Colchester (17), (see Gazetteer for all these sites).

There are only two kilns from Britain with double parallel main flues, they are at Holt, Clwyd and Horton, Surrey (pages 188 and 164). This is in marked contrast to Italy where many more double flue kilns have been recorded (see Chapter 5). Other than having two flues, the design of the combustion chamber of these kilns is similar to the rest under discussion.

Resting on the cross walls was the oven floor, which however built, had holes or VENTS to allow circulation within the kiln. Different craftsmen used different methods of forming the oven floor. Some were apparently made solely of clay, which must have presented difficulties, for this is a vital part of the kiln on which the load would rest during firing, and any instability in the floor would prejudice the efficient functioning of the kiln. A common technique was to bridge the gap between the cross walls with tile or stone slabs, so forming a continuous layer of tile or stone as a base for the oven floor. Gaps in the tile were created by cutting semi-circles or triangles out of each end of the tile so that when placed together circular or square holes were formed. Clay was then laid on top of the tile leaving gaps to coincide with those in the base. With so much raw clay and green tile used for these kilns, it would have been necessary for the clay to harden before a heavy load was placed on the oven floor. The most convenient, and probably most successful, way of doing this would have been to fire the kiln when empty and to carry out any repairs made necessary by shrinkage or collapsing before the first load was placed in it. Such a sequence was detected in the Carnarvon kiln (page 186).

How the kiln was constructed above the oven floor is uncertain. Rarely have the combustion chamber walls survived much above the oven floor, and where they have, no sign of an entrance has been noted, with the possible exception of Eccles, Kent (page 158). Presumably, therefore, the kilns were loaded by passing the tile and brick over the side walls of the kiln, as is the case with many contemporary examples of the simple brick-kilns. The majority of kilns were built into the slope of a hill, making the outside ground level higher than parts of the oven floor. This clearly makes loading the kiln easier, for a two metre high kiln-wall would not be such an obstacle if the ground level outside was higher than the kiln floor. The only example from Britain which provides evidence for the kiln superstructure is from Muncaster (page 121), where Bellhouse described the beginnings of corbelling 900 mm above the level of the oven floor. Whether this is genuine corbelling, and therefore some form of roofing, or merely a collapsing side wall, is not clear from the published account (see also Quernmore, page 159). A study of primitive brick-kilns from other parts of the world, and of recent kilns in Britain, indicates that a formal roof of tile clay or stone would not have been necessary. The kiln load could have been covered, if required, by using kiln wasters, or even turf. It looks as though the kiln walls were taken up vertically to a height of 1-2 m which would enable the tile and brick to be passed over to a person stacking them inside without difficulty. The construction of side walls to such a height would also have had the added effect of creating the necessary draught of air through the kiln which was required to reach a sufficiently high temperature to fire the tile and brick.

When kilns are built into the hillside, the stokehole requires little digging as it appears more or less close to the ground surface. On other occasions quite sizeable and elaborate pits are dug, for example at South Shields, where the stokehole is as big as the kiln (page 171). The purpose of the stokehole is to provide access to the main central flue via a fire tunnel, if one existed, in order to stoke the kiln during the firing. Assuming the kiln was used more than once, the main flue could also be cleaned out from the stokehole in between firings. The length of the flue or fire tunnel, from the stokehole to the combustion chamber varies considerably from one kiln to another. Experience indicated to the tiler what particular design was appropriate. One of the main factors he would have taken into account was the nature of the fuel he was using, whether it be wood, charcoal or straw (see page 91). There is a suggestion from Gellygear that coal was used. This, along with other factors, determined the length of flue he needed in the kiln in order to reach the firing temperature.

In some stokeholes it appears that attempts were made to provide shelter for the stoker. At Canterbury (page 154) a series of post holes and a possible timber slot may have indicated such a shelter, and at Eccles (page 158) a short stretch of wall could be the same, although Mr. Deticas believes it was built to prevent a sudden draught from entering into the flue or the kiln.

Many kilns and their stokeholes were dug into clay, and it comes as no surprise to learn that special steps had to be taken to drain the combustion chamber and stokehole. Several published photographs of excavated tile-kilns shows how easily they filled with water during the course of excavation. The most intricate arrangement for draining a kiln can be seen at Muncaster (page 121). Here a line of <u>imbrices</u> ran along the main central flue and ended at a box-like arrangement of tiles 1.5 m outside the kiln. Another box-like structure existed at a lower level a further 0.5 m away, but nothing was found to connect the two, although one suspects that they were linked in some way. A similar line of <u>imbrices</u> was found at Blackboy Pits near St. Albans, Hertfordshire, (page 142) along the centre of the main flue and about 150 mm below its floor.

At Potters Bar, Herts (page 149), an unusual structure which, judging from the large number of wasters found, must have been connected with tile-making, had a tile drain along what looked like a central flue. The drain consisted of eleven box-flue tiles set end to end, a total length of 3.81 m, and ran from the mouth of the flue across the stoke pit and beyond. Perhaps the best example and certainly the largest, is the drain found leading from the combustion chamber of the kiln at Wykehurst Farm, Cranleigh, Surrey, to a stream 8.83 m away (page 167). A trench 1 m deep was dug from the kiln to the stream and a continuous line of <u>imbrices</u> placed along the bottom, surely removing all doubt that it was a drain. The excavators at Crookhorn Farm, Hants, noted a trench dug into the subsoil leading from the stokehole for a distance of at least 10 m and more likely to a clay pit some 18 m south-west of the kiln. However, one should note the comments made on page 91 concerning the use of such devices for introducing air into the combustion chamber particularly when straw was used as a fuel.

Apart from the drains the features so far described are typical of a tile-kiln, but occasionally one finds little touches of design reflecting the different approaches used by individual craftsmen. For example, the kiln at Moat Farm, Lexden, near Colchester, excavated by Holbert in 1970 (page 129), revealed two such idiosyncratic features. The cross walls each had three tiles set vertically between them, presumably to prevent them moving sideways during the firing of the kiln, although it is difficult to see how just one set of three tiles between two walls could have had much effect when the kiln was loaded. Possibly more such spacers originally existed. In the same kiln, which in fact appeared to have been built inside the shell of an earlier kiln, was an imbrex tile which may have been the remnants of a flue. A similar feature was noted in the Messingham kiln No. 1 (page 150).

So far, no mention has been made of the materials used in constructing kilns. The most common was tile, occasionally wasters being used, but not always. Tegulae were frequently used and so were green tiles, i.e. tiles which had not been previously fired in a kiln. Clay was used for the oven floor and for covering the various surfaces. Sometimes local stone was used in building kilns, either along with tile or completely on its own, and it has been argued that in a group of tile and stone built kilns, the stone variety would be the earliest, the reason being that until the stone kilns had been built there would be no tile available . Such a sequence is suggested at Brampton (page 111).

There are a group of structures usually identified as kilns which lack some of the basic features described above, in particular the cross walls and associated cross flues, and consequently the oven floor. It is not always possible to be sure that these did not once exist, and have subsequently been removed, but sufficient examples can now be cited to show that this form of a kiln must be considered a possibility. In most of these examples what remains is a central flue, and it is argued that the tile and brick was stacked either side of the flue, and in some cases even across the flue, in such a way as to allow the hot air to circulate around the tile.

These structures have been described by some excavators as 'clamp-type kilns'. They are not true kilns where the combustion chamber is divided from the oven by a floor, and they appear a little sophisticated to be called clamps, but the absence of an oven floor separating the fuel from the kiln-charge makes clamp perhaps a better description. Some of the structures which may fall into this group include Mount Bures (page 132), Theydon Garnon (page 133), Braxell's Farm (page 136), Scotland Farm (page 137), Heckington (page 159), Messingham 2 (page 150) and possibly others.

Structures related to kilns are hard to find. At Arbury, Warks, a group of post holes were found and thought to have been associated with one of the kilns (fig. 6.30). The most impressive is perhaps from Crookhorn Farm, Hants, where a series of post holes were found surrounding the kiln representing two phases of a timber structure. (Fig. 6.13).

The reasons for the siting of brickworks have been touched upon by other contributors to this volume as have possible modes of organisation and until further work has been done on the distribution of kiln-products discussion on the siting of tile and brick kilns is perhaps best left until another occasion.

Dr. Peacock has drawn attention to the apparent lack of kilns close to towns (page 8) and there are also very few civilian examples of groups of kilns clustered together such as those found at Holt and Brampton which are to be associated with the army. The most dramatic illustration of a large civilian tilery can be seen in the ploughsoil at Minety where at least ten kilns have been located (Fig. 6.1 and page 181), along with stone buildings which may be workshops connected with the production of tile, or living quarters. Minety offers the greatest potential of any site in the country in helping to understand the organisation of a Romano-British brickworks.

Acknowledgements

The compilation of the Gazetteer would not have been possible without the help of many people. They include, C. J. Baddeley, B. Barr, D. J. Breeze, D. T-D. Clarke, D. B. Connah, M. D. Cra'ster, M. Darling, A. P. Detsicas, J. Dool, A. G. Down, J. A. Greenaway, J. P. Gillam, M. Green, E. Hartley, T. M. Jeffrey, F. Jenkins, G. M. Leather, T. G. Manby, E. J. Owles, C. Partridge, B. F. Rawlins, S. J. Read, W. J. Rodwell, C. Saunders, A. J. Scammell, K. Scott, B. B. Simmons, G. Soffe, D. J. Stagg, V. Swan, P. L. Waters, G. A. Webster, R. White and S. White.

Bibliography

Corder, P., 1957, 'The Structure of Romano-British Pottery Kilns'. Arch. Journ. 94 (1957), 10-27.

Grimes, W. F., 1930. 'Holt, Denbighshire. The Twentieth Legion at Castle Lynons'. Y Cymmrodor, 41 (1930).

Knowles, A. K., 1977. 'The Roman Settlement at Brampton, Norfolk: Interim Report'. Britannia 8 (1977), 208-221.

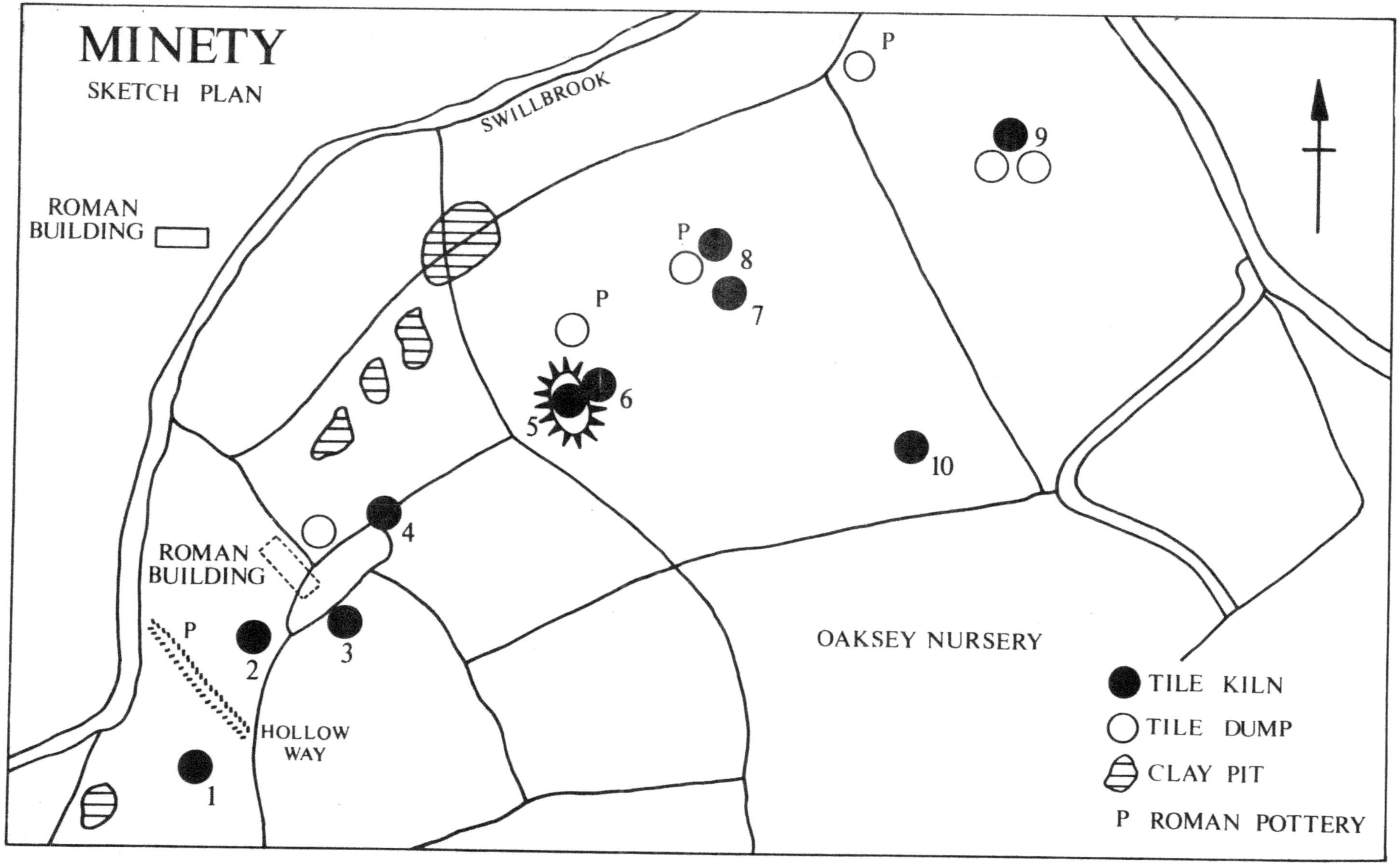
MINETY
SKETCH PLAN
ROMAN BUILDING
SWILLBROOK
P
9
P
8
7
P
6
5
10
4
ROMAN BUILDING
P
2
3
HOLLOW WAY
1
OAKSEY NURSERY
TILE KILN
TILE DUMP
CLAY PIT
P ROMAN POTTERY

Fig. 6.1

Table 6.1 Tile-Kilns: A summary of details

Site	Internal Size (M)	Number of cross walls	Classification	Comments
Cumbria				
Brampton 1	3.6 x 3.6	4	I	An auxiliary tilery and pottery
Brampton 2	3.0 x 3.0	6	III	
Brampton 4	2.5 x 1.8	4	III	
Brampton 6	2.6 x 2.6	5	III	
Brampton 7	Rectangular	?	III	
Brampton 8	Rectangular	?	?	
Muncaster A	2.0 x 2.0	3	I	An auxiliary tilery
Muncaster B	1.8 x 1.8	4?	?	
Essex				
Alphamstone	-	-	-	
Ashdon	3.5 x 2.6	7	III	
Colchester 7	1.8 x 1.4	2	II	Probably a pottery kiln
Colchester 17	1.4 x 1.4	?	IV?	Probably a pottery kiln
Colchester 31	2.1 x 1.7	5	III	Probably a pottery kiln
Moat Farm, Lexden 1	4.0 x 4.0	-	-	
Moat Farm, Lexden 2	3.0 x 3.0	7	III	
Mount Bures	-	-	-	
Theydon Garnon	-	-	-	
Hampshire				
Braxell's Farm	-	-	-	
Crookhorn Farm	c. 4.1 x 3.5	6	III	
Scotland Farm	-	-	-	

Table 6.1 cont.

Site	Internal Size (M)	Number of cross walls	Classification	Comments
Hereford and Worcester				
Upper Sandlin Farm	3.2 x 2.9	6	III	
Hertfordshire				
Black Boy Pits	1.0+ x 2.0	2+	III	Only part recovered
Netherwild Farm	c. 4.0 x 3.1	-	-	
Park Street	2.7 x 2.1	6	III	
Potters Bar	-	-	-	
Humberside				
Messingham 1	3.2 x 2.1	5?	I	
Messingham 2	3.5 x ?	-	-	
Kent				
Canterbury A	-	-	-	
Canterbury B	2.4 x 1.8	3	III	
Eccles	4.9 x 4.7	9	III?	
Lancashire				
Quernmore		2?	?	An auxiliary tilery
Lincolnshire				
Heckington 1	-	-	-	
Heckington 2	c. 3.3 x 2.7	-	-	
Heighington	3.3 x 3.3	5	III	

Table 6.1 cont.

Site	Internal Size (M)	Number of cross walls	Classification	Comments
Surrey				
Horton	3.0 x 2.3	7	III	
Wykehurst Farm	2.8 x 2.8	6	III	
Sussex				
Wiston	3.8 x 3.7	4	II?	
Tyne and Wear				
South Shields 1	3.2 x 2.1	8		An auxiliary tilery
South Shields 2	c..3.0 x 2.7	7		
Warwickshire				
Arbury 1	3.2 x 2.9	6	II	
Arbury 2	1.6 x 1.6	-	IV	
Arbury 3	c. 3.0 x 3.0	6/7?	III?	
Griff Hill Farm	2.5 x 1.6	4/6?	III	
Kenilworth	c. 3.4 x ?	-	-	
Lapworth	1.8 x 1.8	4	III	
Wiltshire				
Minety A	3.4 x 3.3	-	-	
Minety B	-	-	-	

Table 6.1 cont.

Site	Internal Size (M)	Number of cross walls	Classification	Comments
Yorkshire				
Grimescar	4.8 x 4.0	-	-	This may be an overall measurement
Scotland				
Mumrills	2.0 x 2.0	2	I	An auxiliary tilery
Wales				
Caernarvon		-	-	An auxiliary tilery
Gellygaer	2.4 x 2.4	5	III	An auxiliary tilery
Holt - Double flue kiln	5.4 x 5.4	12	III	Described as a pottery kiln
Holt - Kiln plant - kiln 1	2.4 x 2.1	7	III	Described as a pottery kiln
Holt - Kiln plant - kiln 2	5.2 x 2.6	6	III	Described as a pottery kiln
Holt - Kiln plant - kiln 3	4.3 x 4.3	7	III	Described as a tile kiln
Holt - Kiln plant - kiln 4	5.3 x 4.1	8	III	Described as a tile kiln
Holt - Kiln plant - kiln 5	5.3 x 4.4	8	III	Described as a tile kiln
Holt - Kiln plant - kiln 6	4.4 x 3.7	7	III	Described as a pottery kiln
Holt - Kiln plant - kiln 7	2.7 x 2.4	4	III	Described as a pottery kiln

GAZETTEER

Introduction

The Gazetteer which follows is an attempt to bring together all structures which may have been connected with the firing of tile or brick, drawn at the same scale to make comparison easier and with standard symbols. The plans have all been redrawn by Nick Griffiths without whose meticulous work this survey would not have been possible. In a number of cases the published plans have not always given sufficient detail to enable them to be redrawn with a guarantee of complete accuracy and a 'degree of interpretation' has had to be adopted. There is also still some uncertainty as to whether all the kilns were used for tile brick, but where there is doubt they have been included.

The following structures have not been included as they are no longer considered to be for firing tile or brick (for grid references and other details see McWhirr and Viner 1978).

Altwood Bailey, Berks	Probably a corn-drier
Hampsted Marshall Park, Berks	Probably a corn-drier
Birdoswald, Cumbria	Mr. Gillam now considers this be a bread oven.
Iwade, Kent	A pottery kiln producing poppy-beakers inter alia. Information from Mrs. V. Swan.
Corbridge, Northumberland	Large quantity of pottery found, but no evidence of tile or brick (?)
Elmsell, Suffolk	Pottery Kiln; Proc. Suffolk Inst. Arch. 30 (1964), 117; JRS 55 (1965), 214.
Philips Norton, Somerset	No evidence of tile having been made.

To the Britannia list (McWhirr and Viner, 1979, 372) should be added the following sites which were either inadvertently omitted or have been discovered since the list was compiled in 1977. The same letters have been used to distinguish between sites where kilns have been found and those where tile production is suspected.

Avon

S Tracey Park, W of Bath	ST 7171	Information from Dr. A. J. Parker and R. Iles

Cheshire

S Wilderspool	SJ 6186	F. H. Thompson, Roman Cheshire, Chester, 1965. 78-80.

Derbyshire		
S Carsington	SK 2451	Information from T. W. Courtney
Gloucestershire		
S Gloucester, St. Oswalds	SO 8318	Glevensis, 10 (1976), 8
S Little Britain, Minchinhampton	SO 8402	Glos. Cotswolds, RCHM 1976
Leicestershire		
S Newbold Verdon	SK 4504	Information from M. Harding and K. Scott
Norfolk		
K? Shouldham	TF 6809	CBA GpVII Bulletin for 1971 1975 and 1977. Information from Mrs. V. Swan
Suffolk		
S Stanton	TL 9774	CBA Gp VII Bulletin 3 (1956), 6 Information from Mrs. V. Swan

Caernarvon should, of course, have appeared under Wales!

Tile-Kilns from the following sites are included in the Gazetteer

Cumbria

Brampton
Muncaster

Essex

Alphamstone
Ashdon
Colchester
Mount Bures
Theydon Garnon

Hampshire

Braxell's Farm
Crookhorn Farm
Scotland Farm

Hereford and Worcester

Upper Sandlin Farm

Hertfordshire

Black Boy Pits
Little Hadham
Netherwild Farm
Park Street
Potters Bar

Humberside

Messingham

Kent

Canterbury
Eccles

Lancashire

Quernmore

Lincolnshire

Heckington
Heighington

Surrey

Horton
Wykehurst Farm

Sussex

Wiston

Tyne and Wear

South Shields

Warwickshire

Arbury
Griff Hill Farm
Kenilworth
Lapworth

Wiltshire

Minety

Yorkshire

Grimescar

Scotland

Mumrills

Wales

Caernarvon
Gellygaer
Holt

Kilns plans are all drawn to a standard scale with the exception of the kiln-plant at Holt (Fig. 6.37). Some of the large plans do not contain scales.

BRAMPTON NY 5261

A tilery was found by chance in May 1963 during surface levelling for an extension to the recreation grounds of Irthing Valley School and excavation took place during 1963 under the direction of R. Hogg, then Keeper of Archaeology, Carlisle Museum. Eight kilns were found, two of which may have been used for pottery. They lay in an area of heavy clay with the wooded valley of the River Irthing 0.8-1.2 km to the north and west. Hadrian's Wall, at its closest, is 2.2 km from the kiln site and The Stanegate 0.8 km away. The nearest military settlement to the kiln is the fort at Old Church, Brampton, which lies 1.2 km west of the tilery. Water was available from the Brampton Beck 60 m south-west of the site, but some 8 m below the level of the kilns. The orientation of all the kilns was found to conform to the slope of the ground which meant that the combustion chamber of the kiln was dug into the 'hillside' with the stokehole pointing downhill.

KILN 1 Fig. 6.2

This was a rectangular kiln complete with oven floor and foundation walls orientated north-south with the stokehole to the south. The overall dimensions of the kiln were approximately 4.5 m square externally and 3.6 m internally. A flue 1.4 m long projected from the south wall into a stokehole. The north and east walls of the kiln were described as being merely the clay sides of the pit hardened by heat, although the published plan indicates the remnants of a stone wall on the east and an outline of a stone wall on the north.

The west wall was faced with roughly-coursed red sandstone one stone thick. Built against the east wall was a series of sandstone pillars 250 mm square which acted as supports for the sandstone slabs of the oven floor. The south wall was similar to that on the east, but had been pierced for the flue. The floor of the kiln was supported in an unusual way. There were four free-standing semicircular arches 760 mm in internal and 1.2 m in external diameter which aligned with the main flue and on either side of the arch was a sandstone column about 400 mm square. The columns and arches were linked by larger sandstone slabs as shown in the cross section of the kiln. The floor of the combustion chamber, once the columns and arches had been built, was roughly paved with sandstone slabs which were continuous with the floor of the flue from the stokehole. The floor of the kiln which rested on the columns and arches was 350 mm thick and was made of sandstone slabs some 100 mm thick, over which was spread clay. The sandstone slabs had been drilled to form vent-holes which were continued through the layer of clay. These vent-holes were on average 80 mm in diameter and corresponded to the gaps between the arches and columns beneath the floor. They were irregularly spaced and completely absent from the southern end of the kiln floor. There was no evidence from the fill of the kiln or from around it to show how the kiln was built above ground level.

BRAMPTON 1

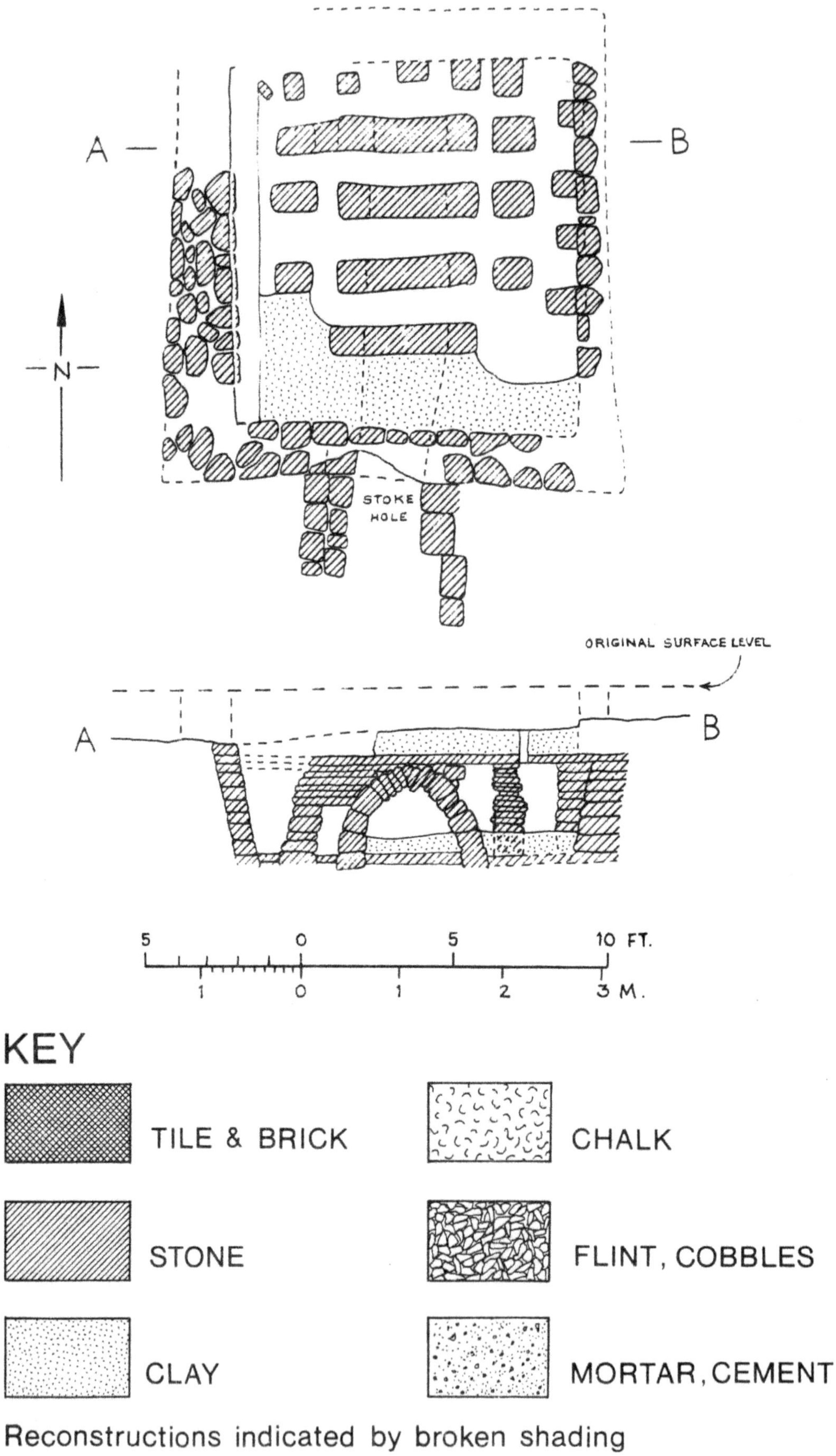

Fig. 6.2

BRAMPTON 2

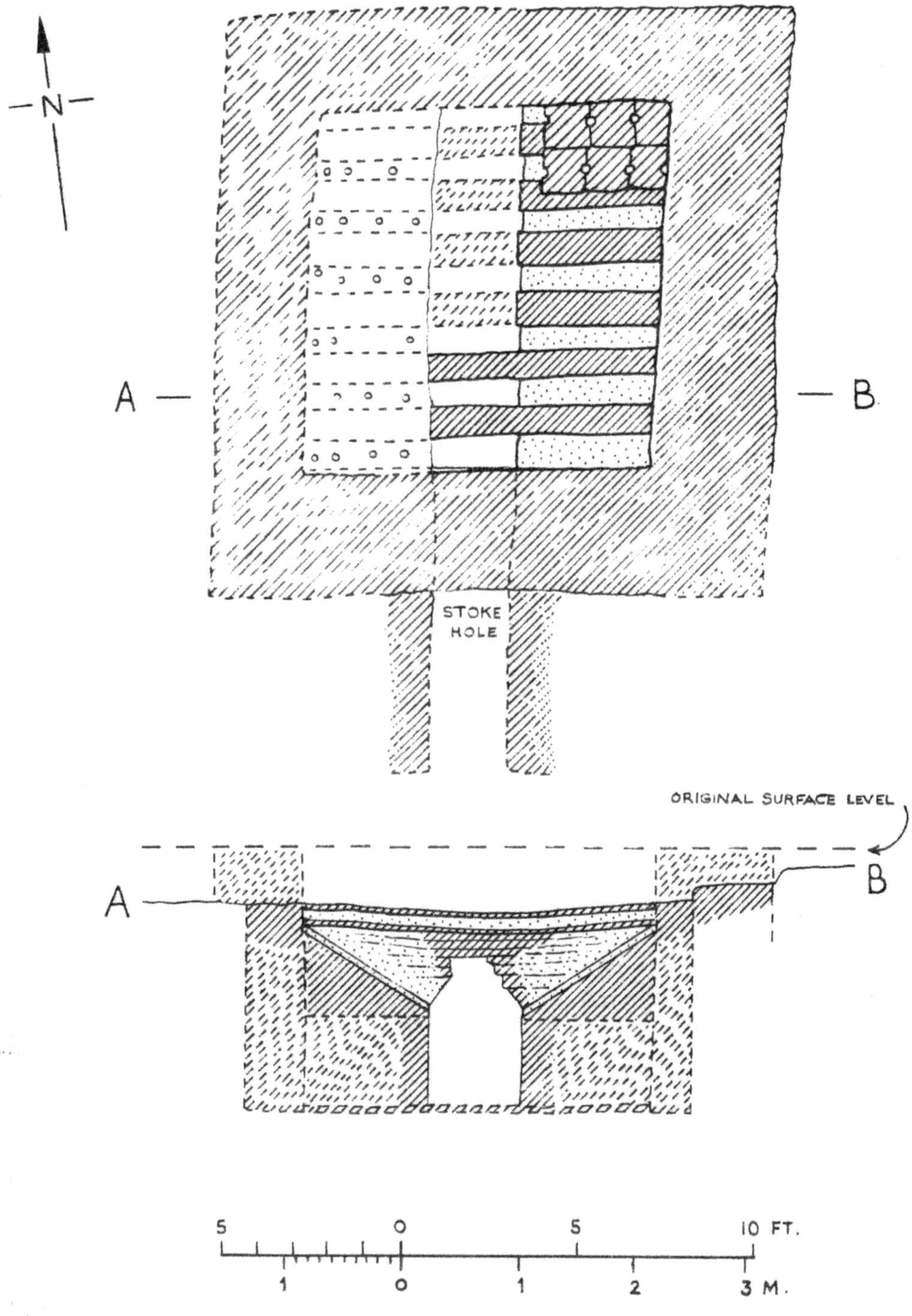

Fig. 6.3

The flue leading into the combustion chamber projected from the southern side of the kiln for a distance of some 1.4 m and was 680 mm wide and 910 mm high. Seven hundred and fifty millimetres from the kiln the walls of the flue, which were burnt, were well constructed with sandstone ashlar and the roof arch was of a flattish shape, but for the most part in a collapsed condition.

The flue lead into a large pit or stokehole which was not completely excavated, but was shown to be 7.6 m long and to have sloping sides. Close to the flue arch leading into the kiln was found a lamp stamped FORTIS which it was suggested served as an inspection lamp for examining the interior of the kiln. Both Hogg and Bellhouse suggest that this unusually-designed kiln, Hogg says 'faulty-designed', was, along with kiln 2, the earliest of the kilns to be built out of this group of eight and was made out of local material as no brick was yet available.

KILN 2 Fig. 6.3

This kiln was approximately 3 m square internally and orientated north-south, the stokehole being to the south. The inner walls of the kiln, which were lined with sandstone slabs laid end on, were almost entirely concealed by the substructure for the kiln floor. The walls of the main flue were 780 mm high and the flue itself 830 mm wide. There were six cross walls supporting the kiln floor, eech 280 mm thick and 200 mm apart and carried across the main flue by means of sandstone slabs set in clay forming corbelled arches. The floors between the cross flues sloped upwards from the main central flue.

On the floor of the main flue 300 mm from the rear wall of the kiln stood a stack of fourteen tiles each 380 mm square and 50-60 mm thick. It is suggested by the excavator that this pillar represented an extra support for the floor of the kiln showing a structural weakness in the original design.

Lying immediately on the cross walls was a layer of sandstone slabs each about 380 mm square and with semi-circular holes cut into the centre of the edges spanning the cross flues, so that when placed together they joined to form a vent-hole 60 mm in diameter. A 120 mm layer of clay was placed directly on the slabs followed by another layer of sandstone slabs 50 mm thick forming the kiln floor. The outer foundation walls of the kiln, which were 890 mm thick and in places stood three courses high, were faced on the inside with coursed slabs 450 mm long and about 120 mm thick. The main flue was 830 mm wide and 1.2 m in overall height and extended 1.5 m beyond the south wall of the kiln.

The stokehole pit was found to contain alternating layers of wood ash and burnt clay suggesting to the excavator not only flue cleanings and discarded oven superstructure, but also that the pits were used for dumping waste material from other kilns which were still in use. Evidence of the high temperature and continuous firing to which the kiln fabric had been submitted was visible in both the sandstone and the clay. The sandstone became purple with the surface forming a highly vitreous green glaze and the continuous heating and cooling of the kiln had caused complete shattering of the sandstone structures. The clay was fired evenly to a dark purple colour.

BRAMPTON 4

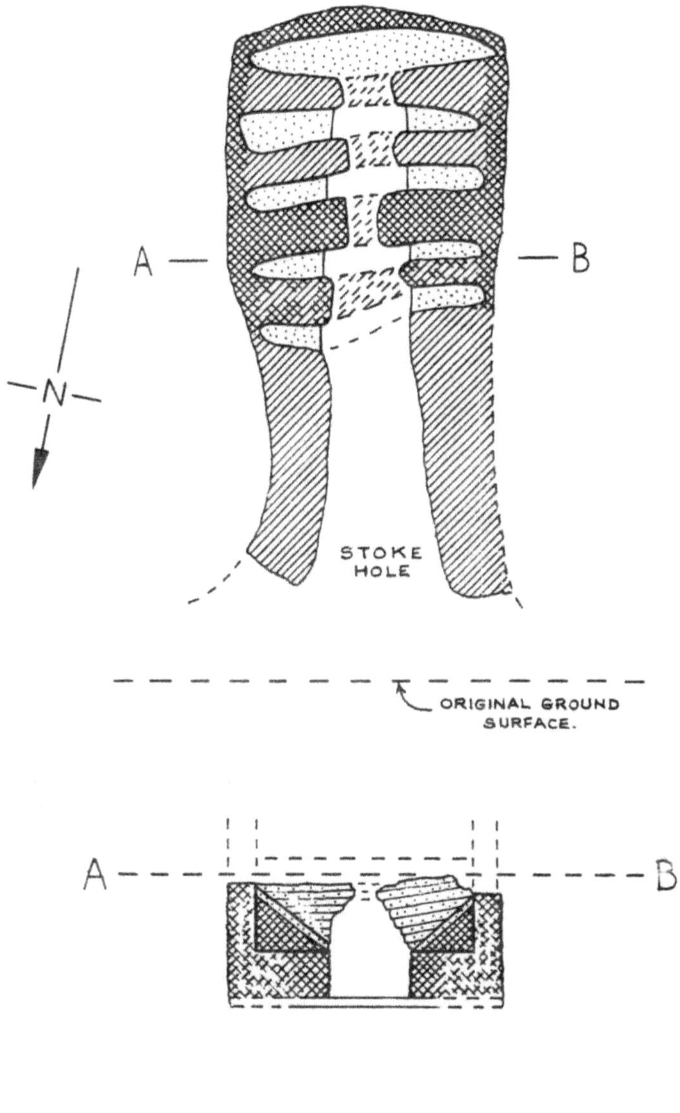

Fig. 6.4

KILN 3

A circular updraught kiln 1.14 m in diameter containing a number of fragments of dark grey pottery which were thought to be the main products of the kiln. It seems most likely that this was a pottery kiln.

KILN 4 Fig. 6.4

A rectangular kiln built of tile and sandstone, 2.5 m x 1.8 m with a 2 m projecting flue of similar width to the central flue of the kiln. The outer walls were composed of tiles and the platforms supporting the cross walls (i.e. the walls of the main flue) were constructed of large tiles. There were four cross walls with arches which went across the main flue and the floors between the cross walls sloped upwards from the central flue. The first cross wall was built of stone and tile, the second wholly of tile whilst the third and fourth were constructed only of sandstone. The tiles used in construction were mainly wasters. This kiln used the same stokehole as kiln 5.

KILN 5

A circular updraught kiln approximately 1.27 m in diameter with projecting tongue attached at the rear of the kiln. It was built of tile and sandstone and seems likely to have been for the production of pottery.

KILN 6 Fig. 6.5

An almost square kiln orientated north-south with five cross walls supporting the kiln floor and a stokehole to the north. The excavator commented upon the lack of symmetry and planning of what remained and noted that the kiln was not truely square. It was 2.6 m long internally and 2.6 m wide at the rear of the kiln, but this narrowed down to 2.4 m at the front or stokehole end. The walls of the kiln and flue were made of sandstone and the floor of the flue and combustion chamber were paved with sandstone flags.

The five cross walls were built of sandstone except for the west side of the first and the east side of the fourth wall which were made of tile. The fifth wall appeared to be different from the rest in having no joining arch across the central flue. The floor between this cross wall and the rear of the kiln was not inclined, but packed with solid clay implying that it had never been used as a flue. The same feature was noted with kiln 4. The remainder of the floors between the cross walls were made of broken tile with a capping of clay (as with kiln 2) and were inclined from the centre of the kiln upwards towards the outside .

In the centre of the main flue and standing on the flagged floor in line with cross walls 3 and 4 were two pillars of sandstone blocks, each about 300 mm square which were interpreted as extra supports required when the effects of the heat on the corbelled cross arches had caused them to sag.

The main flue leading from the stokehole, which was built of sandstone, was 2 m long from the inside face of the kiln, and 1 m wide narrowing down to 800 mm at its junction with the main flue of the kiln.

BRAMPTON 6

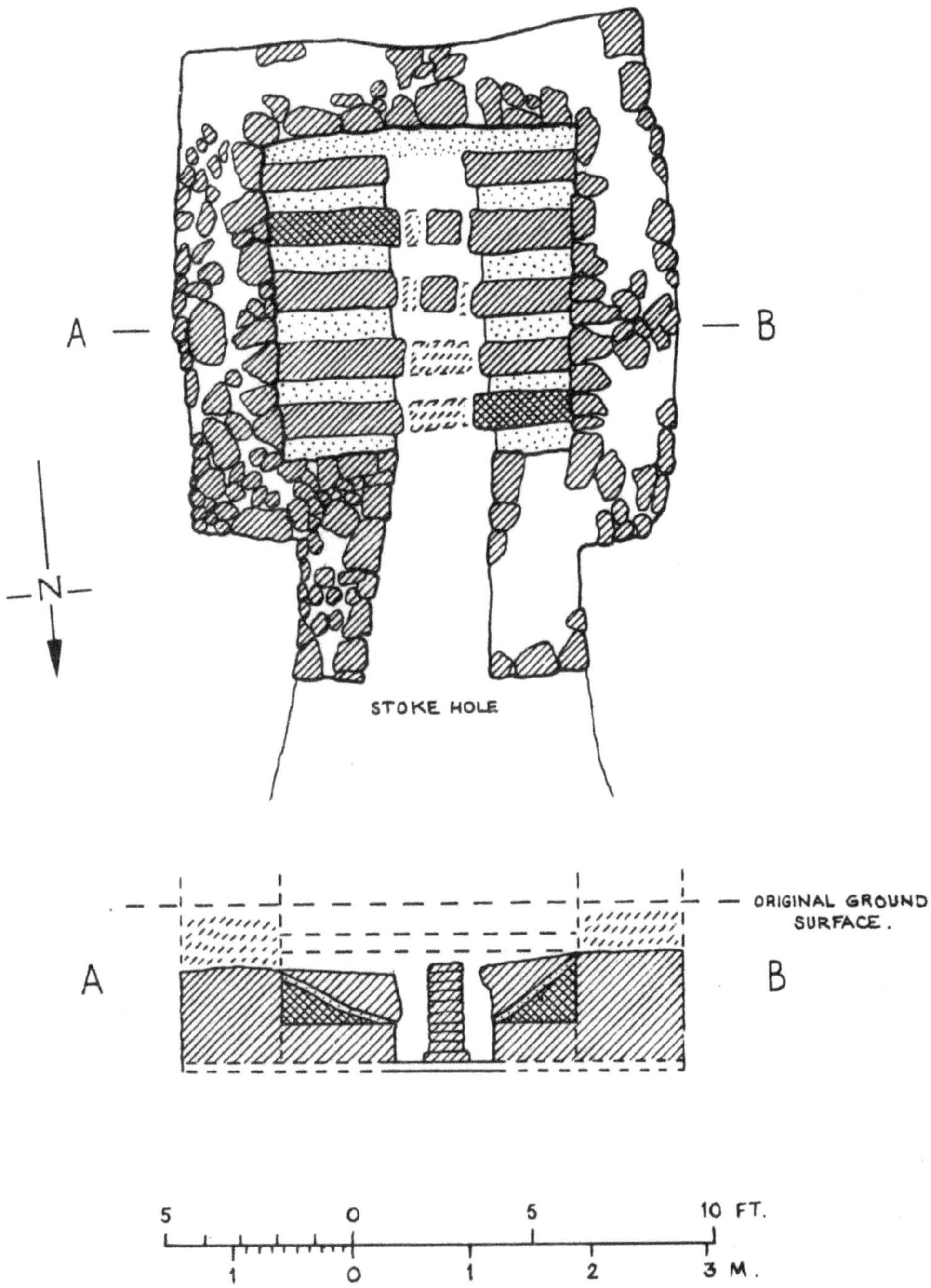

Fig. 6.5

KILN 7

Very little survived of this kiln of which there is no published plan. It appeared to be rectangular and equated with Grimes' type III, i.e. with upward-sloping flue bottoms between the cross walls. The main flue was 760 mm wide. The kiln differed from the rest in that the central flue was paved with cobbles which may have been a late feature of kiln development, when it was found that the earlier kilns with sandstone slabs were prone to excessive wear. Tiles were used in the platforms which formed the side walls of the main flue.

KILN 8

Only fragmentary foundation walls survived and there is no published plan. The kiln was rectangular and said to be similar in plan and size to kilns 2 and 6.

Date

The pottery, over 800 fragments, was studied by J. P. Gillam who then dated it to between A.D. 100-125. The 'inspection' lamp is a product of L.AEMILIUS FORTIS who worked at Mutina (now Modena) in north Italy and the bulk of his lamps was found in Flavian deposits with a few extending into the Hadrianic period. So a date of c. A.D. 70-110 was suggested for the lamp although in so remote an area from the source of production a later date is possible. The tilery appears to have been working during the early part of the second century, probably before work started of Hadrian's Wall.

Products

Few details of the products of the kilns are given in the published account other than 'finds of tegulae, imbrices, flat flooring tiles and box-tiles'.

Pottery wasters were found supporting the suggestion that pottery was being produced in this area and it seems likely that kilns 3 and 5 were pottery kilns.

Other Finds

A hoard of iron work was found in 1964 consisting of 58 pieces of Roman ironwork in a pit 600 mm in diameter and at least 910 mm deep, it being difficult to ascertain the actual depth as the modern ground level has been removed. None of the objects appear to be connected with tile making.

Discussion

Hogg concluded that although there was no direct proof, the Brampton tilery was most probably a military establishment operated by an auxiliary unit. He thought that the kilns were too simple and crude to be of legionary origin and the absence of any legionary stamps further convinced Hogg that they were auxiliary, probably used for producing tiles of the fort at Old Church, Brampton, about 1.6 km west of the tilery, when the Stanegate was consolidated in the early years of the second century. Bellhouse noted various clay pits 9 m in diameter into which had been dumped charcoal and kiln debris and it was from one of these pits that the iron hoard could have come.

It is worth noting that other military tileries also produced pottery as well. At Holt, for example, both pottery and tiles were being produced for the XXth

Legion and it is possible that on occasions tile kilns could be used for firing pottery, especially heavy and bulky articles like mortaria.

References

Hogg, R., 1965. 'Excavations of the Roman Auxiliary Tilery at Brampton'. CW[2] 65, 133-168.

Manning, W. H., 1966. 'A Hoard of Romano-British Ironwork from Brampton', CW[2] 66, 1-36.

Bellhouse, R. L., 1971. 'The Roman Tileries at Scalesceugh and Brampton'. CW[2] 71, 35-44.

MUNCASTER (ESKDALE) SD 1398

The first record of a kiln in the shadow of Muncaster Fell was made in 1884 and since then various notes appear in print reporting on the activities of Miss M. C. Fair and R. L. Bellhouse, the latter spending much time unravelling the unpublished accounts, including photographs, of Miss Fair. The notes which now follow are taken from the published account of Bellhouse and no attempt has been made by the writer to consult the originals which are in Carlisle Museum.

Bellhouse concludes that the sequence of events was:-

i. Kiln found in 1884.

ii. Miss Fair identified two kilns, A and B, during trial excavations in 1922 and 1923. The site seems to have been visited regularly before 1922, for example in 1917 when a <u>tegula</u> with grafitto was found.

iii. Miss Fair examined a potter's workshed (?) or supervisor's house and mentions that at least three kilns were found and destroyed.

iv. Miss Fair found three kilns in 1946 forming a plant and planned like a shamrock. They were fully loaded with box-tiles, roof and floor tiles and fire bricks, and had been deserted after being partly fired.

v. Bellhouse excavated a kiln in 1957 and 1959 which he identified as Kiln A of ii. above.

vi. Bellhouse identified another kiln found during the same excavations as Kiln B and possibly the same one as found in 1884.

KILN A Fig. 6.6

First identified by Miss Fair in 1922/3 when it was partly examined. Notes and photographs made at the time have led Bellhouse to the conclusion that this was the same kiln which he excavated in 1957 iand 1959.

The kiln, which was approximately 2 m square, had been built into the hillside and had a flue built of rough granite blocks set in red clay, part of which had collapsed in antiquity. The axis of the kiln was approximately north-west to south-east and the main outer walls survived to varying heights. The west wall survived to a height of 760 mm above the kiln floor and was coursed in brick and tile. The east wall was ruinous. The back wall, above the level of the kiln floor, was of coursed brick partly smeared with clay, but below the

MUNCASTER A

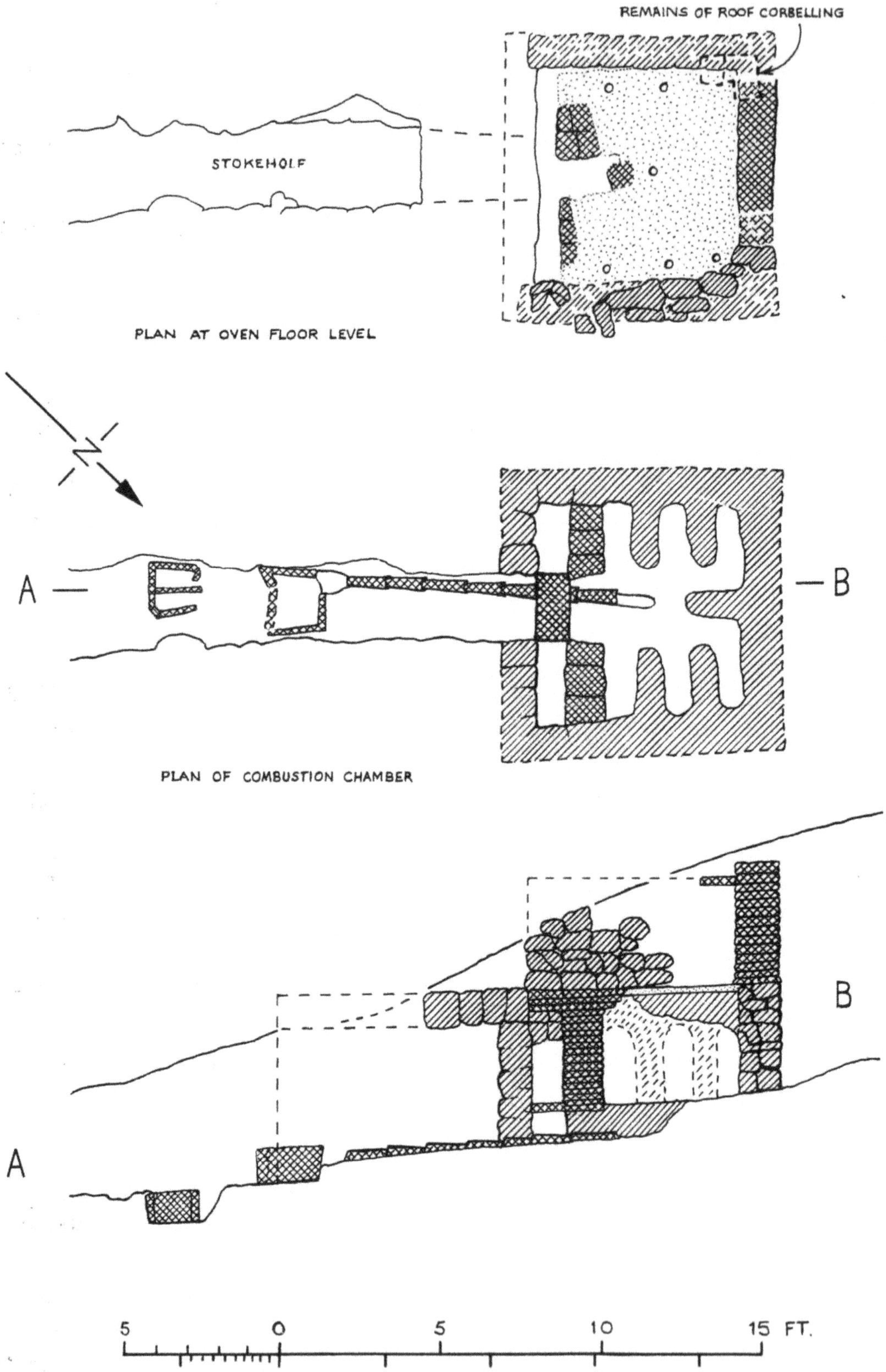

Fig. 6.6

kiln floor, in the combustion chamber, it was of the same build as the granite side walls. The southern wall was said to be a substantial wall of brickwork pierced by a single central entrance.

An important and possibly unique feature of this kiln was observed at the junction between the north and west walls where the excavator reported the beginnings of corbelling some 900 mm above the kiln floor, interpreted as the beginnings of a corbelled roof.

The kiln floor, which had a smooth surface of powdered brick and clay, was pierced with vents about 60 mm in diameter. There was a great deal of broken brick and tile lying on it.

This kiln contained another unusual feature which consisted of a line of imbrices on the floor of the flue leading into the combustion chamber of the kiln. The line started from a box-like arrangement made from heavy tiles and enclosing a space of about 430 mm by 480 mm. The first four imbrices were described as being inverted and covered with a mass of burnt clay and broken tile. The remaining three imbrices were the normal way up and led into the centre of the furnace where they rested in a trough out into the solid rock floor. Bellhouse at first interpreted this feature as a system for regulating an extra supply of air into the heart of the kiln so as to maintain an oxidising atmosphere. In a subsequent paper he drew attention to the fact that Kiln A was on a wet site and that the line of imbrices may have been a drain and that the tile boxes were collecting tanks for water. The fact that the rock floor of the kiln had to be cut out to take the tiles may suggest that care had to be taken with their position, which would be the case if they were to drain the floor of the combustion chamber. On the other hand the size of the 'collecting boxes', for there were two, would not have allowed much water to accumulate and be drained off. Similar features have been found at other kilns (see page 100).

Products

The usual group of products appear to have come from this kiln plus a 'rebated voussoir'. There were a few sherds of pottery from the stokehole, but none from the kiln.

Date

There was said to be no datable material from this kiln.

KILN B

Located by Miss Fair in 1922/3 and subsequently excavated by Bellhouse in 1960 from which it was possible to suggest that this was the kiln found in 1884.

The kiln was approximately square measuring 1.8 m internally. During the early stages of excavation it became apparent that there had been extensive robbing of both tile and stone, both side walls being almost compltetely robbed and it was only possible to determine their positions be areas of burnt clay. The axis of the kiln was shown to be north-west to south-east. The kiln floor was described as being supported on piers and found to be quite well preserved. It had only three vents. Four brick piers, based on large granite blocks, were found to be continuous across the kiln with their courses sloping steeply inwards towards the middle. The combustion chamber was not entered and excavated and so details of the sub-floor structure are not available. It was suggested that constant use at high temperature had caused the kiln to be reconstructed and several features were connected with this rebuilding. New piers were built on the remains of the collapsed kiln floor and even these eventually sagged inwards causing a gap to develop between the piers and the outer wall of the kiln. This gap was filled to make level the firing floor by using tegulae and box-tile wasters set in clay. The earilest kiln floor was made of large red tile and the piers of the second floor were largely of tile wasters apart from the granite blocks. It is not easy to construct a plan from the published information.

Products

When several kilns are found together it is difficult to say whether one particular kiln was used for one kind of tile. It is likely that mixed loads were fired and that different loads were fired in the same kiln if it was in use for any length of time.

Wasters found in Kiln B could have come from Kiln A, for example, but it is worth noting that among the wasters found were tegulae, box-tiles and notched or rebated voussoirs. However, Bellhouse suggested either that the presence of box-tiles and tegulae wasters in the stokehole might indicate the contents of the last firing, or that they could easily have been dumped into a disused stokehole from a nearby kiln depending upon the position of the wasters found. Both kilns are rather small when compared with other tile-kilns.

Date

Little pottery was found. From Kiln B a rim was found in one of the floor vents and reported to be Gillam type 120. The rest of the pottery came from the threshold of the stokehole of Kiln A and dates to the Hadrianic-Antonine period.

References

Early references to the site are summarized in Bellhouse 1960.

Bellhouse, R. L. 1960. 'Excavations in Eskdale, the Muncaster Roman Kilns'. CW^2 60, 1-12.

Bellhouse, R. L. 1961. 'Excavations in Eskdale, the Muncaster Roman Kilns'. CW^2 61, 47-56.

O.S. Map of Roman Britain 4th Edition. 1978, 29, where the site is listed as Parkhouse.

ALPHAMSTONE Fig. 6.7 TL 8735

A tile and clay structure found in a sand and gravel pit near Moat Farm in 1928 has been interpreted as a tile-kiln in VCH. After the initial observation of an arch, subsequent digging brought to light a flue 4.0 m long and 480 mm wide internally with the wall surviving in places to a height on 760 mm. At one point two tiles of an arch remained in position. The walls of the flue were said to have been built of unfired tiles very few of which were perfect. Behind the walls the sand was burnt purple for several centimetres confirming the interpretation as a flue, as does perhaps the thick layer of charcoal noted at the time.

No floor was found and little of the superstructure remained save for one piece of wall 'above the west wall and eight inches outside it', and of similar construction to the flue.

Finds

Fired tiles recovered from the site include pilae tiles 230 mm square and 38 mm thick, and portions of many flanged tegulae. All the remains were removed immediately the investigations were completed, presumably destroyed by the removal of sand and gravel. There are no finds in Colchester Museum.

References

Colchester Museum Report for 1929, 24-28 (including a plan and photograph).

VCH, Essex III (1963), 35-6 (including a redrawn plan).

ASHDON Fig. 6.8 TL 5838

The discovery of numerous tile fragments and pottery on the surface of Oak Field, the property of Lord Braybrooke, lead R. C. Neville to carry out excavations on the site in 1852, which resulted in the discovery of a square kiln. It measured 5.48 m square externally including the 0.91 m outer walls although the plan which accompanies Neville's text shows the side walls to be about 1.11 m wide. The combustion chamber was about 3.5 by 2.6 m. There were seven cross walls which separated eight lateral flues each 180 mm wide and having sloping floors 300 mm above the level of the main flue and therefore of Grimes type III. The main flue was 760 mm wide at the entrance to the kiln, but along most of its length it was only 600 wide.

The construction of the walls was not described in detail. The north-east wall of the kiln was said to be 'carefully constructed of Roman tiles' and there is mention in the text of flanged tiles, presumably tegulae, in the walls. Clay appears to have been employed in bonding the tiles and in sealing joints.

Date

Very little pottery was said to have been found during the excavations and the provenance of the coins, 'three or four brass coins of the Constantine family' was not stated.

ALPHAMSTONE

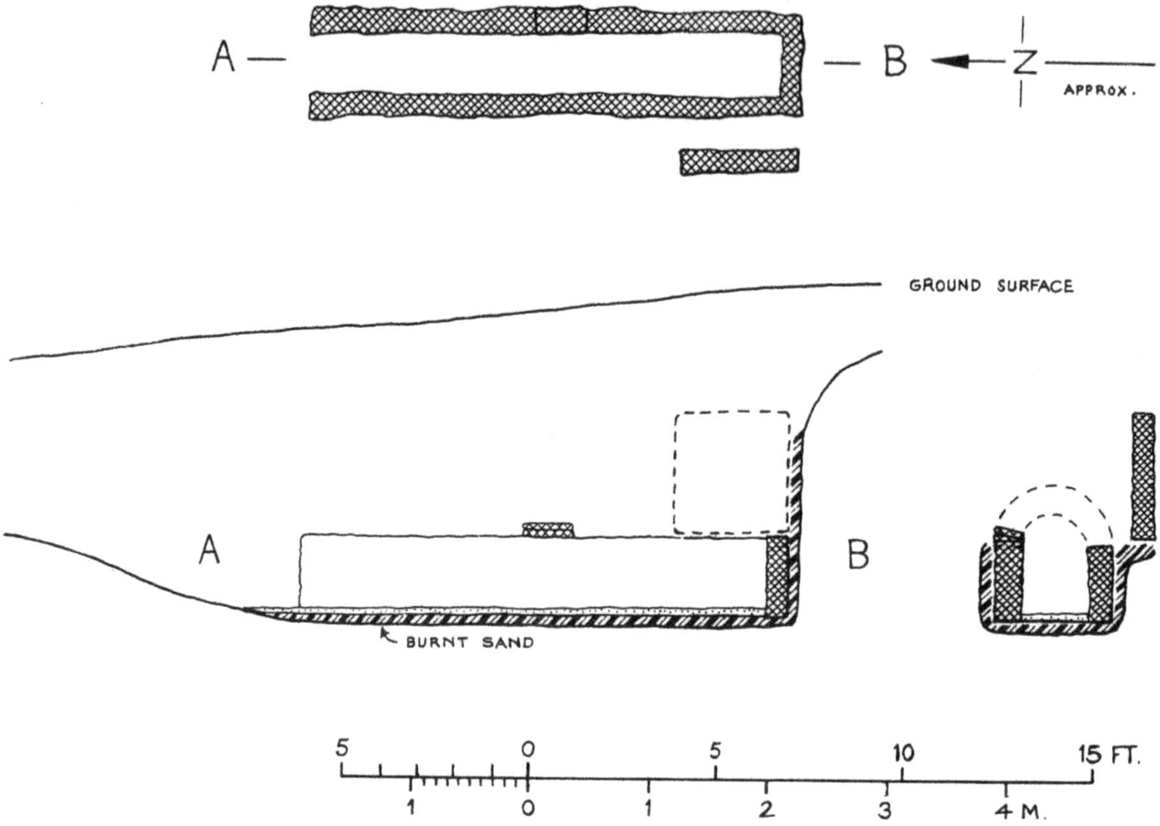

Fig. 6.7

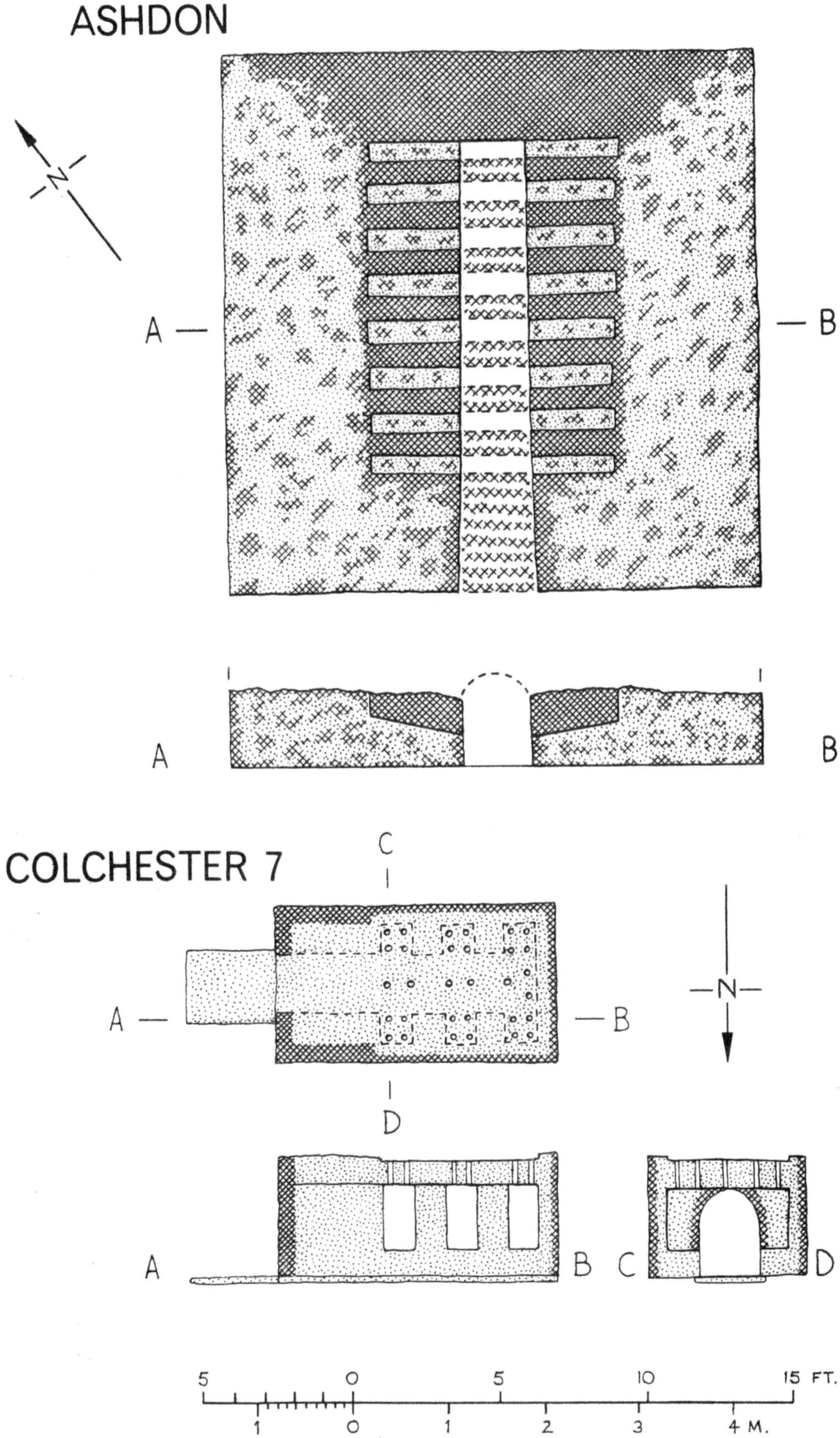

Fig. 6.8 -

Discussion

The measurements given on a drawing in the Braybrooke papers in the University Museum Cambridge enable a plan and cross section to be drawn. When this plan is compared with the published version there are slight discrepancies (Neville, 1853. 22). The plan illustrated here (fig. 6.8) is based on the measurements and is not a copy of the plan in Archaeological Journal.

References

Neville, R. C., 1853. 'Account of an Ancient Kiln, Excavated in the Parish of Ashdon, March 22, 1852', Arch. Journ. 10, 21-4.

VCH Essex III (1963), 45.

Plan and water colour in the Braybrooke (Neville) papers in the University Museum Cambridge.

COLCHESTER

The kiln numbers used by M. R. Hull in his report (Hull, 1963) have been retained for those kilns discovered before 1963.

KILN 7 Fig. 6.8

Found in 1877 in the south-west corner of field 496 (see Hull, 1963 for site plan), this kiln is approximately 1 km west of the colonia defences. Attempts were made to preserve the kiln, but eventually it appears to have been destroyed. From the plan published by Hull the kiln chamber measured 1.8 m by 1.4 m internally; the overall measurement of the rectangular structure, including the flue, is 2.9 m by 1.7 m. When found the kiln floor was intact and about 50 mm thick with 31 vent-holes 50 mm in diameter arranged regularly as shown on the accompanying plan. The side walls of the kiln are described as 220-300 mm thick and built of clay bricks of various sizes.

The main central flue, arched with tiles, was 600 mm wide and 2.7 m long although its floor extended another 900 mm outside the kiln proper. The flue-arch leading into the stokehole and the walls of the rectangle were built of tiles laid in clay. The piers dividing the lateral flues were built of rectangular blocks of clay and were separate from the main kiln wall. In the published account the difference between the various words used is not clear, as for example between blocks of clay and tiles /bricks. It is assumed that the first are unfired whereas the latter are fired. The piers divided the combustion chamber to provide three cross flues which had flat floors 300 mm higher than the main flue and therefore of Grimes type II.

Products

From the structure itself came pottery including mortaria and tile, whereas near the entrance to the kiln were large quantities of mortaria and 'pinched' vessels.

Date

All the material from kilns 7-11 is discussed together where a date of late third to early fourth century is suggested.

Discussion

The presence of the pottery near to the entrance of this kiln lead T. May to suggest that the kiln was used for producing mortaria. This may have been so and perhaps some of the smaller rectangular kilns were used for producing heavy, large vessels like mortaria, although Hull seems to think that it may have also been used for tiles.

References

The early works are cited in Hull in

Hull, M. R., 1963. The Roman Potters' Kilns of Colchester, London, 3 and figs. 3 and 4.

KILN 17 Fig. 6.9

A twin kiln, the larger of which was described as 'bottle-shaped in plan with a central rib supporting the clay floor on arched flues of hand-puddled clay'. The kiln chamber was about 1.4 m square and large quantities of pottery were found associated with it making it almost certainly a pottery-kiln.

Hull, 1963 16-7 and fig. 11.

KILN 31 Fig. 6.9

During the excavations of 1959 on the southern slope of the hill south of Sheepen Farm, that is in the eastern end of field number 1074, a rectangular kiln was found associated with a 'pear-shaped' pottery kiln, both being served from a common stoke pit and having a complicated history of phases.

The plan produced by Hull is not very detailed, but it does show a main central flue 3.4 m long (from stokepit to rear of kiln) and about 680 m wide. The flue was lined with 'crude' clay blocks. Six cross flues are suggested from what remains and their floors are flat in the eastern half of the kiln and some 150-200 mm above the floor of the central flue, whereas in the western part of the kiln they are sloping at an angle of about 30°. This may represent the different phases of the kiln when one half was possibly rebuilt.

Hull suggests that soon after the kiln was built the cheeks of the mouth appeared to have been wearing badly; to remedy this thin grooves 80 mm deep were cut in the floor against each wall, and in them a tile was set vertically and luted into position with clay.

Part of the kiln floor survived in the north-west corner and had 50 mm diameter vents in it as well as vents at the ends of some of the cross flues. The kiln floor was repaired several times with thin layers of soft clay which hardened when the kiln was fired. This clay carried the impressions of several 'herring-bone' mortarium stamps as a result of placing mortaria upside down in the kiln during firing, although it is difficult to believe that the oven floor was still wet when the mortaria were loaded.

The oven floor was supported by cross walls carried across the main flue by arches which were described as a 'double row of clay voussoirs'. Over much of the interior of the kiln was a rendering of clay applied by hand leaving various finger-marks.

COLCHESTER 17

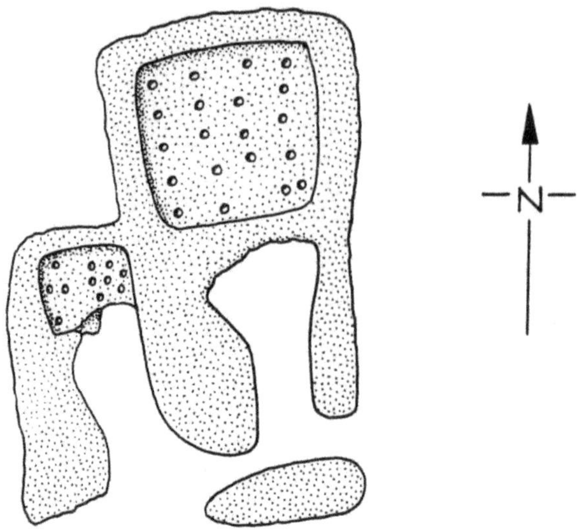

COLCHESTER 31

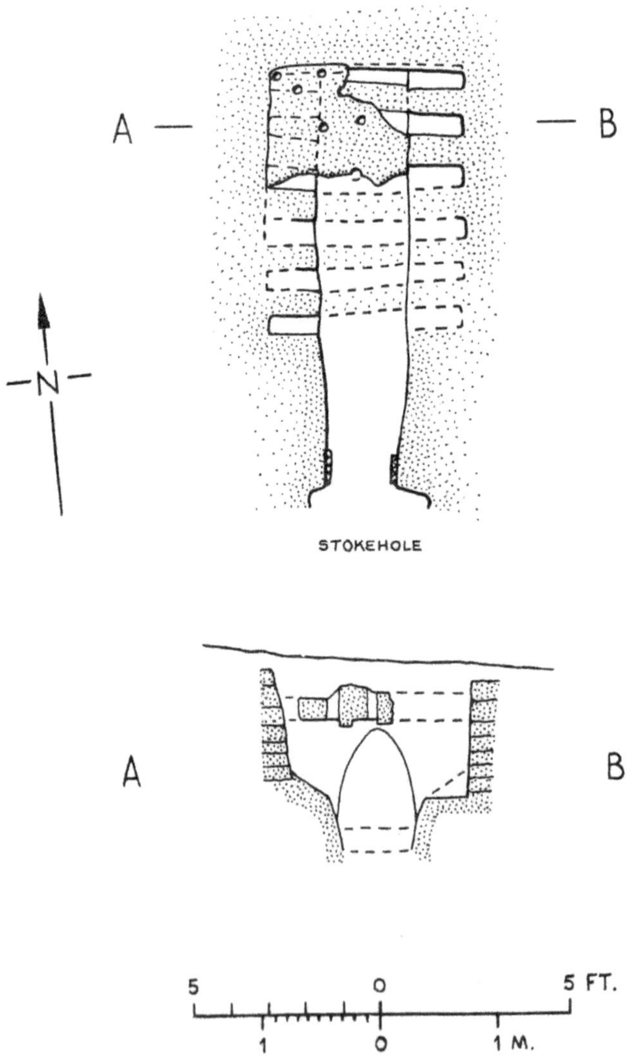

Fig. 6.9

Products

In view of the mortaria stamps on the kiln floor it is assumed that this kiln was used for firing mortaria at some stage in its life. There appears to be no evidence that it was ever used for tiles.

Date

Hull dates the kiln to about A.D. 175-210.

Reference

Hull, M. R. The Roman Potters' Kilns at Colchester, London 1963. In particular see pages 39-43, figs. 10, 15 and 16 and several plates.

MOAT FARM, LEXDEN TL 9826

The site lies on the northern bank of the River Colne directly opposite the Sheepen site to the south of the river. Two kilns, one built inside the other, were found during ploughing and subsequently excavated by P. R. Holbert in 1969-70.

KILN 1 Fig. 6.10

Only the outer wall of the kiln survived, the rest having been destroyed when kiln II was constructed. The outer walls were about 7-800 mm thick and enclosed a kiln chamber approximately 4 m square. They were built with clay blocks 80 mm thick, 280 mm wide and of varying lengths. On the inner face of the kiln walls was a 70 mm shelf or ledge presumably designed to carry the kiln floor. Various markings on the inside surface of these walls suggested to the excavator that it had cross flues with sloping bottoms and that the kiln was of Grimes type III. It was also clear to the excavator that the kiln had actually been used.

KILN II Fig. 6.10

Inside and against the outer walls of kiln I was constructed a second kiln resulting in it effectively having outer walls 1.2 m thick. These newly-constructed walls were built of unfired clay blocks of about 70 mm thickness. The firing chamber was approximately 3 m square. The main flue of the kiln projected beyond the outer walls and was of one build with kiln II.

There were seven cross walls in the combustion chamber mainly built of tegulae broken roughly to size by removing the flanges. The walls were carried across the central flue by arches formed with voussoir tiles 380 mm by 250 mm, tapering from 57 mm to 38 mm in thickness. The cross flues formed by these walls were about 170 mm wide on average and had sloping bottoms built up with clay. Between each wall, although not in a line, were placed three tiles set vertically to act, according to the excavator, as spacers to stabilise the sub-floor structure and to prevent distortion. Nearly all the walls and arches were given a thick coating of clay which became vitrified by heat during the operation of the kiln.

Resting on the cross walls and arches was the kiln floor which consisted of a layer of tiles each about 330 mm by 260 mm and 70 mm thick. These

MOAT FARM, LEXDEN

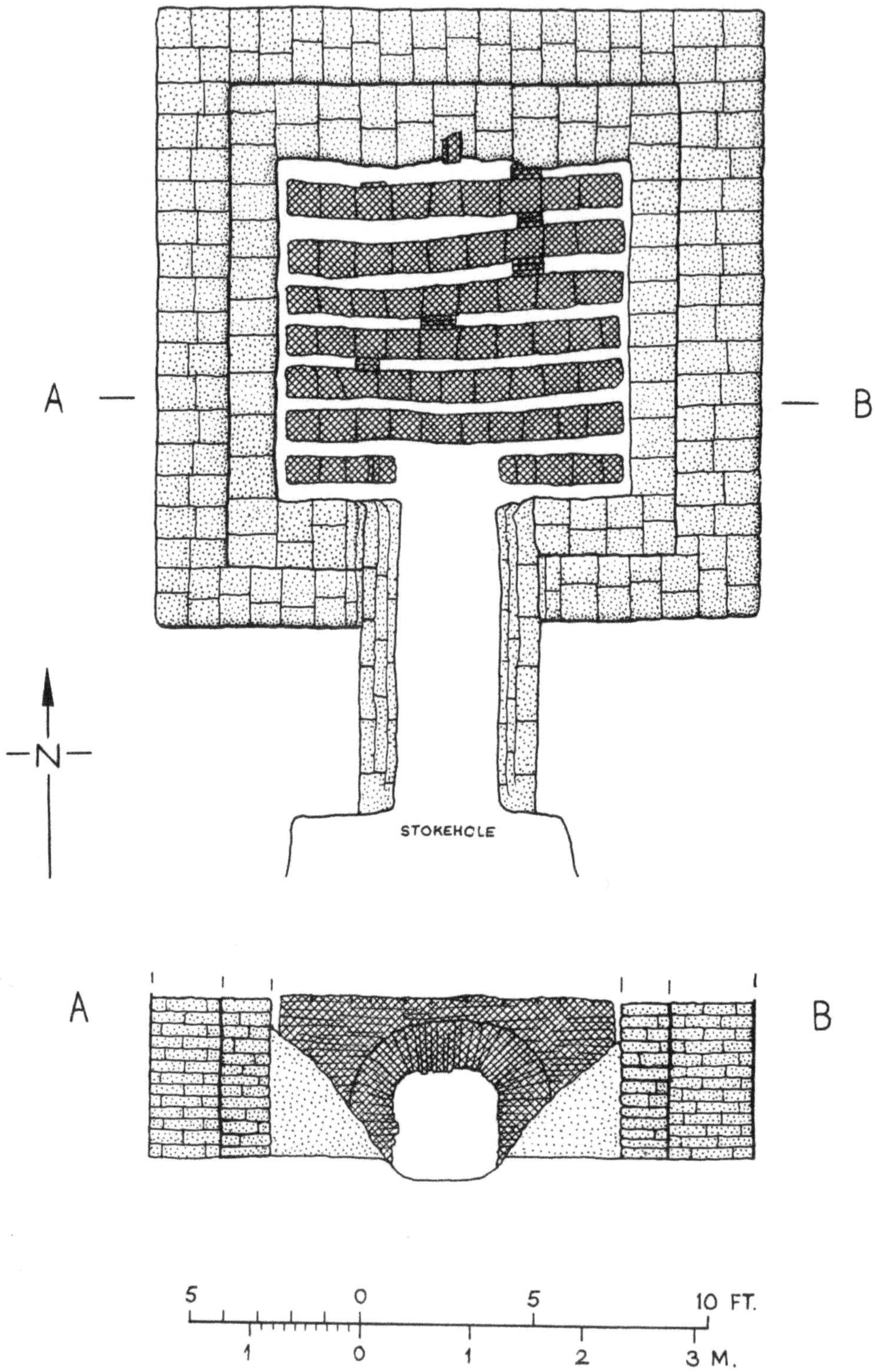

Fig. 6.10

MOUNT BURES

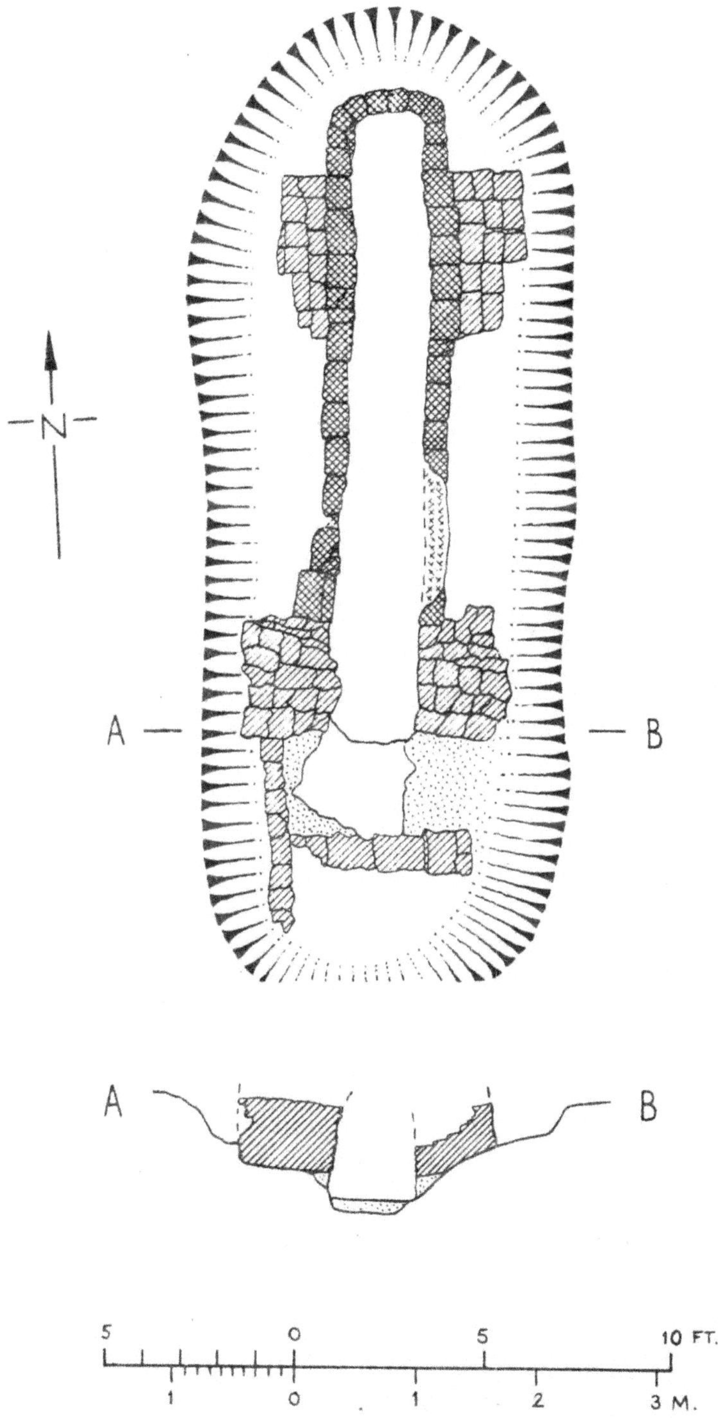

Fig. 6.11

tiles had a 'V' shaped notched cut out of each long side so that when laid across the lateral walls the 'V' notches came together forming a series of square holes above the cross flues. The tiles in turn were covered with several layers of clay, the holes being kept clear and rounded off.

In the centre of the northern wall of this kiln was found an upturned imbrex tile which, together with another laid on top, the excavator suggests could have formed a vent connecting the main flue with the outside. From the section it appears that the imbrex tile is above the floor level of the kiln and so may have been a vent for the firing chamber rather than the flue.

The stokehole which projected southwards downhill was not excavated to any extent, but in it was found a thick deposit of black wood-ash containing pottery and tile fragments.

Products

There is no evidence from the published account of any indication that it was possible to identify the products of either kiln.

Date

The excavator dates the kilns to the latter half of the first century A.D.

Discussion

Although the building of one kiln inside another is an unusual form of construction, this would be a logical step if the construction of the original kiln was found to be faulty. The building blocks of Kiln I were described as 'cindery' in places and thought to have been made from peat rather than clay which could have been the result of using clay taken from too near the surface.

References

Britannia, 2 (1971), 273.

Holbert, P. R., 1971. 'Excavation of Roman Tile Kilns at Moat Farm, Lexden, Colchester, 1969-70', Colchester Group Annual Bulletin Vol. 14 (1971), 22-34.

MOUNT BURES Fig. 6.11 TL 9132

M. R. Hull when writing the third volume of the VCH for Essex drew attention to the discovery in 1955 of roof tiles near Elms Farm and in particular one fragment of tile he considered to be a waster. The absence of mortar and pottery led Hull to suggest that the site was that of a tile-kiln. Excavation by P. R. Holbert in 1971 followed a proton magnetometer survey by Dr. Tite.

The surviving remains consisted of a flat-bottomed tile-lined flue 5 m long and 0.6 m wide. The walls of the flue were made mainly from tegulae laid in clay which showed signs of considerable use, the tiles being burned bluish-black in colour. Just short of the rounded north end of the flue were two buttress-like masonry blocks, one on each side of the flue. A similar pair were found at the southern end and were probably connected with the superstructure to the flue, whatever that may have been. Beyond these two blocks in the south a low stone wall extending a further 1.5 m may, with a wall at right angles, have formed the rear of a small stokehole area. The whole of this tile structure had been constructed in a trench cut into the natural gravel, the bottom of which was about 1.6 m below modern ground level.

Products

Four types of tile found on the site, which however could not be shown to have been made in the 'kiln', were, tegulae, imbrices, flat pilae or building tiles and box-flue tiles. It is not clear how much pottery was found; only five sherds are listed in the report.

Date

The excavator assigns a second-century date to the structure although his reasons for doing so are not clear.

Discussion

It is far from clear what function this structure served. It seems most unlikely that the surviving flue formed part of the more conventional tile-kiln although in the absence of any superstructure even this is not certain. Perhaps it is to be associated with a drying shed for either pottery or tiles. One is reminded of the so-called tile kiln from Scotland Farm, Hants., (see page 137), and other similar structures discussed earlier (page 101).

References

VCH Essex 3 (1963), 60.

Britannia 3 (1972), 334.

Holbert, P. R. 1972. 'A Romano-British Tile Kiln at Mount Bures, Essex', Colchester Archaeological Group Annual Bulletin, 15 (1972), 19-22.

THEYDON GARNON Fig. 6.12 TL 4703

A possible tile-kiln was found in 1891 on Coppersale Common in a field called Solomon's Hoppet. It consisted of two walls about six metres long, 600 mm high and 600 mm apart, built with roof-tiles, flue-tiles and paving-bricks. The space between the walls was filled with ashes and wet clay. The plan is reconstructed from the measurements given.

References

Essex Arch. Soc. 4 (1893), 222.

Essex Naturalist 4 (1890), 79.

Essex Naturalist 15 (1901), 48.

RCHM Essex, 1921, 61.

VCH Essex iii (1963), 188.

O. S. Map of Roman Britain, 4th Edition 1978, 29, where the site is listed as Epping.

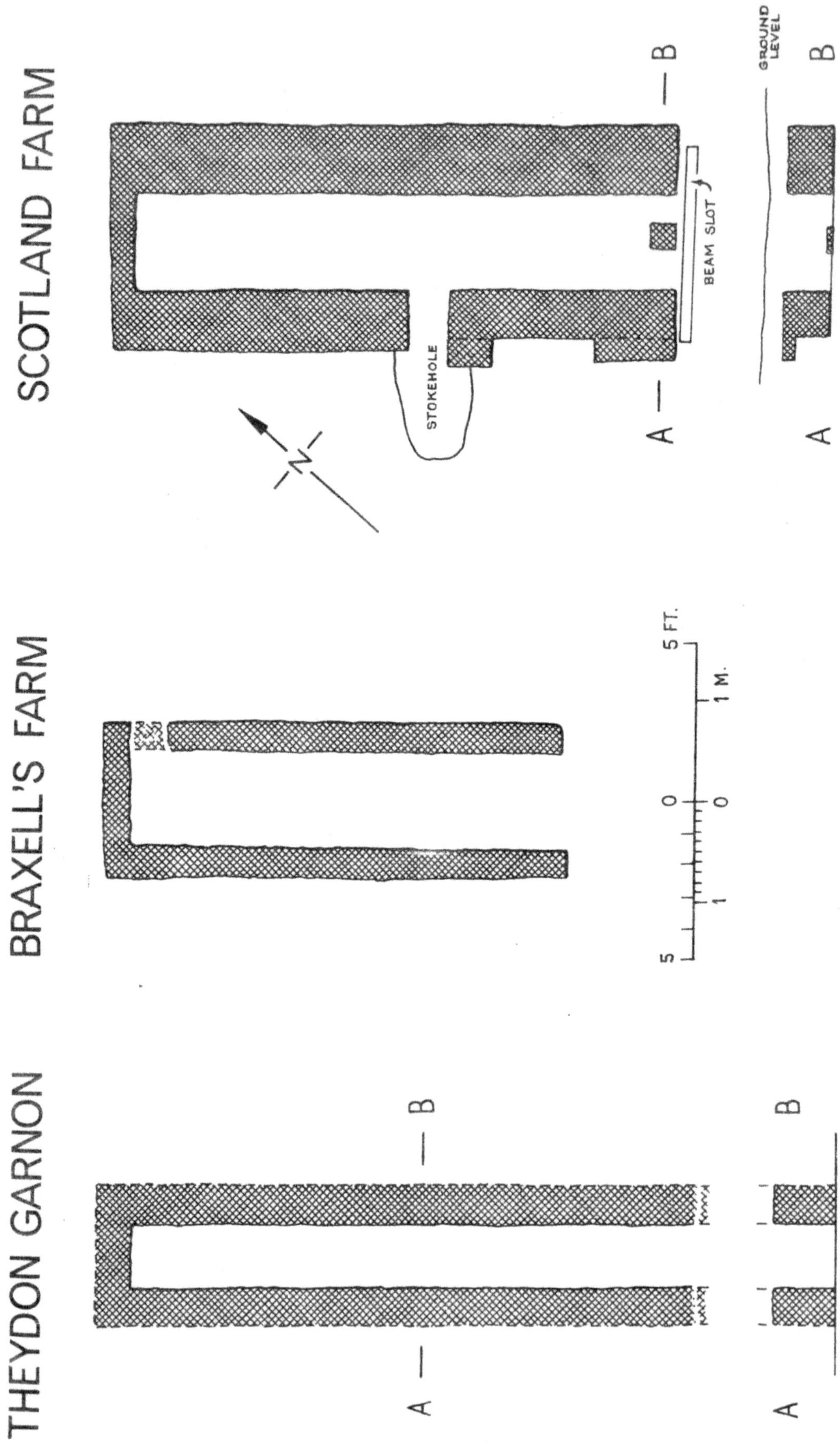
THEYDON GARNON
A
B
BRAXELL'S FARM
5
0
1
5 FT.
1 M.
SCOTLAND FARM
STOKEHOLE
BEAM SLOT
GROUND LEVEL
N

Fig. 6.12

CROOKHORN FARM

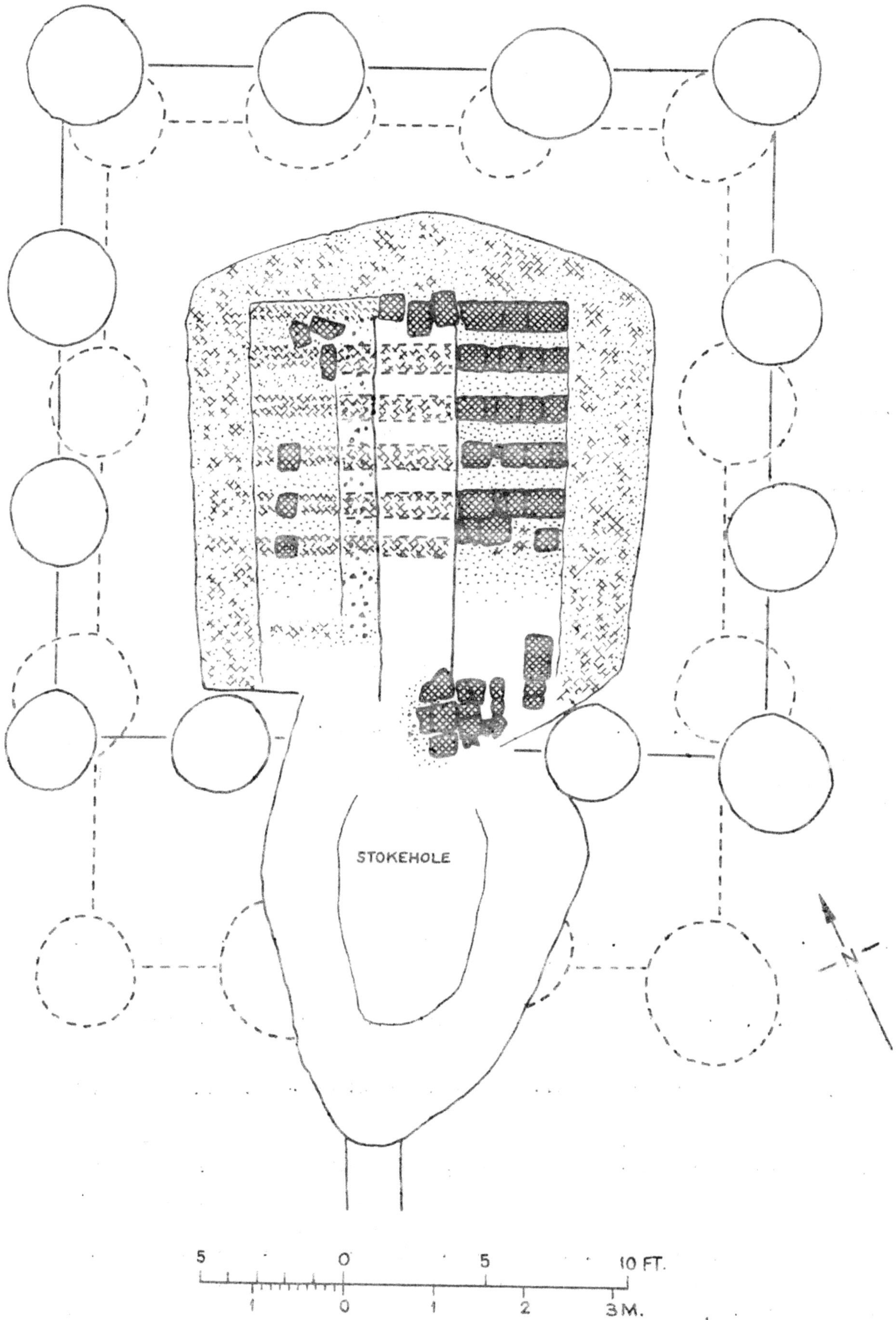

Fig. 6.13

BRAXELL'S FARM, BOTLEY Fig. 6.12 SU 5115

During 1956 land on Braxell's Farm was being prepared by bulldozer for cultivation when Roman tile was found and reported to Winchester City Museums. Under difficult circumstances Mr. F. Cottrill was able to recover the plan of the structure which had been disturbed and to make a few notes which are now in the Winchester City Museums.

The rectangular structure was 4.4 m by 0.9 m internally and walled on three sides by five or six courses of tegulae. The 220 mm wide walls were made of roofing tiles 'laid in the direction of the walls' and set in clayey-sand which was discoloured by heat. The tiles in the wall were described by Cottrill as soft and pliable from imperfect firing. The fourth side was open and Cottrill's notes clearly indicate this to be an original feature and not the result of the bulldozers' activities. However, a break of 380 mm at the junction of two of the walls may well have been caused by the contractors.

Products

Associated with the debris of the tile structure were tegulae, box-tiles and flat tiles.

Date

To one side of the structure were found a few sherds of pottery of late first-century date.

Discussion

The plan is not unlike the tile structure found at Mount Bures, Essex, and Scotland Farm, Hants. For further discussion see later page.

References

Proc. of the Hants. Field Club, 22 part 1 (1961) 22-4.

CROOKHORN FARM Fig. 6.13 SU 6807

During 1974/5 excavations in advance of development uncovered an industrial and agricultural site in which there was a tile-kiln and associated timber structure. South-west of the kiln was a large Roman clay-pit dated to the second and third centuries into which fed a drain from the stokehole of the kiln. North-east of the kiln were indications of pits filled with wasters.

The kiln consisted of a central flue with side flues at a higher level and with sloping floors, i.e. type III. Sub-floor walls and flues had been dismantled in the western part of the kiln and there was no trace of the oven floor. Internally the combustion chamber measured c. 4.1 m by 3.5 m with six cross walls. The whole structure was built of tile and brick, but where these bricks were fired is not clear. The outer wall was built on foundations of gravel and tile wasters set in clay again indicating that tile was being made before this kiln was constructed. From the stokehole a drainage ditch lead towards the clay pit 18 m away.

Surrounding the kiln were two periods of post-holes representing two post-built rectangular structures which are interpreted as some form of covering to protect the kiln during loading. There is no other kiln from Britain which has such clear indications of a surrounding structure although the post-holes found at Arbury (page 175) may be similar. Possible parallels exist in other provinces and there is evidence for them in the medieval period (information from Graham Soffe). The first phase of post-holes were packed with flint and contained no tile waste suggesting that the timber structure was erected at an early stage and before tile was being produced, whereas the post-holes of the second phase had a packing of stiff clay and wasters. Perhaps the first timber structure covered s simple clamp which produced tile and brick for the kiln itself.

The plan of the timber structure in each of the two phases was similar and probably represents a rectangular open-sided building, i.e. basically just a roof supported on posts. Whether any of the other buildings excavated belong to the tile-works is uncertain, they do seem to have a more agricultural look about them.

Products

The usual range of tile and brick seems to have been made some of which was stamped TIFR. One of these stamps was found in 1926 in the villa at Crookhorn and another came from a Roman building at Langstone 4 km east of the tilery indicating that it supplied its products to both places.

Date

Archaeomagnetic dating samples were taken from the kiln giving a date of A.D. 330 ± 20.

Reference

Britannia, 7 (1976), 366.

Information from Graham Soffe.

SCOTLAND FARM, ODIHAM Fig. 6.12 SU 7453

The site is on a layer of valley gravel with outcrops of London Clay only a few hundred feet away, just over 1.6 km to the south of a Roman villa at Lodge Farm. Flue tiles, imbrices and tegulae wasters were found and have been deposited in the Willis Museum, Basingstoke. The rectangular structure found during excavations by K. D. Graham, measured 5.7 m by 2.3 m overall externally and had tile walls of various widths enclosing an area 5.4 m by 1.0 m. The walls were built of different types of tile bonded together with yellow clay and gave the impression that they had been salvaged from an earlier structure. The whole structure was bedded on the natural gravel though the bottom of the foundation trenches had been covered with a layer of yellow clay. The floor between the walls consisted of the natural clay/gravel mix which had hardened during firings to form a solid floor. In the middle of the south-western side was a 460 mm break in the wall and on one side of the break the wall widened.

A pit, interpreted as the stokehole, extended for about 0.9 m and was 600 m wide, the bottom being delineated by a layer of ash which rose sharply up away from the kiln. There was no wall at the south-eastern end; instead there was a single tile 250 mm square in the centre of the opening, resting like the walls, upon a foundation of yellow clay. A few centimetres away was a beam slot 1.8 m long, 130 mm wide and 100 mm deep cut into the gravel. The impression gained by the excavator was that this end could be opened and closed when required, in particular for the removal of ash, a thick layer of which was found extending for several metres away from the structure.

The ash layer was also found over most of the floor inside the walls and in it were preserved traces of burnt logs measuring 450 mm by 80 mm in diameter. These logs were laid side by side parallel to the long axis of the 'kiln' on either side of the stokehole on the south-western side.

The excavator thought that when in use the 'kiln' had served the locality for a relatively short period and then been deliberately dismantled.

Products

Many tegulae, imbrices and building tiles were found. The tegulae all appeared to be of similar size and measured 380 mm by 220 mm, the majority having a semi-circular mark onthe underside. No complete imbrex was found and so no dimensions were given in the report. The average size of the building tiles, which appeared to be the commonest product of the 'kiln', was 180 mm by 180 mm by 30 mm.

Date

No dating evidence for the use of the structure was found. A coin of c. A.D. 330-335 was discovered in the rubble infilling and the excavator drew attention to the fact that the major building period of the villa at Lodge Farm was about A.D. 306-381. A link between the two is possible.

Discussion

Although at the beginning of the report there is reference to the discovery of flue tiles on the site none are mentioned in the appendix on 'Finds and Products'. One would like to know whether they were combed or whether they had a relief pattern made with a roller die and if so, of which type. Further discussion of this type of structure will be found on page 101, but it can be stated here that from the quantity of tile and the presence of wasters it was connected with tile-making.

References

Britannia 2, (1971), 282-3.

Graham, K. D. 1971 'The Roman Tile-Kiln at Scotland Farm, Near Hook, Hants.', The Journal of the Farnham Museum Society I (1971), 22-29.

O.S. Map of Roman Britain, 4th Edition, 1978, 29, where the site is listed as Potbridge Farm.

UPPER SANDLIN FARM, LEIGH SINTON Fig. 6.14 SO 7551

Roofing tile and part of a chimney were found at Upper Sandlin Farm and reported by G. H. Jack in the Antiquaries Journal for 1925. An excavation was carried out by G. R. Nicoll in 1955 and published by P. L. Waters in 1963 (see below).

The kiln which had stone outer walls was built entirely of tile internally. It was dug some 1.5 m into the ground and rested on the underlying rock. The combustion chamber was approximately 3.2 m x 2 9 m. There was a central flue, 1 m wide and the platform either side of this flue was raised about 350 mm to form a platform for the six cross walls. These cross walls were built of tegulae and brick; the floors between them were sloping upwards towards the outside of the kiln. The main flue projected 1.5 m into the stokehole which was covered by a thick layer of charcoal containing pottery. Several Roman ditches, about 1 m wide, were found, and interpreted as drains.

Date

No stratified pottery was found to date the construction of the kiln but fragments of samian in the stokehole, together with a stamped mortarium from a nearby ditch led Waters to postulate a date in the latter half of the second century for the active operation of the kiln, and perhaps continuing into the third century.

Products

Numerous pieces of tile were recovered from in, and around, the kiln and surrounding ditches. These included tegulae, imbrices, bricks and box-flue tiles.

References

Antiquaries Journal 5 (1925), 285.

Waters, P. L., 'A Romano-British Tile-Kiln at Upper Sandlin Farm, Leigh Sinton, Worcs.', Transactions Worcestershire Archaeological Society, New Series 40 (1963), 1-5.

UPPER SANDLIN FARM

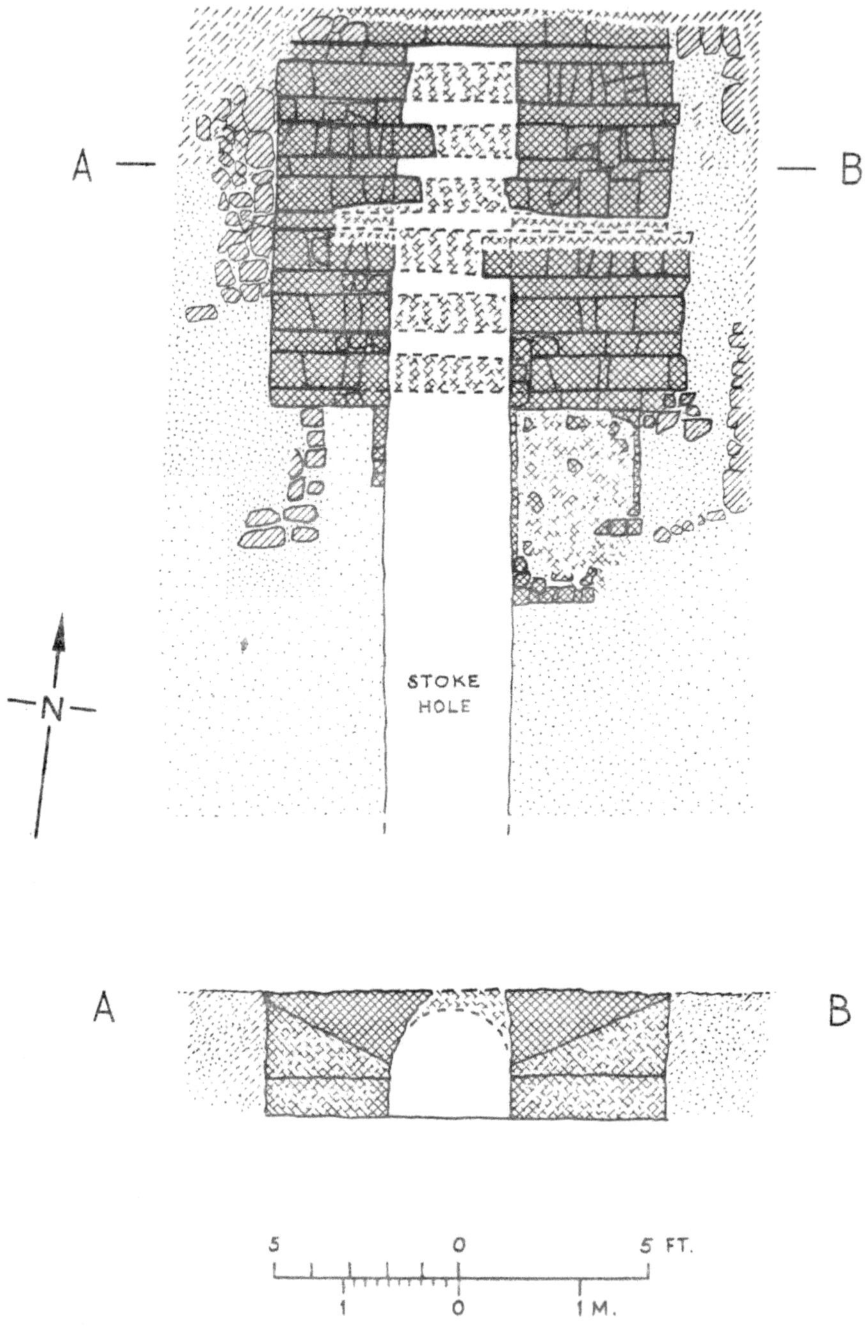

Fig. 6.14

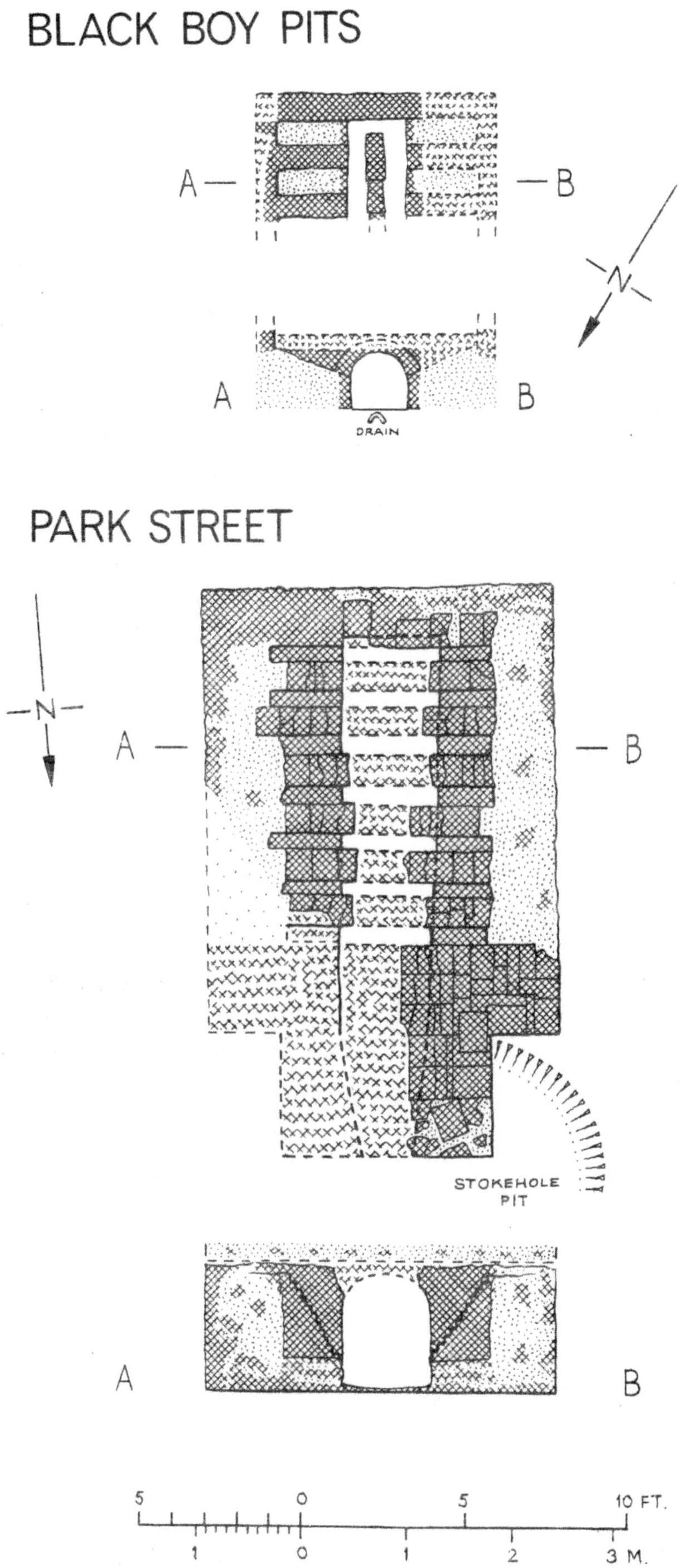
BLACK BOY PITS
A
B
N
DRAIN
PARK STREET
N
STOKEHOLE PIT
5
0
5
10 FT.
1
0
1
2
3 M.

Fig. 6.15

BLACK BOY PITS, NEAR ST. ALBANS Fig. 6.15 TL 1202

During the extraction of gravel in 'Black Boy' pit in 1932 Dr. Norman Davey noticed a quantity of tile which on closer inspection proved to be the remains of a kiln. A metre or so of the kiln had been destroyed before the site was discovered and so precise dimensions are unavailable. Working under extremely difficult circumstances (the kiln was situated at the top of a 8 m vertical face!!) Dr. Davey was able to excavate what remained and to produce an informative plan.

The surviving remains comprised a main central flue and two cross walls and flues. The floor of the main flue was made of clay and was 1.4 m below modern ground level. The kiln, which was 2 m wide internally, was built of tile and clay blocks, varying in thickness up to 70 mm, bonded together with clay daub, _tegulae_ being frequently used. Two cross walls each 220 mm thick and arched across the main flue were found and others must have existed originally. The cross flues were 220 mm wide and began 300 mm above the level of the main flue then sloped upwards at an angle of 20° towards the outside of the kiln.

Perhaps the most interesting aspect of this kiln was a line of _imbrices_ found along the centre of the main flue which Davey interprets as a drain to remove water from the combustion chamber which had been dug in a seam of clay and was therefore likely to retain water. These tiles were about 150 mm below the floor of the flue and Davey calculates would have 'met the surface of the ground several metres to the front, or stoking end of the kiln'. The kiln was constructed on a slope and dug into the hillside.

Products

Large quantities of tile were found in varying stages of firing and included roofing tiles, walling tiles and box-flue tiles, leaving no doubt in Davey's mind that the primary use of the kiln was for tile. Fragments of only two pots were found.

Date

The pottery which was found buried in the debris filling the main flue was dated to the first half of the second century.

Discussion

Although Davey entitles his paper the 'Roman Tile-and Pottery-Kiln at ...' there seems little doubt that the main purpose of the kiln was for firing tiles; indeed there is no firm evidence to suggest that pottery was made at all. The small size of the kiln might suggest to some that it was planned for pottery, but as can be seen from other sites, small tile-kilns are not uncommon.

References

Davey, N., 1933. 'Roman Tile-and Pottery-Kiln at Black Boy Pits, St. Stephens, near St. Albans'. _Trans. of St. Albans Architectural and Archaeological Society_ for 1932. 212-214.

LITTLE HADHAM

Several sites claiming to be tileries have been found in the parish. Extensive pottery and tile works are known from Bromely Hall Farm in an area centred on two fields known as Wickham Spring and Barley Hill about 2.4 km south-west of the village of Little Hadham. Five kilns are suggested from surface finds at Westland Green, 1.5 km west-south-west of Little Hadham. None of these kilns has been extensively written up and it is not altogether clear how many kilns have actually been found. No details of a tile-kiln, said to have been excavated by Hadham Hall School in the grounds of the School, can be traced.

Barley Hill and Wickham Spring

The results of the excavations conducted in the 1960s have not been published although a brief account has appeared in the Stort Valley Group Newsletter. At least one tile-kiln was said to have been found and excavated in Wickham Spring and one in Barley Hill. There may have been more.

BARLEY HILL c. T_ 419215 c. TL 419215

This kiln, 'consisted of a stokehole and tile-lined flue opening upwards into a clay-lined firing chamber containing the remains of one cross wall'. Few other details are available. Mr. Partridge writes that 'the kiln was much larger than Wickham Spring with a rectangular chamber and two opposing flues' (letter dated 6.vi.77).

Found on the site were fragments of *tegulae*, *pilae* and bonding tiles, but only two pieces of an *imbrex*. If activity in the adjacent pottery works was contemporary with the tilery then the tile kiln would be of fourth-century date.

WICKHAM SPRING TL 418216

No details of this kiln are available although Mr. C. Partridge believes it to have been 'quite small with a single flue'.

WESTLAND GREEN TL 425224

Five tile-kilns have been identified from surface finds by Mr. F. Cowbar. The site is only 1 km away from the kilns at Bromley Hall Farm and indeed may be a continuation of the same works.

Discussion

Clearly this is a very large industrial complex with both tile- and pottery-making taking place and it is unfortunate that no plans apparently survive of the tile-kilns from Bromley Hall Farm. The destination of the kiln products is uncertain, but the proximity of the site to Stane Street, makes it relatively easy for them to be marketed over a fairly wide area.

References

JRS 55 (1965), 211; 58 (1968), 194; 59 (1969), 221.

Britannia 1 (1970), 289; 3 (1972), 330.

Stort Valley Archaeological Group Newsletter 14.

NETHERWILD FARM

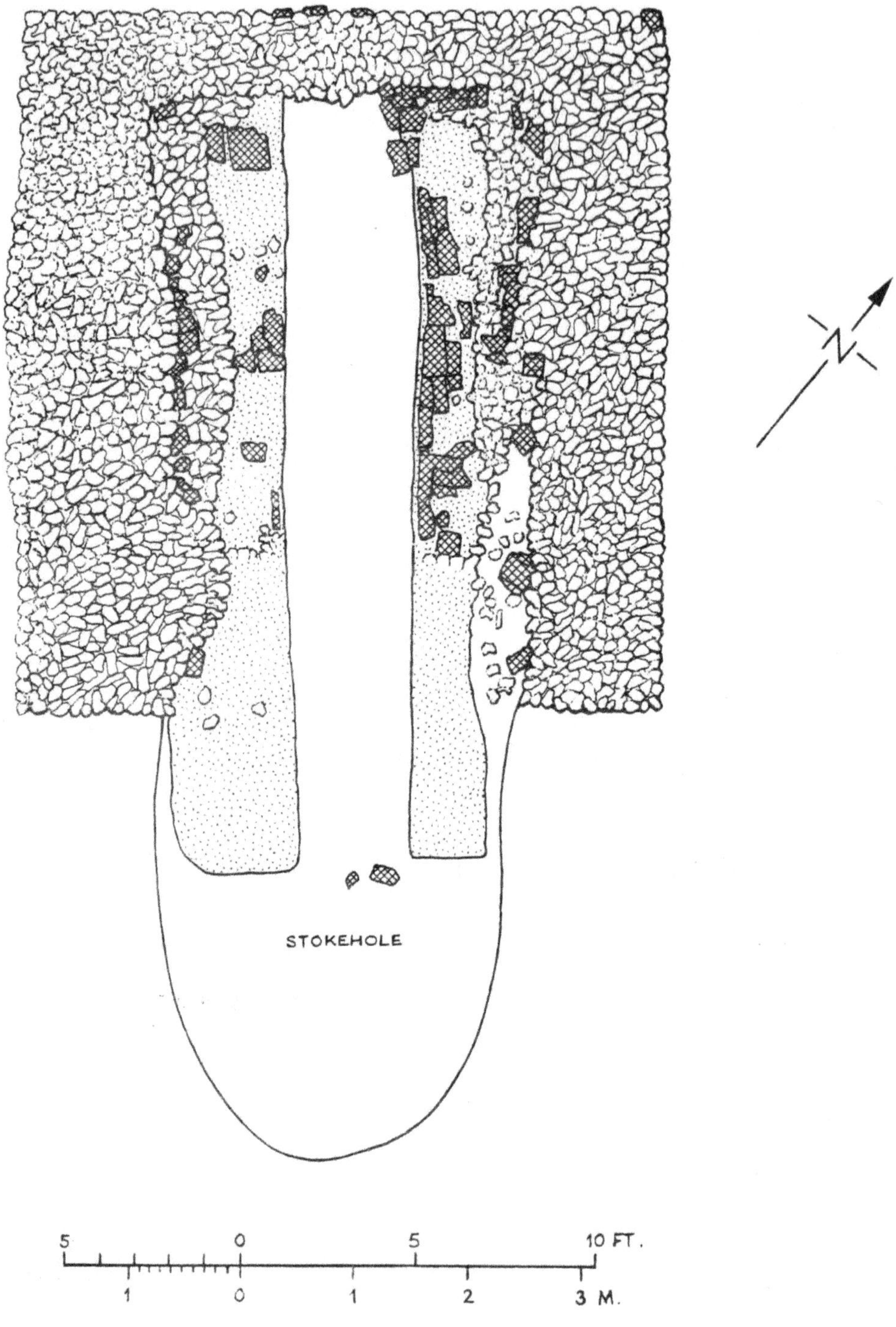

Fig. 6.16

NETHERWILD FARM, COLNEY STREET Fig. 6.16 TL 1401

Ploughing in 1964 in a large field adjoining the water meadows beside the River Colne revealed four areas of red-burnt clay and tile debris, the most northerly of which was excavated by B. F. Rawlins in 1965. An almost square kiln was found aligned north-west to south-east with its stokehole to the south-east. The foundations of the side walls varied in width, the inner edge being ill-defined except in the east corner where the wall was 1.21 m wide. Elsewhere the foundations were as much as 1.83 m across. It is difficult to understand which walls are being described in the published account of this kiln and no further attempt is made here to describe them.

The main central flue was 1.07 m wide with a floor of natural clay which had been burnt to a depth of 380 mm. The excavator noted that 'a cutting through the side walls of the flue revealed that originally the central portion had been 1.52 m wide, but had been reduced to 1.07 m by the addition of further tile courses'. The main flue projected beyond the surviving foundations on thesouth-east, side of the kiln for a further 1.09 m, but because of damage to this part of the site one cannot be sure of the original design. Nothing remained of any cross walls or flues, or of the kiln floor which would have rested upon them. The stokehole pit covered an area of about 3.20 m by 3.66 m and trodden into its bottom was a scatter of charcoal, mid-second century pottery and a whetstone. A crescent-shaped bank of clay, not shown on the excavator's plan, had been built around the south side of the stokehole pit to prevent it from filling with water.

Products

Bonding or flat tiles were produced in the kiln along with both types of roofing tiles. From the site came some badly-fired pottery suggesting to Rawlins that pottery was also made nearby.

Date

Rawlins dates the kiln to the mid-second century A.D. and suggests, from its build and degree of burning, that it had a fairly long life, probably until the end of that century, being finally demolished in the late third century.

Discussion

The details given in the preliminary report do not show how the dates given above were arrived at. East of the kiln was a fourth-century bath house and south-west of the bath house the very damaged remains of a dwelling house of uncertain date. Any connection between these buildings and the living quarters of a tilemaker cannot be established. The presence of houses close to kilns can be interpreted as the owners' or workers' living accommodation or alternatively the presence of a kiln may have been due to the need to make tiles close to the site of the building being constructed . Other kilns are thought to lie 60 m to the south of this kiln, which might lend some weight to the idea that the buildings belong to the tiler.

References

Rawlins, B. F., 1966. 'Preliminary Report on Excavations of Roman Tile-Kilns'. Bulletin No. 3 Watford and S.W. Herts. Arch. Soc. April 1966.

JRS, 56 (1966), 209.

PARK STREET Fig. 6.15 TL 1303

The site lies 2.81 km south-east of Verulamium on boulder clay. A small stream runs close by the kilns which is situated on a gentle southwards-facing slope. It was discovered during the removal of a tree in the garden of 67 Mayflower Road, Park Street and subsequently excavated by B. F. Rawlins in 1969 before the construction of a garden pond which now seals the kiln.

The kiln was aligned north south with the stokehole to the north. It had been built in a pit dug 1.21 m into the natural clay and the side walls of the kiln had been built against the vertical face of this pit. These walls were 630-680 mm thick and built of bonding tiles, many of which were wasters, the inner face using complete or large pieces of tile, while the outer face used small pieces all bonded with puddle clay. The south wall was still standing 15 courses high, i.e. 1.09 m, and was 530 mm thick widening to 630 mm at its topmost two courses. It was built mainly of complete bonding tile with clay bonding 3.8-5 cm thick. Each course slightly overhung that below giving a forward tilt of about 80°. The north wall is stated by Rawlins to have been only 410 mm thick, but it is far from clear what this measurement refers to on the plan.

The overall outer dimensions of the kiln given by the excavator were 4.3 m by 3.6 m with the flue projecting 1.2 m beyond the kiln's north face. Measurements taken from the plan using the bold line as the limits of the kiln do not quite agree with the figures just quoted, being 4.1 m by 3.3 m. The internal measurement of the kiln chamber are approximately 2.7 m by 2.1m.

The main central flue was 4.11 m long and 850 mm wide at the northern end and 950 mm at the south; there was a rise in the floor level of the flue from north to south of 38 mm. The flue walls were made of three courses of complete bonding tiles, the two lowest of which were laid as stretchers and the third as headers, and clay bonding courses 64 mm thick, making the height of the flue side walls about 280 mm. The flue was paved with 95 square pilae tiles, 17 of which were stamped with the letter M, stopping 580 mm short of the south wall. The flue was said to project into the stokehole pit 1.2 m beyond the kiln's north wall, although the published plan makes it difficult to interpret this part of the kiln.

There were six cross walls each remaining to a height of about 1.12 m and built of complete bonding tile giving an overall width of about 279-304 mm. The tile was bonded with puddled clay which bore the horizontal impressions of finger marks. The arches which carried these six cross walls over the main flue had been destroyed, but sufficient remained to show that the spring of the arch began 533 mm above the flue bottom. Between the cross walls were seven cross flues each 150 mm wide, the floors of which began 250 mm above the tiled floor of the central flue and sloped upwards towards the outside of the kiln at an angle of about 60°. The best preserved cross flue was 1.93 m long from one side of the kiln to the other and rose in 14 tile-built steps to meet the outside wall.

The main and cross flues were found to contain much red-burnt clay, many tile fragments and yellow clay. From the lower levels of this fill was found early fourth-century pottery most of which was in a red under-fired state according to the excavator. The stokehole pit was not completely excavated, but it measured at least 4.26 m by 3.35 m and was 1.28 m deep. The excavator recognised four distinct layers of charcoal each 25-50 mm thick separated by clean yellow clay which he suggested indicated four seasons' use of the kiln.

Products

Bonding tiles	(280 mm x 350 mm)	3 diagonal grooves made with fingers across the full extent of the tile.
Tegulae		3 grooves diagonally across the tile or as a semi-circle at the lower rebated end.
Imbrices		Triple groove along the crest combed with an 8 toothed comb 30 mm across.
Pilae tiles	(170 mm square)	3 diagonal grooves across the tile; 17 also stamped with the letter M 25 mm high.

Date

Rawlins concluded that the building of the kiln was linked with building activity at Verulamium in the early fourth century, basing his assumption on the discovery of fourth-century pottery in the fill of the kiln.

Discussion

Doubt must be cast on the date given for the construction of the kiln as pottery from the fill of the flues only gives a date at which that fill was accumulating. In an area producing both pottery and tiles, as this seems to be, the presence of under-fired pottery of fourth century date merely indicates that pottery was being made at that time and has no bearing on the date of construction of the tile-kiln.

References

Rawlins, B. F. 'A Roman Tile-Kiln at Park Street near St. Albans'. Hertfordshire Archaeology 2 (1970), 62-5.

O. S. Map of Roman Britain, 4th Edition, 1978, 29, where the site is listed as How Wood.

POTTERS BAR (OR PARKIELD) Fig. 6.17 TL 2501

During 1950 Mr. G. R. Gillam's attention was drawn to a small ledge in a field known as Parkfield at Potters Bar. The field is on a gentle slope facing north-west and is essentially on London clay with a capping of pebble gravel at the summit. Excavations were conducted in 1953-4 revealing a brick built flue 1.21 m below modern ground level. It was 3.58 m long internally and built of unmortared brick which in places survived to a height of 457 mm comprising 8 courses of brick. The side walls of the flue leaned inwards, being 609 mm wide at the bottom and 431 mm at the top; however, at times the flue narrowed down to only 355 mm. The floor of the flue was made of clay 50 mm thick which sloped downwards towards the stokepit. The clay floor and material beneath were affected by heat to a depth of 203 mm and the whole flue was chocked with ash. No other structural features were found associated with this flue.

The stokepit was 3.65 m long and 3.42 m wid

POTTERS BAR

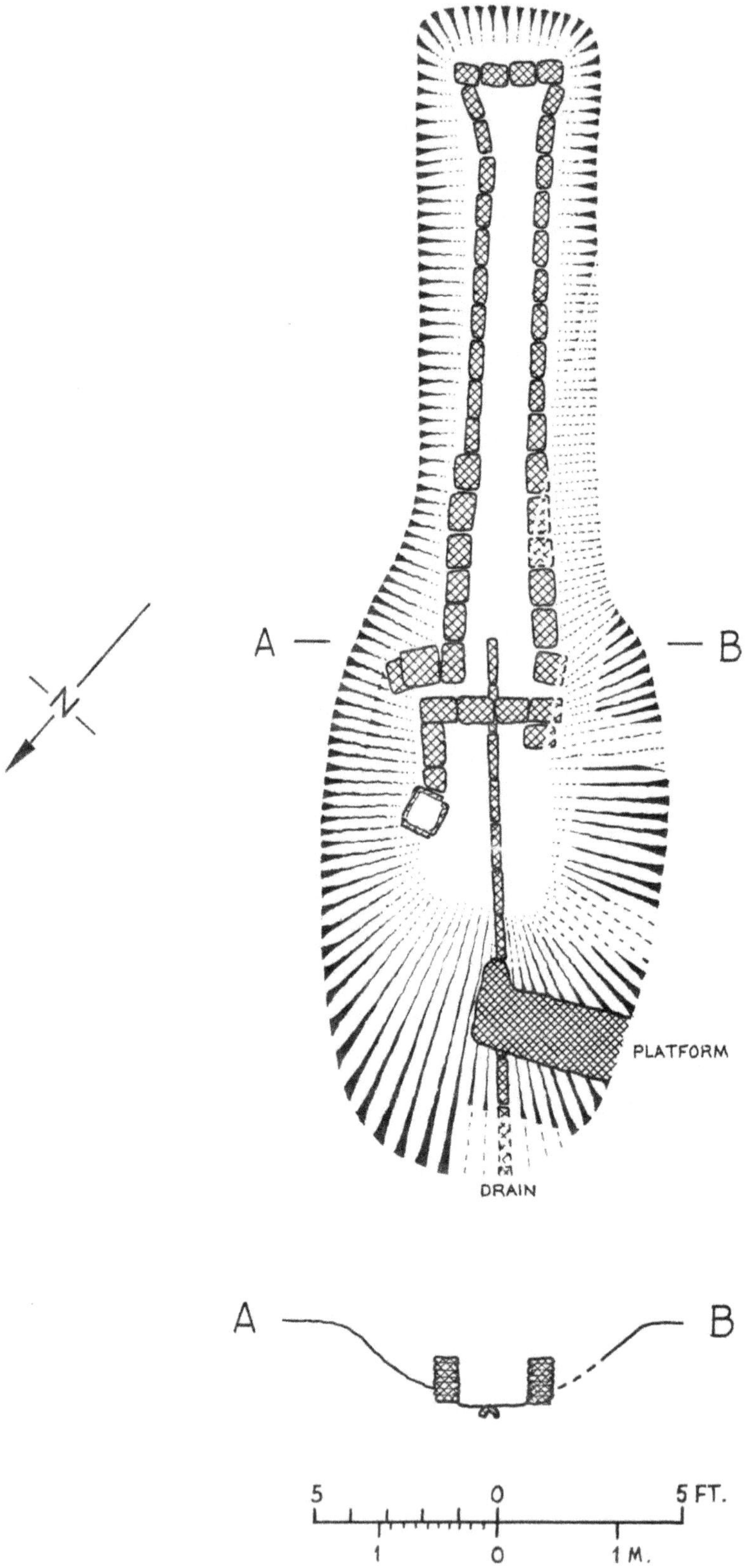

Fig. 6.17

The stokepit was 3.65 m long and 2.43 m wide. Astride the entrance to the flue were two brick piers and tiles had been used to pave an area 1.21 m by 457 mm on the floor of the pit. Along the north-east side of the stokehole four tegulae had been set on edge to form a box the top of which was level with the paving just described. Running from the mouth of the flue and across the stokehole was a line of box-flue tiles which was interpreted as a drain.

Products

Bricks, tegulae and imbrices were found along with three pieces of roller-pattened flue tiles. Mr. A. W. G. Lowther stated that the latter were pattened with die No. 32 (compare Lowther, 1948, 32) and, like other tiles which had been decorated in this way, had come from voussoir-shaped box tiles. Also found and of some interest was a scatter of large and roughly-cut tesserae and tile chips found in the stokehole.

Date

Evidence from first-century pottery found on the tile paving of the stokehole and the pattened flue tiles dated by Lowther to the late first or early second century points to the kiln, if that is what it was, being in operation at that time.

Discussion

Destruction of the upper structure of the flue made interpretation difficult, but the excavator was in no doubt, from the large number of tile wasters, that it was connected with tile-making and suggested either that the flue was used to heat a drying floor or that it was the main central flue of a tile-kiln, the superstructure of which has long disappeared.

References

Gillam, G. R. 'A Romano-British site at Potters Bar, Middlesex'. Bulletin of Barnet and District Record Society. No. 9, 1956, not paginated.

Lowther, A. W. G. A Study of the Patterns on RomanFlue-Tiles and their Distribution. Research Papers of the Surrey Archaeological Society No. 1. 1. Farnham 1948.

MESSINGHAM SE 8904

Two structures of rectangular shape were excavated in 1964 at Messingham following the discovery of pottery and tile in the 1946 and 1949 and a proton magnetometer survey in 1953. One of these was most probably a tile-kiln; the function of the other is less obvious and it may well have been used for another purpose.

KILN 1 Fig. 6.18

The outer walls of this kiln were unusual in being made only of clay, which was on average 450 mm thick. These walls enclosed an area measuring 3.2 by 2.1 m. At the eastern end, where the heat from the fire would have been most intense, the clay walls were strengthened by the inclusion of stone slabs. Along the sides of this rectangular chamber were 'five piers of tiles up to 760 mm long and between 220 mm and 350 mm apart'. These piers only survived to a height of 4-5 courses of tile and so there are no details of how they supported the oven floor across the centre of the combustion chamber. They could have been the remains of cross walls which once supported the oven floor.

This sub-floor structure is one of the few examples of a type I kiln, i.e. where the floor of the central and side flues are on the same level. The excavator, however, drew attention to the fact that 'between the piers the clay walls of the kiln sloped inwards at an angle of about 45°, but towards the centre of the kiln the side flues seem to have been at the same level as the main flue'. Even so there had been no attempt to raise the floor level between the piers/walls and so on the Grimes classification it would be classed as type I.

KILN 2 Fig. 6.18

Kiln 2 was situated 5 m south of kiln 1 and was also built of clay. It consisted of a channel or flue 3.5 m long and between 670 to 900 mm wide at its east end. The sides of the channel, which were reddened by fire, sloped inwards from the base. The excavator was of the opinion that as tiles were not used in its construction it would hot have been strong enough for firing tile or brick.

Products

The usual range of tile and brick was found in the area of the two structures, but none can be shown with certainty to have been made there.

Date

No material was found which helps to date the construction or use of the kilns. Pottery found roundabout is dated to the first half of the fourth century A.D.

Discussion

The presence of wasters and the large quantity of tile found strongly points to the making and firing of tile somewhere in the area and at least one of the

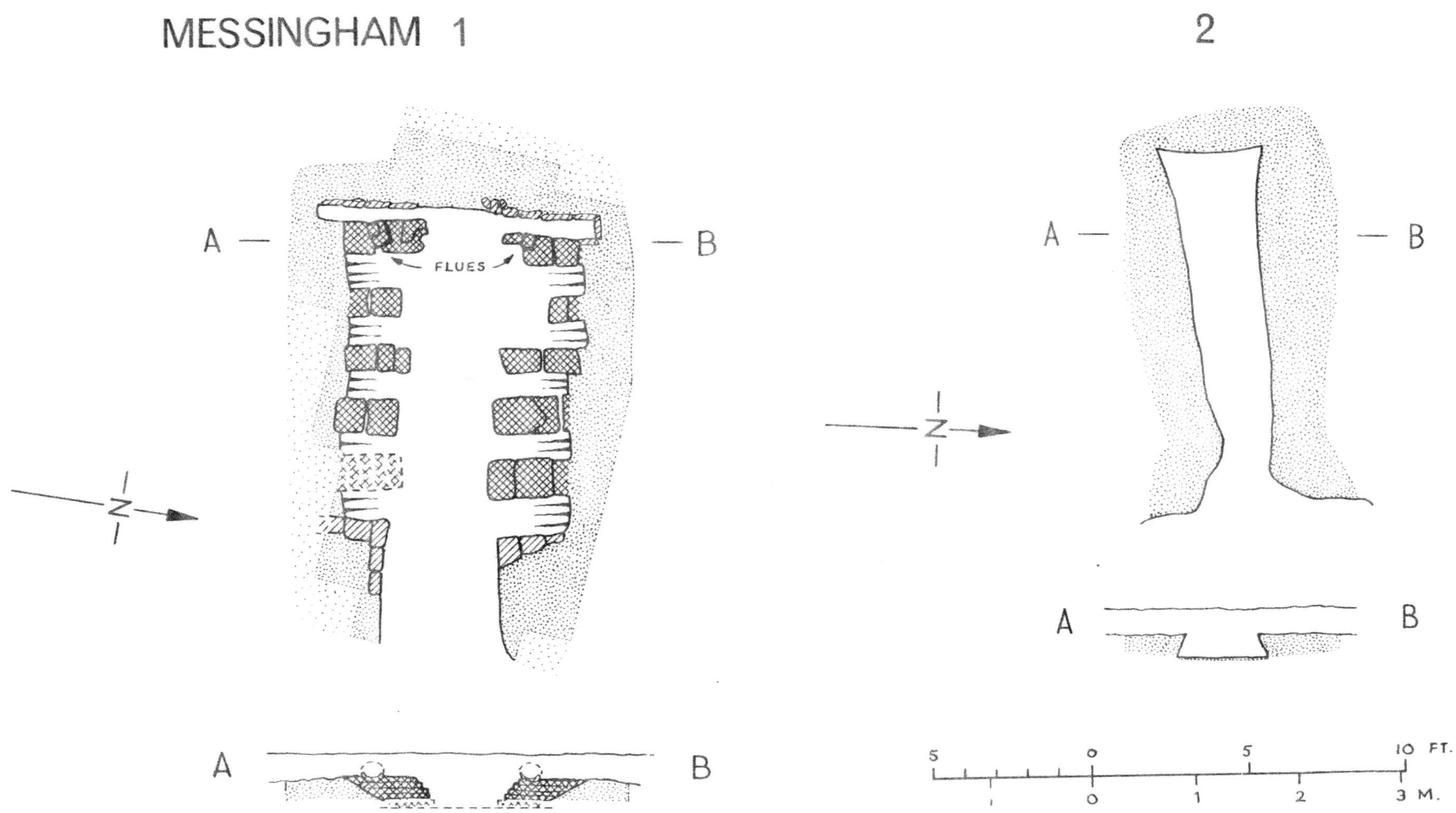

Fig. 6.18

structures looks convincingly like a kiln. Kiln 2 could be the remnants of a more conventional kiln with internal features removed. Alternatively, there are a number of similar structures now recorded and they may represent some form of forced drying or a cross between a kiln and a clamp (see page 101).

References

Stead, I. A. Excavations at Winterton Roman Villa, 1958-1967. London 1976, 98-101.

KENT

CANTERBURY

KILN A Fig. 6.19 TQ 1558

This kiln was discovered in 1952 when levelling was taking place for Market Way, the approach road to the Cattle Market at St Stephen's Road. The Canterbury Excavation Committee carried out excavations under the direction of Mr. F. Jenkins.

The site, which lies on brick earth, is c. 530 m due north of the Roman town wall of Durovernum and situated on a gentle slope on the north side of the River Stour in an area where brickmaking was still being carried out in the 1950s. The plan, which as Mr Jenkins pointed out, is unusual for a tile-kiln, is described as 'T-shaped consisting of a large stokehole pit, from which a tile-built flue channel led into an open vertical cross-vent, which acted as a chimney'. The main flue was 5.94 m long and 910 mm wide and consited of two tile-built walls each 600 mm thick with the tiles being set in puddled clay. The trench into which the flue had been dug was 2.13 m wide at the original ground level. The floor of the flue stopped slightly downwards from the stokehole to the vents. The two cross-vents were each 228 mm wide and 15 mm long and their floors sloped upwards away from the main flue. Mr. Jenkins interpreted these vents as 'chimneys to control the draught through the kiln, but not apparently to conduct the heat into the oven'.

As there was no evidence of cross walls or arches to support the kiln floor, Jenkins concluded that 'some other system of firing was employed'. He discussed the problem with local brick burners and concluded that the structure was a tile kiln and that it worked as follows:-

"The layers of 'green' tiles were stacked with fuel in the horizontal flue channel, the lowest standing on the bottom. Each tile would stand edgewise, those in the first layer at right angles to the long axis of the flue, and those in the next, in the other direction, and so on with alternate layers until the requisite number was stacked up. From the impressions found on some of the larger pieces of clay found in the stokehole pit, it seems that pads of puddled clay were used to support the tiles and act as spacers.

The flue was thus filled with tiles and fuel for most of its length with the exception of a space, perhaps one metre long, at the stokehole end. The stack of tiles in its final form could have been built up to a height equal to six layers,

CANTERBURY A

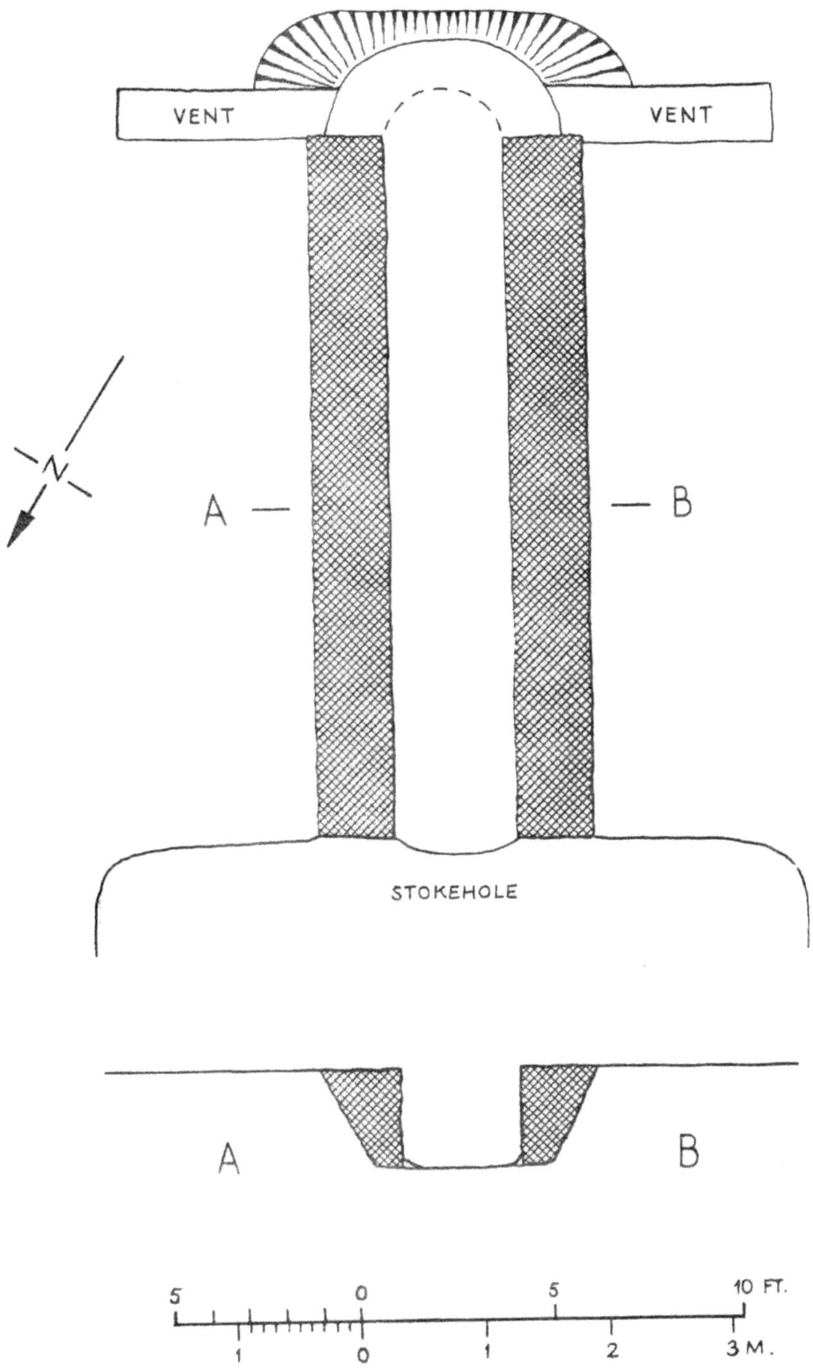

Fig. 6.19

that is, about one metre above the top of the side walls of the flue. Fuel was now placed on the stack, by this time covered by soil and turves, and then ignited".

The stokehole pit which was cut into the natural brick-earth Jenkins concluded had been deliberately filled in with tips of debris consisting of burnt daub and many fragments of tile. At the bottom of the pit was a layer of 'puddled brick-earth' mixed with black ash containing many tile wasters and a small quantity of pottery. To the rear of the stokehole four post holes placed at right angles to a narrow channel were found cut into natural brick-earth. Jenkins suggested that they were the remains of a 'rough lean-to timber hut used by the kiln attendants, the narrow slot being where a wattled wall had once rested'.

Products

Many fragments of *imbrices* and *tegulae* with semi-cricular grooves on the lower edge were found and uniquely, a possible ridge tile. In cross section it was like an *imbrex* although more triangular and had slots cut into the lower edge, two per side, to fit over the two adjacent flanges of the uppermost *tegulae* on the roof. All were scored with a lattice pattern suggesting that some form of mortar covering was used to make the joints waterproof.

Bricks of various size were found, but few were complete.

Date

On the north-east side of the stokehole was a pottery kiln which seems to have been in use at the same time as the tile-kiln, both being dated to A.D. 130-140. After lying derelict for some years they appear to have been deliberately filled in and levelled off in *c*. A.D. 200.

Discussion

The kiln is unusual in plan for a tile-kiln, but judging from the large quantity of tile including wasters, found there can be little doubt that it was connected with tile-making in some way. Perhaps it was used for drying tile before firing in a kiln.

It is interesting to note that another pottery-kiln of mid-first century date was found about 65 m east-north-east and itself contained tile in its structure. Assuming that the date of the kiln is correct, then the tile used in its construction must have been made at some time around the middle of the first century A.D.

References

Jenkins, F. 'A Roman Tilery and Two Pottery-Kilns at *Durovernum* (Canterbury).' *Antiq. Journ*. 36 (1956), 40-56.

KILN B Fig. 6.20 TQ 1458

Three kilns were discovered during the development of a building estate at Whitehall Gardens, two in 1956 and one in 1959. Kiln II of this group was described by F. Jenkins as a tile-kiln of the 'rectangular updraught type'. The other two kilns were used for pottery.

The kiln is situated in the valley of the River Stour on gently rising ground 'to the north-west of the left bank of the river' some 137 m from the Roman town wall of Durovernum Cantiacorum and only 160 m from the Roman road which links Canterbury with London. The kiln was dug into the natural brick-earth which covers the valley floor to a considerable depth.

The combustion chamber of this rectangular kiln measured 2.4 m by 1.8 m with the main flue projecting beyond the south-east side for a further 1.8 m to link with the stokehole pit. There were three cross walls in the combustion chamber and the cross flues between these walls had sloping floors (i.e. type III). The bottom of the main flue was one metre below the original Roman ground surface. No evidence of flooring above the combustion chamber survivied, however there was sufficient tiling left to show that the extended main flue was originally arched over.

The rear and front walls of the combustion chamber bore the vertical imprints of an implement thought by Jenkins to have been an iron- hod spade. The two back corners had been 'hollowed out', possibly to act as flues.

There was a stokehole south-east of the extended main flue, but it could not be fully examined. It was a large pit one metre deep and contained burnt clay, ash and many fragments of tile all of which showed signs of faulty firing.

Products

Jenkins thought that the main product of this kiln was the 'usual flanged tile' (tegula). Some combed tiles showed that box-tiles were also made.

Date

From the fill of compacted burnt clay in one of the cross flues came a coin of Geta (c. A.D. 211-12) which does not help in determining when the kiln was in use. Two sherds of pottery, said to have been found in the ash layer at the bottom of the central flue, seem to date from the late second to early third century, and as the two pottery kilns are dated to the second century it seems likely that the tile-kiln was also in use at that time.

Discussion

The kiln is small for making tiles, but comparable to some other tile-kilns. The large number of wasters found clearly indicates that tile-making was taking place in the vicinity and that the tiles were probably fired in this kiln. Flagons feature in the list of pottery found in the area and it is possible that this kiln could have been used to fire them.

References

Jenkins, F. 'Two Pottery-Kilns and a Tilery of the Roman Period at Canterbury (Durovernum Cantiacorum)'. Archaeologia Cantiana. 74 (1960), 151-161.

CANTERBURY B

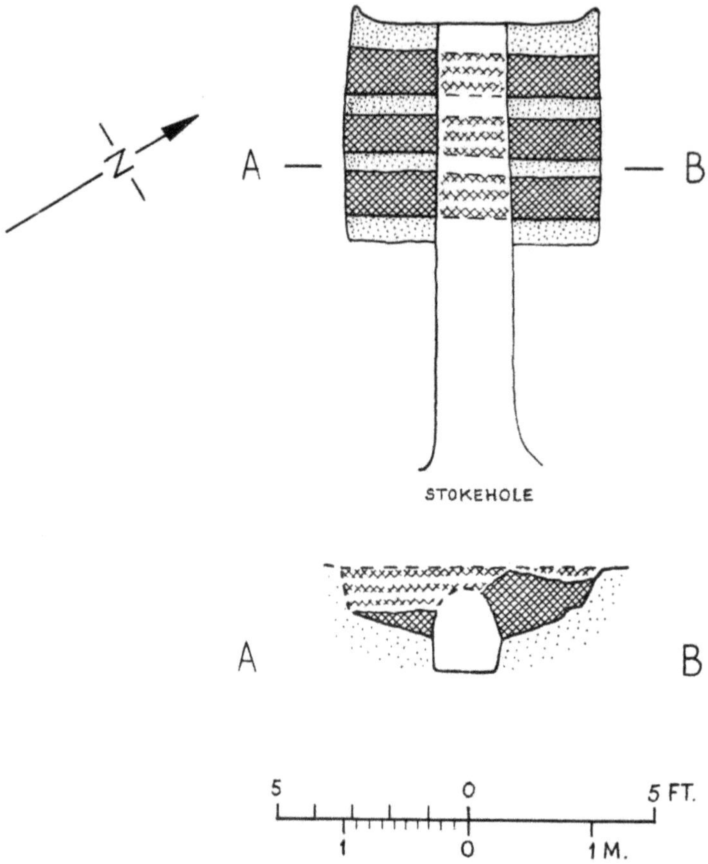

Fig. 6.20

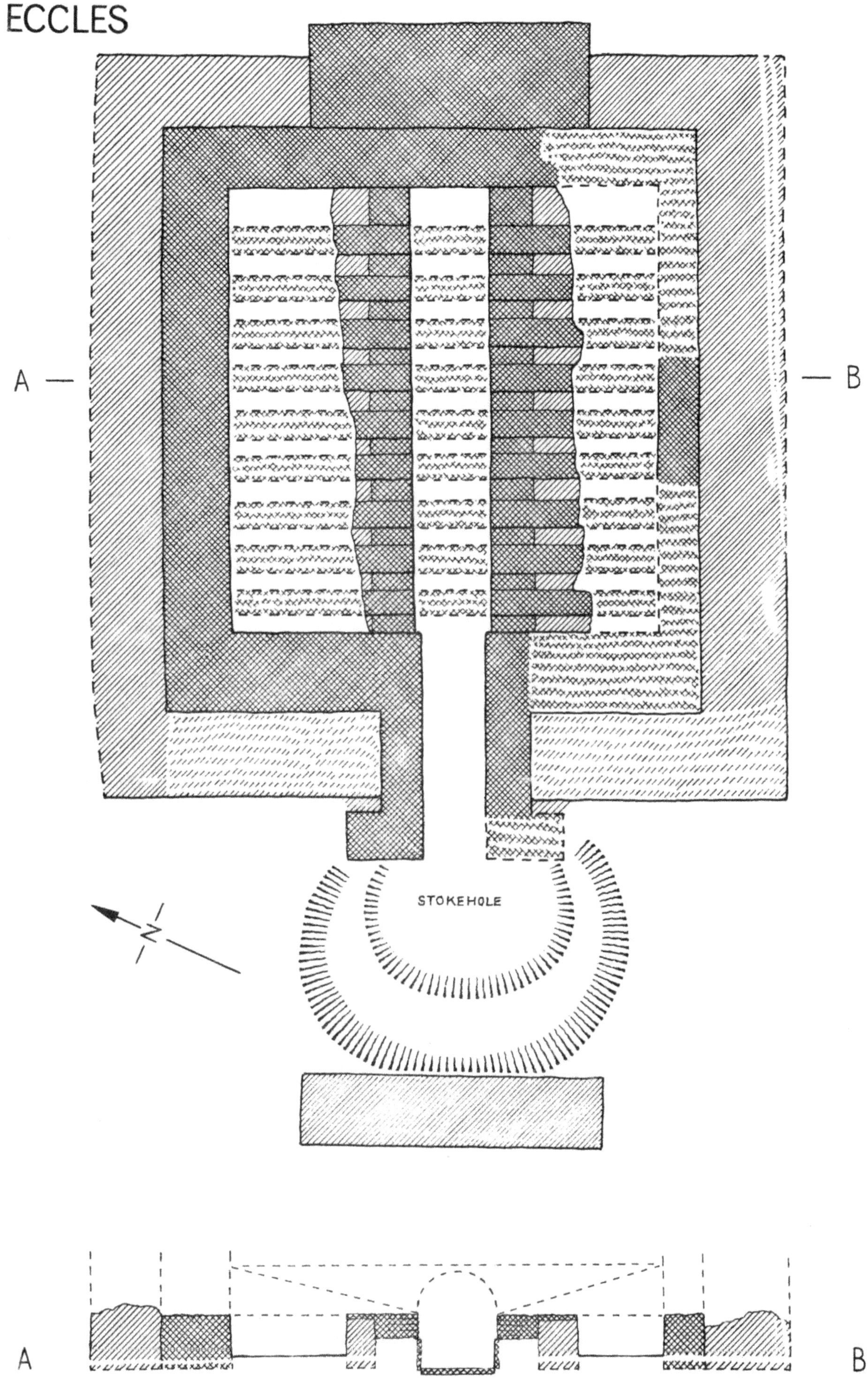

Fig. 6.21

ECCLES Fig. 6.21 TQ 7160

This kiln was found during extensions to a waste-paper storage yard in 1966; it was only discovered after one metre of soil had been removed by bulldozer and it is likely that much more of the superstructure originally survived. What is known of the kiln, which was badly damaged by the bulldozer, derives from only two weekends' work on the site when the structure was cleared and planned.

The kiln is situated at the junction of the Gault clay and terrace gravel, some 450 m to the south-west of the villa complex. The firing chamber which was approximately square, measuring 4.9 m by 4.7 m, is the largest civilian tile-kiln so far recorded in Britain. The outer walls of the kiln were built of ragstone and yellow mortar whilst tile was used for the nine cross walls and lining of the kiln. The sub-floor structure was Grimes type III.

The central flue was 830 mm wide and its floor consisted of three layers of clay-bonded tiles, some of which were <u>tegulae</u>. In the east wall of the kiln there was a gap 3.2 m wide which Detsicas interpreted as an entrance for loading and unloading the kiln. 2.7 m beyond this eastern wall were the remains of another wall half a metre wide of which a stretch 2.9 m survived. The conjectural continuation of this wall resulted from close questioning of the bulldozer driver who was quite certain that it continued as shown on the published plan.

The stokehole pit was about 3.6 m in diameter and on its western side was a length of wall 760 mm wide and 3.4 m long, built of ragstone and yellow mortar, which Detsicas interpreted as being 'similar to a modern baffle-plate to prevent a direct draught into the firing tunnel and so to minimise the risk of an uncontrolled surge of flame into the flue'.

<u>Products</u>

It is not clear what the kiln produced. Four different types of tile were found on site, probably originating from the kiln structure. They were, <u>pila</u> tiles, bonding tile, bridging tiles and one <u>imbrex</u>.

<u>Date</u>

Some stratified pottery was found below the lining of the kiln which dated its construction to the late second century A.D.

<u>Discussion</u>

Detsicas presumed that the kiln was built to provide tiles for period VI of the villa; in this case other kilns must have existed in the area to provide tiles for its earlier phases and for the construction of the tilery itself. This is the only kiln which has produced evidence of an entrance, if that is what it is, and of a stone structure around the stokehole. Its size and these unusual features make this a particularly interesting kiln.

<u>References</u>

Detsicas, A. P. 'Excavations at Eccles, 1966'. <u>Archaeologia Cantiana</u>. 82 (1967), 162-178 (The tilery is on pages 170-178).

LANCASHIRE

QUERNMORE Fig. 6.22 SD 5259

From the area of Quernmore Park 5 km east of Lancaster has come a variety of structures indicating an industrial site possibly connected with the fort at Lancaster. A tile found in the late eighteenth century bore the stamp ALE SEBVSIA. A series of kilns of uncertain use have been found and details of these along with other discoveries have been reported upon by G. M. Leather. It is not altogether clear which structures were used for tile and brick. The kiln at Q1, Lythe Brow (SD 526622) may have been for tile (there are suggestions in the text that it was) and kiln 1 from Q2, Low Pleasant (SD 521592). is labelled as a tile-kiln on the published plan. The latter is reproduced in this Gazetteer but further details, because of the uncertainty of the kiln's function, have been omitted.

Reference

Leather, G. M. Roman Lancaster Preston. 1972/3.

LINCOLNSHIRE

HECKINGTON FEN Fig. 6.23 TF 1745

The kilns are situated 90 metres west of the Car Dyke and were first noted in 1968, but it was not until 1970 when the site was first deep ploughed that rescue excavation became necessary. The work was undertaken by the Car Dyke research group.

KILN 1

This kiln was badly damaged when kiln 2 was constructed, but even so some of its plan was recovered. What survived seemed to be the main central flue of a typical tile kiln and although not traced fully it was at least 3.8 m long internally and 0.7 m wide. Bricks and clay were used in the construction of the walls of the flue and its fill contained well-carbonised wood ash which was thickest at the northern end close to the presumed stokehole for the kiln.

KILN 2

Kiln 1 was partly cut away for the construction of kiln 2, and the flue packed soldily with clay and tile over which was laid a 'working floor' made from rejected tegulae. A shelter or store place was then formed using the shell of the earlier kiln, the southern wall of which had been partly rebuilt, with the addition of a wall across the northern end of the flue of kiln 1. In addition to the main central flue of the kiln, which measured 6.2 m by 1 m there was some evidence for the way in which the tiles were stacked in the kiln. The excavator believes that the impressions left on the platforms either side of the flue result from the stacking of tiles which had been so arranged as to leave vents between the stacks. An alternative suggestion based on comparison with other tile-kilns might be that the impressions that were recognised were in fact left by cross walls which had subsequently been removed. However, the excavator noted no signs of cross walls or of any oven floor

QUERNMORE - LOW PLEASANT

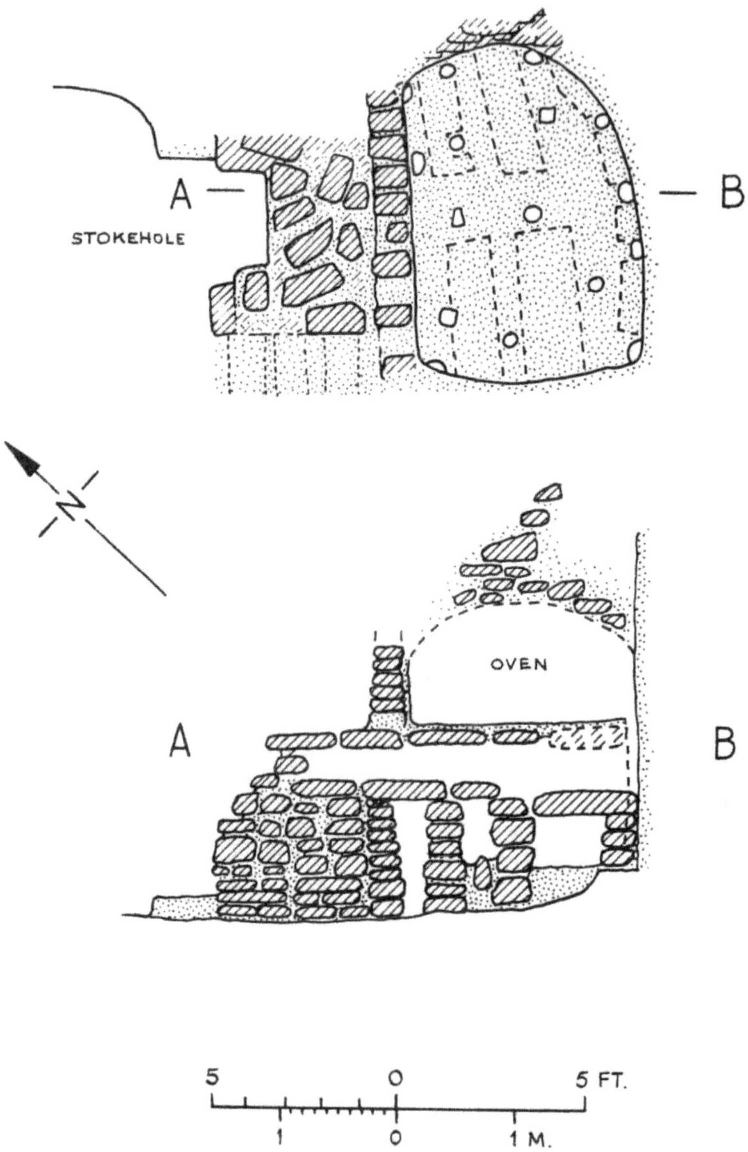

Fig. 6.22

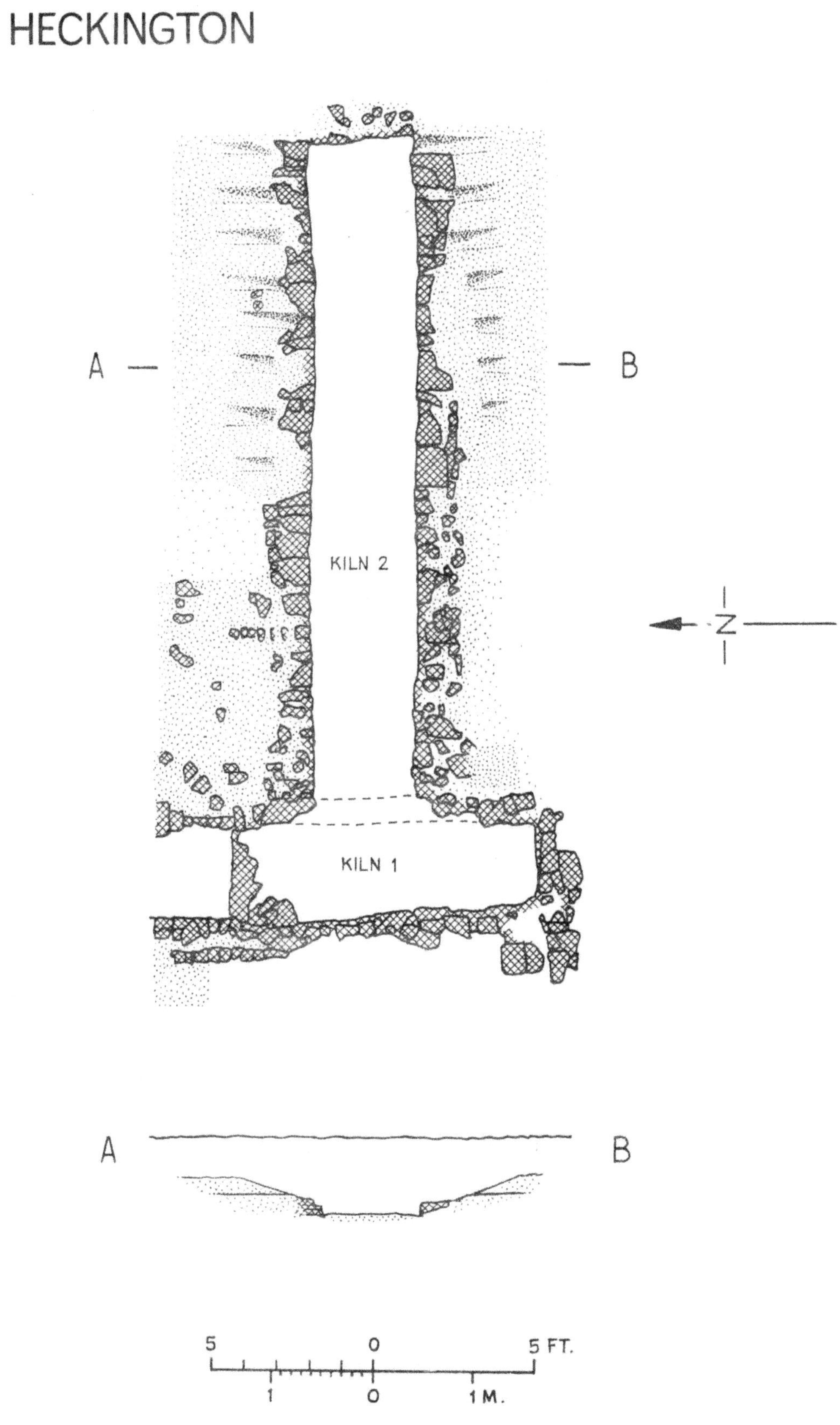

Fig. 6.23

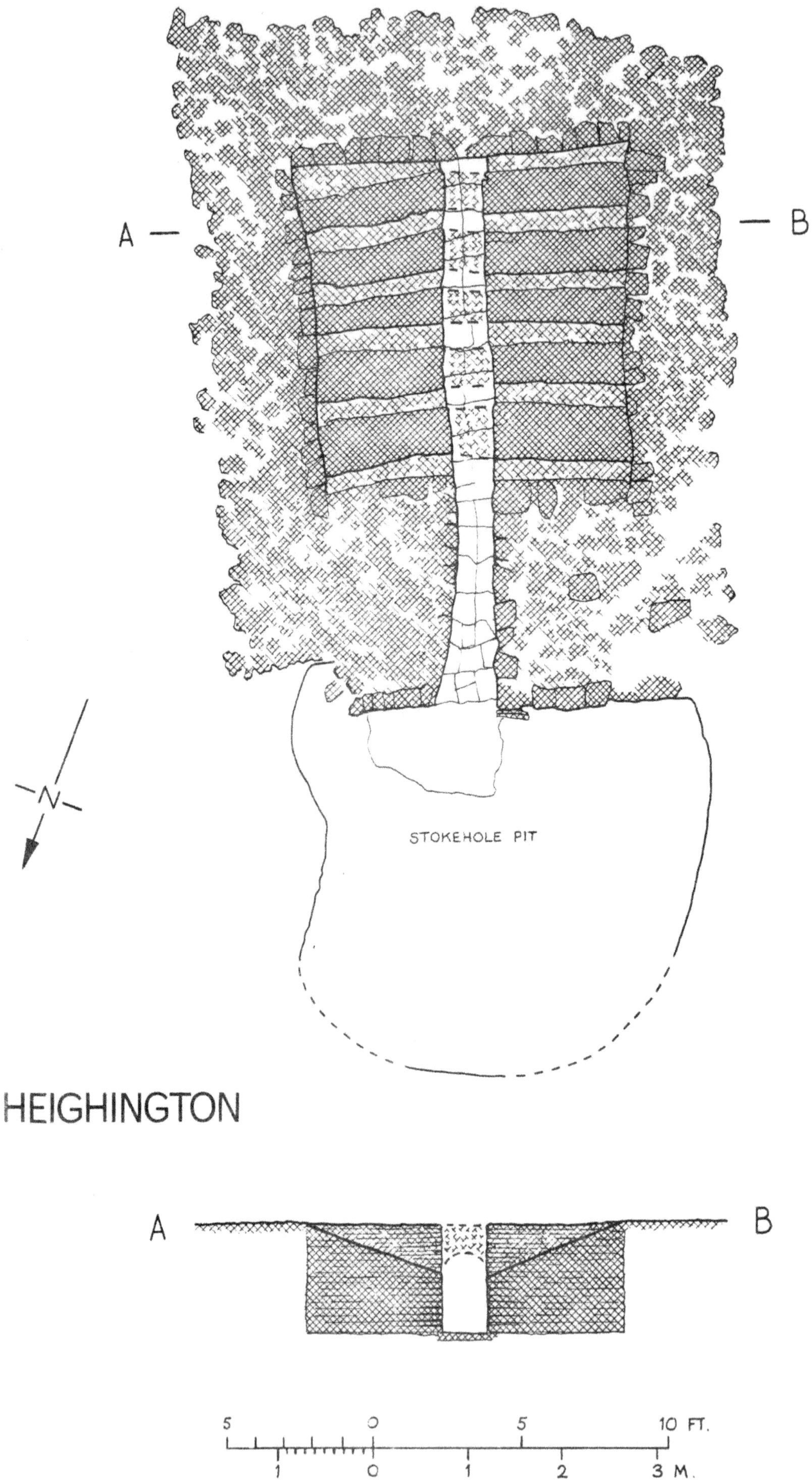

Fig. 6.24

and so concluded that the structure was a clamp-type kiln which originally had no cross walls or floor to support the kiln load, but that air-dried bricks were stacked on either side of the central flue with gaps being left between the stacks so as to allow the gases to circulate within the kiln.

Products

It was difficult to decide what was made in either of the kilns, but five different types of tile were noted from the excavations. The most common were tegulae and the rarest pilae, and it is thought that kiln 2 was used for brick rather than tile.

Date

A rubbish pit, partly beneath kiln 1, gave a terminus post quem for the construction of kiln 1. Pottery from the pit is dated to the third century and along with pottery found in the destruction layer of kiln 2 a date of around A.D. 300 is postulated for the use of the two kilns.

References

Simmons, B. B. Roman Tile-Kilns at Heckington, Lincs. Car Dyke Publication No. 3 1977.

HEIGHINGTON Fig. 6.24 TF 0569

The Lincoln Archaeological Trust and the Society for Lincolnshire History and Archaeology jointly excavated the kiln in 1976, the most recent example to be investigated in Britain and which in some earlier publications was wrongly attributed to the parish of Washingborough. It is situated between Washingborough and Branston Booths about 9.6 km east of Lincoln and only 200 m from the Car Dyke. The firing chamber of the kiln was approximately square (3.3 m), but with the long flue or fire tunnel the overall shape, including the tile walls of the kilns, was rectangular being 7.1 by 5.4 m. The kiln was built of tile, some rectangular and in good condition, others waste fragments. Also used in its construction were 'green' or unbaked clay bricks, which hardened when the kiln was first fired.

There were five cross walls and hence six cross flues, the distance between these walls being on average 200-250 mm. The floors of the cross flues were at a higher level than the tiled main flue and were sloping towards the outside of the kiln, i.e. type III. No trace of the oven floor remained. Linking the combustion chamber with the stokehole was a fire tunnel 2.3 m long. The outline of the stokehole pit was recovered and most of it excavated.

During the life of the kiln various parts showed evidence of collapsing and having been repaired. Eventually, when it was abandoned, it was filled with broken tile and clay.

Products

These included tegulae, imbrices, combed flue-tiles, hypocaust tiles and bonding tile or brick. Of particular interest were fragments of quarter-round tiles and a chimney or ventilator.

HORTON

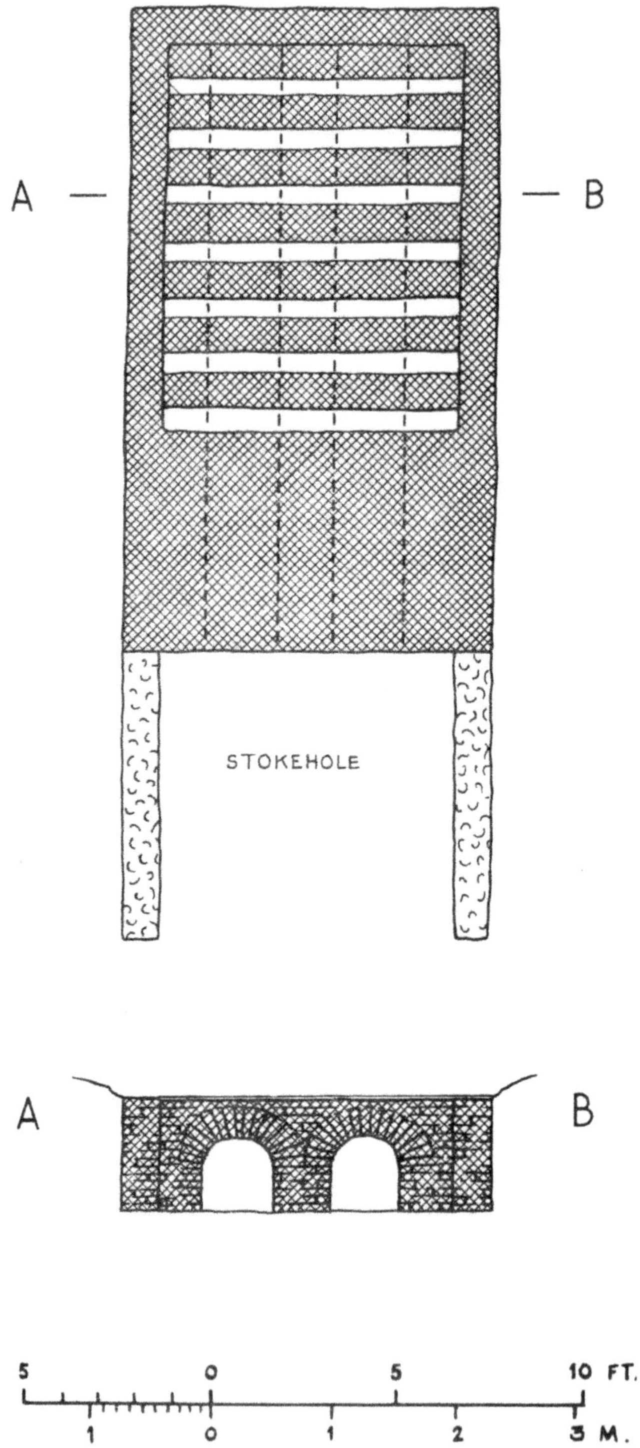

Fig. 6.25

Date

The excavator claims that 'pottery found in the top-soil and in the demolition debris indicates a fourth-century date'. At present it is not clear whether any datable material was found in levels which would date the construction and use of the kiln with any certainty, and so the suggested fourth-century date must be treated with caution.

Discussion

Another tile-kiln found during fieldwork lies about 1 km away from the excavated kiln and a further tile-scatter close to the Car Dyke may be indicative of yet another. Whether these kilns were built to supply tile and brick to Roman buildings in the immediate area or whether they had a wider distribution has not yet been determined.

References

Lincoln Archaeological Trust Fourth Annual Report 1975-76, 22-23.

Britannia 8 (1977), 388.

SURREY

HORTON (EPSOM) Fig. 6.25 TQ 1861

During the construction of the West Park Asylum at Green Man Farm, Horton, near Epsom in 1922, a trench for a drain exposed the combustion chamber of a kiln. Subsequently the whole structure was uncovered, but details were never published until Professor Goodchild came to write up his excavations of the tile-kiln at Wykehurst Farm, Cranleigh. In that report, Goodchild included a brief account of the Horton kiln and a plan based on a drawing made at the time the kiln was discovered.

It was built of tile set in 'pink cement' and measured 5.2 by 3 m overall. There were two walls of chalk blocks on either side of the stokehole thus effectively increasing the overall length of the kiln by a further 2.3 m Internally the combustion chamber measured 3 by 2.3 m and was fired by two parallel flues each 60 mm wide. The two fire tunnels leading from the stokehole to the combustion chamber were each 1.8 m long. The sub-floor structure of the kiln consisted of seven cross walls (including one against the back wall) and seven cross flues. It is not clear from the published drawings how the floors between the cross walls were constructed.

Date

Goodchild states that 'since the kiln was obviously part of the widespread Ashstead industry, its period of activity was fairly certainly that of the latter, i.e. c. A.D. 70-150'. There is no evidence from the site to support this date.

Discussion

This is one of only two examples from Britain of a kiln with double parallel flues, and it is also unusual in having the flanking walls either side of the stokehole. Unfortunately, in the absence of further details especially about the

WYKEHURST FARM

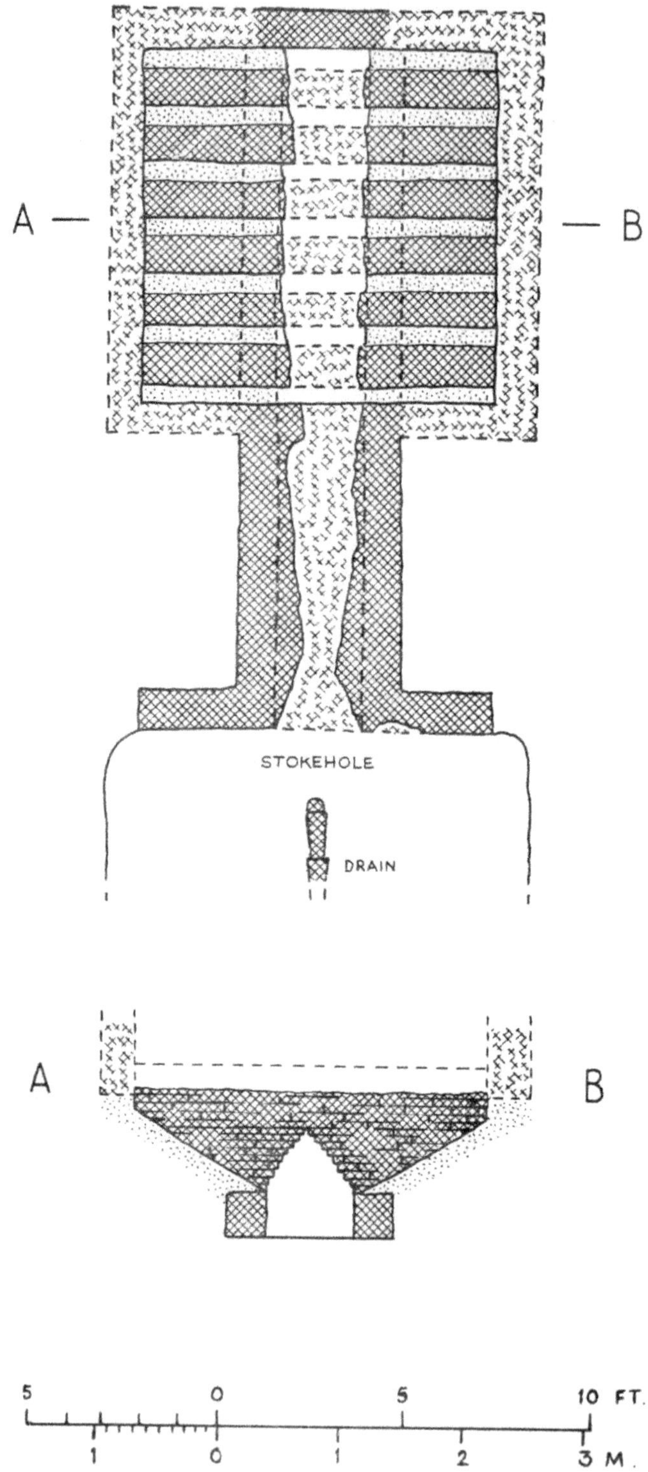

Fig. 6.26

products and their destination, little more can be said about this extremely interesting kiln.

References

Goodchild, R. G. 'The Roman Brickworks at Wykehurst Farm in the Parish of Cranleigh'. Surrey Arch. Collections 45 (1937), Horton, pages 90-92.

WYKEHURST FARM, CRANLEIGH Fig. 6.26 TQ 0840

This kiln is situated on the Wealden Clay close to the Roman road which branches from Stane Street at Rowhook (Sussex) and heads for Farley Heath. The first indication of Roman actitity in the area seems to have been observed by S. E. Winbolt in 1923 and reported in the Surrey Archaeological Collections (see below). Further discoveries by members of Cranleigh School in 1933-34 led to excavations by the School Archaeological Society in 1936 led by R. G. Goodchild.

The kiln was constructed entirely of tile and clay, there being no traces of stone or mortar. The combustion chamber, which was approximately 2.8 m square, does not appear to have been lined with tile, but seems just to have utilised the clay into which it was dug. There were six cross walls each about 300 mm wide with flues between them about 150 mm wide. The flues were at a higher level than the central flue and had sloping floors, i.e. type III. From the combustion chamber there projected an elaborate flue leading to the stokehole, by far the longest such flue noted from rectangular kilns in Britain, being 2.6 m long. Where the flue joined the stokehole its walls turned a right angle on each side forming an almost continuous tile wall on the south side of the stokehole, broken only by the arched entrance to the flue. The first 2.6 m of this flue was arched over forming what Goodchild describes as the 'stoking tunnel', but when it entered the combustion chamber, according to the section reproduced in the report, the cross walls were carried over the flue by corbelling the tile rather than by forming arches.

The oven floor only survived in a few places and 'was made by placing tegulae across the spaces between the cross walls, leaving gaps for the vents and finally covering the tegulae with a thick layer of clay daub'. Goodchild believes that once the tiles had been stacked on the oven floor 'they were enclosed with a wall of clay blocks.....large numbers of these clay blocks, darkened by the smoke from the furnace, were found crammed into the main flue, where they had been thrown after the kiln was dismantled'.

The stokehole was a 3.4 m square pit dug into the natural clay and with sloping slides. In order to keep the pit and the kiln drained a trench one metre deep was dug from the mouth of the kiln to a stream 8.8 m away. In the bottom of the trench was a continuous line of overlapping imbrices.

Trenching to the east of the kiln revealed what Goodchild describes as a 'roughly-paved brickyard'. This consisted of a 60 mm layer of broken tile and brick, the majority of which were wasters, marking out an oval-shaped area 21 m by 17m. No details of any structures which might have been associated with tile and brick-making were found in the very limited area investigated, but they may have existed. One part of this brickyard seems to have been used for making tesserae as large quantities of red-brick cubes were

WISTON

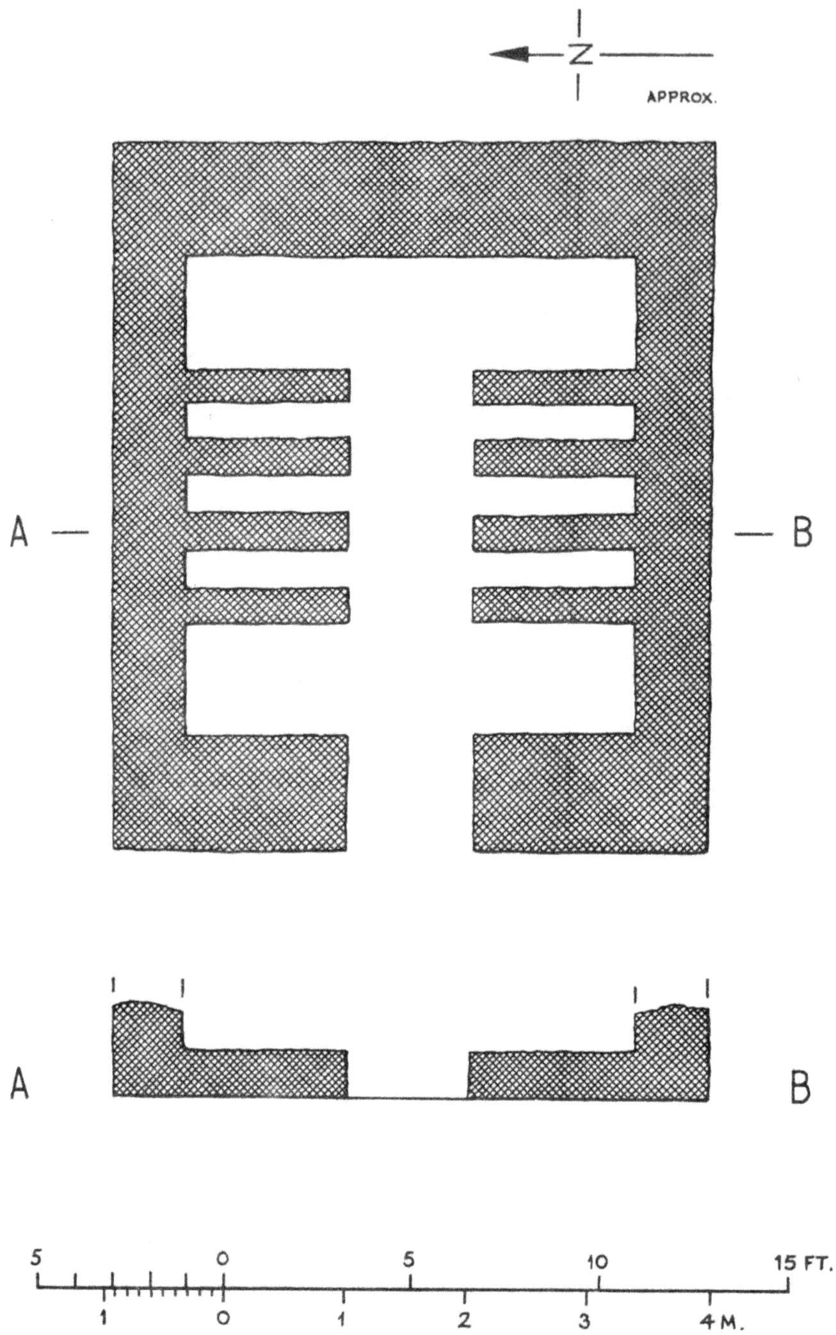

Fig. 6.27

found, many of which were very rough and considered to be rejects. Blue and brown coloured cubes were also noted and an iron object which may have been a chisel used in making the tesserae.

Products

Besides the tesserae referred to above the usual range of tegulae, imbrices and building or pilae bricks was found. Box-flue tiles from the site were all decorated with combs; there were none with roller-pattened designs of which so many were noted at Ashstead close by.

'Pear-shaped' tiles unlike anything found in Britain were identified by Goodchild as roof-tiles. 'From a flat top 100 mm wide they broaden out and terminate in a semicircle of about 280 mm in diameter, being 380 mm long from top to bottom, with a pierced nail-hole at the top'.

Semicircular tiles 230 mm in diameter were also found, these being the sort used in the construction of engaged of half-engaged columns.

Date

The lack of finds which might date the kiln made it difficult for Goodchild to offer a date for its use but by analogy with Holt and Hoheneck and other general considerations about the 'heyday of building in Roman Britain' he suggested a late first- or early second-century date.

Discussion

This kiln has produced a number of interesting features as well as some unusual products. It was built on first class clay and close to the road system, but the distribution range of its products is uncertain although some ideas are contained in the report.

References

Goodchild, R. G. 'The Roman Brickworks at Wykehurst Farm in the Parish of Cranleigh'. Surrey Arch. Collections 45 (1937), 74-96.

SUSSEX

WISTON Fig. 6.27 TQ 1513

Although clearly a tile-kiln, the structure was originally interpreted as part of a Roman building. It was found in 1848 a short distance north-east of the rectory-house at Wiston. The walls were constructed entirely of tile with flint used in the foundations. The combustion chamber was approximately square, 3.8 by 3.7 m with walls on the north and south each 600 mm wide and those on the east and west 900 mm. There were four cross walls with flues between them. The flues at the east and west end of the chamber were much larger than those between the cross walls. The main central flue was one metre wide. Few other details survive and little more can be said about the kiln.

Reference

Figg, W. 'On the Remains of a Roman Building Discovered at Wiston in 1848'. Sussex Archaeological Collections, 2 (1849), 313-315.

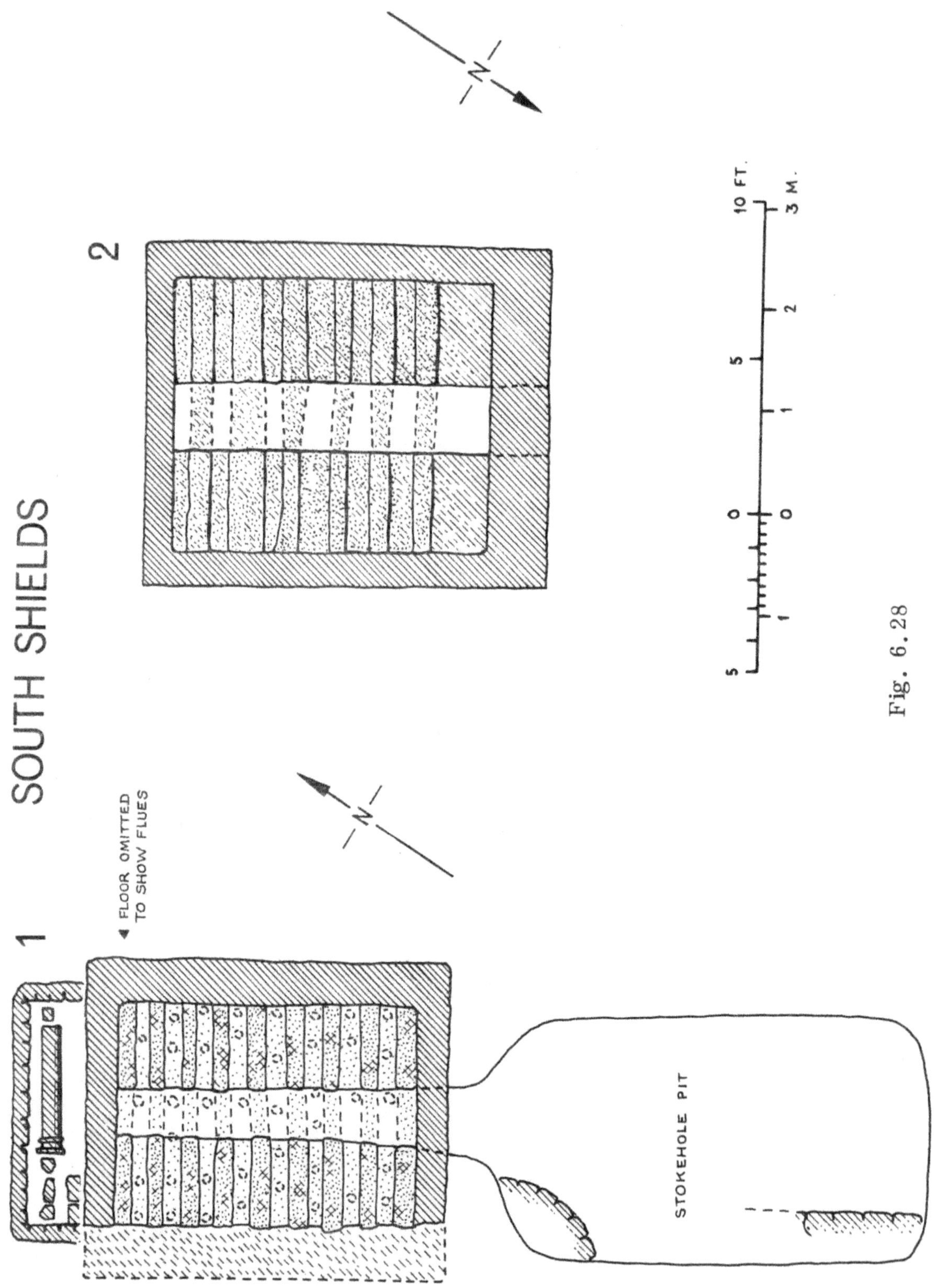
SOUTH SHIELDS
1
2
FLOOR OMITTED TO SHOW FLUES
N
N
STOKEHOLE PIT
5 0 5 10 FT.
1 0 1 2 3 M.

Fig. 6.28

SOUTH SHIELDS NZ 3667

Fourth century alterations to Granary I in the fort included the construction of two tile kilns which were excavated by Professor Sir Ian Richmond in the 1950s. Although unpublished a detailed typescript and plans was prepared by Richmond and are in his papers housed in the Ashmolean Museum.

KILN 1 Fig. 6.28

This kiln lies in the north-west corner of Granary I and to the north-west of kiln 2. Richmond describes the kiln as follows:-

> 'Its chamber is built with sandstone walls and a stokehole at the west end. The stokehole has a low pointed arch and is served from a great stoking-pit. The pit is now unlined, though a quadrant-shaped bench of stone filled its north-west corner, and part of a stone lining appears at the south end of its west side. The kiln is built with a central flue, on each side of which 8 lateral flues are now preserved. This appears to be the full quota of flues related to the existing floor of the kiln and implies a kiln-chamber 3.3 m by 2.2 m. But the chamber was originally not only longer towards the east but deeper, and traces of a terminal scarcement belonging to a lower and earlier system of flues appears at its east end. The scarcement is formed of a small re-used sandstone column...... The flue walls are built of clay reinforced with tile, the whole very heavily burnt. The floor which they carry is also built of clay and tile and has been pierced somewhat irregularly with eight rows of holes varying in number from three to six in a row, set above the eight lateral flues'.

KILN 2 Fig. 6.28

This kiln was built of stone, unfaced except on the inside of the chamber, and was provided with an arched fire-tunnel. The kiln measured 3.2 m by 2.7 m internally, had a central flue and six cross flues. The kiln, however, was never fired and had never been provided with a stokehole.

Products

The use to which kiln 1 was put was illustrated by the material recovered from the stokehole, namely rooftiles and floor tiles. None of the tiles from the kilns were stamped.

Date

A small group of early fourth-century pottery from the stokehole of kiln 1 suggested to Richmond, along with other considerations, a fourth century date for the kilns.

Discussion

This is the only example from Britain of a tile-kiln in a fort, although the recently discovered kiln from Caernarvon may be another example. Tiles from an earlier period of the fort were stamped Cohors V Gallorum.

ARBURY 1

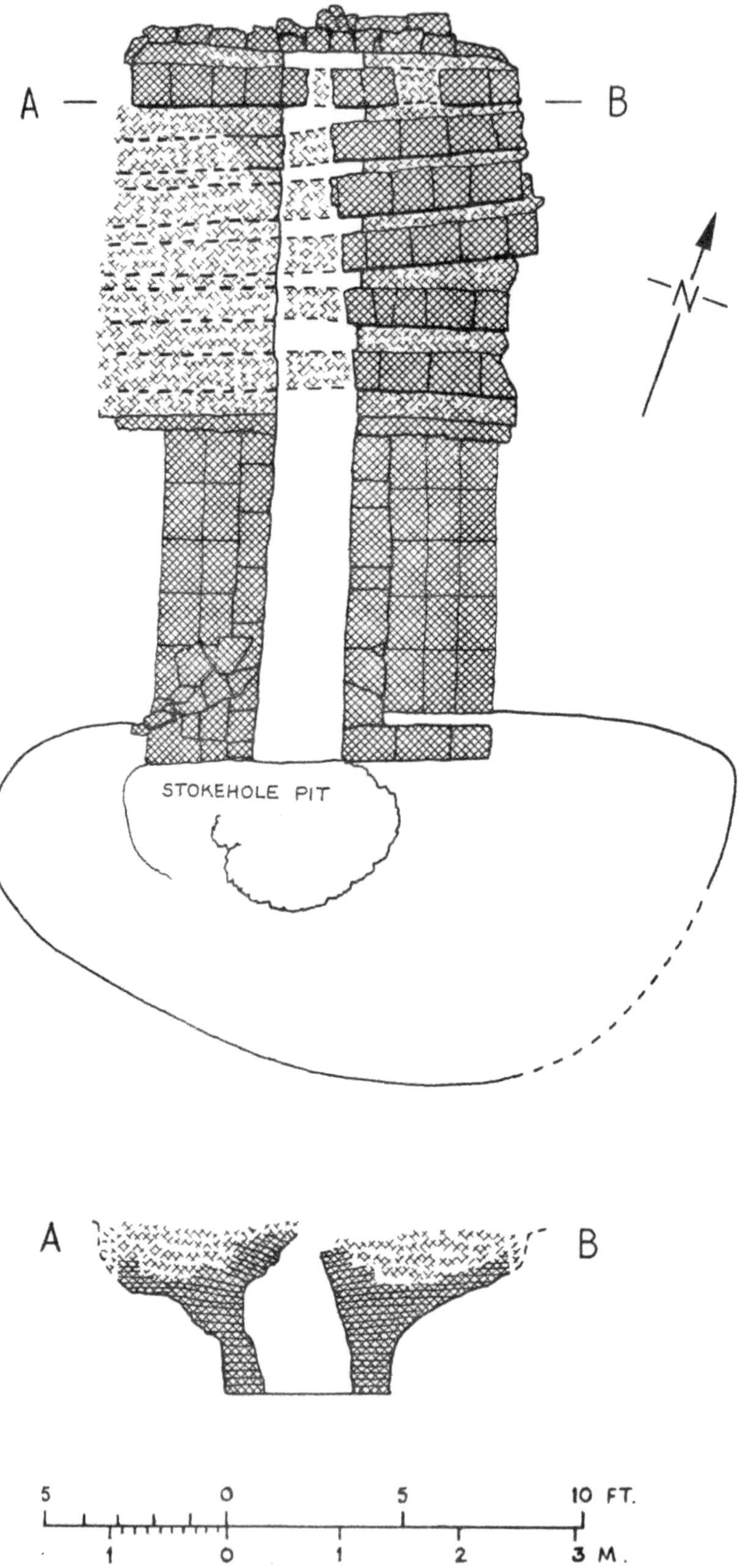

Fig. 6.29

References

Richmond, I. A. The Roman Fort at South Shields - A Guide. Newcastle, 195?

JRS, 40 (1959), 96.

Richmond papers in the Ashmolean Museum, Oxford.

WARWICKSHIRE

CHILVERS COTON

Excavations in 1964 and 1968 revealed three kilns at Arbury and in 1968 another at Griff Hill Farm, 2 km to the south-east.

ARBURY SP 3489

The kilns lie on a sandy outcrop in the Corley Keele which consists mainly of marls, the raw material for tile-and brick-making. There is a stream 90 m south of the kilns.

KILN 1 Fig. 6.29

The combustion chamber was 3.2 by 2.9 m internally and there were six cross walls and seven cross flues. There appear to be no outside walls to the chamber on the east and west sides, but the pit into which the chamber was dug was lined at the back (north) and front (south) with walls built entirely of tile bonded with silty clay. The cross walls which were constructed in the same way were built from a level area of tile 900 mm above the floor level of the central flue and the floors between the walls were described as '...level, and the ends were shaped to the vertical with clay'. This would imply a type II kiln. The main flue which was 760 mm wide projected 2.9 m beyond the southern end of the combustion chamber making it 5.9 m long overall. It was made of brick and tile founded on sandstone. The flue, or firetunnel, was corbelled over and according to the plan, each side wall turned through a right-angle on reaching the stokehole, not unlike the kiln from Cranleigh (page 167). The use of green or unfired tile in the main flue backing was noted.

No oven floor was found, leading the excavator to the conclusion that none need have existed; instead, he suggested that the tiles to be fired were stacked so as to bridge the gap between cross walls.

When the kiln was abandoned the stokehole was filled in. Later a trench 760 mm deep was dug into this fill, which may have been the beginnings of another kiln, never to be completed. On the same level as this trench, and presumably associated with it, were three post holes and a small area of flat tile.

The usual range of tile and brick was found.

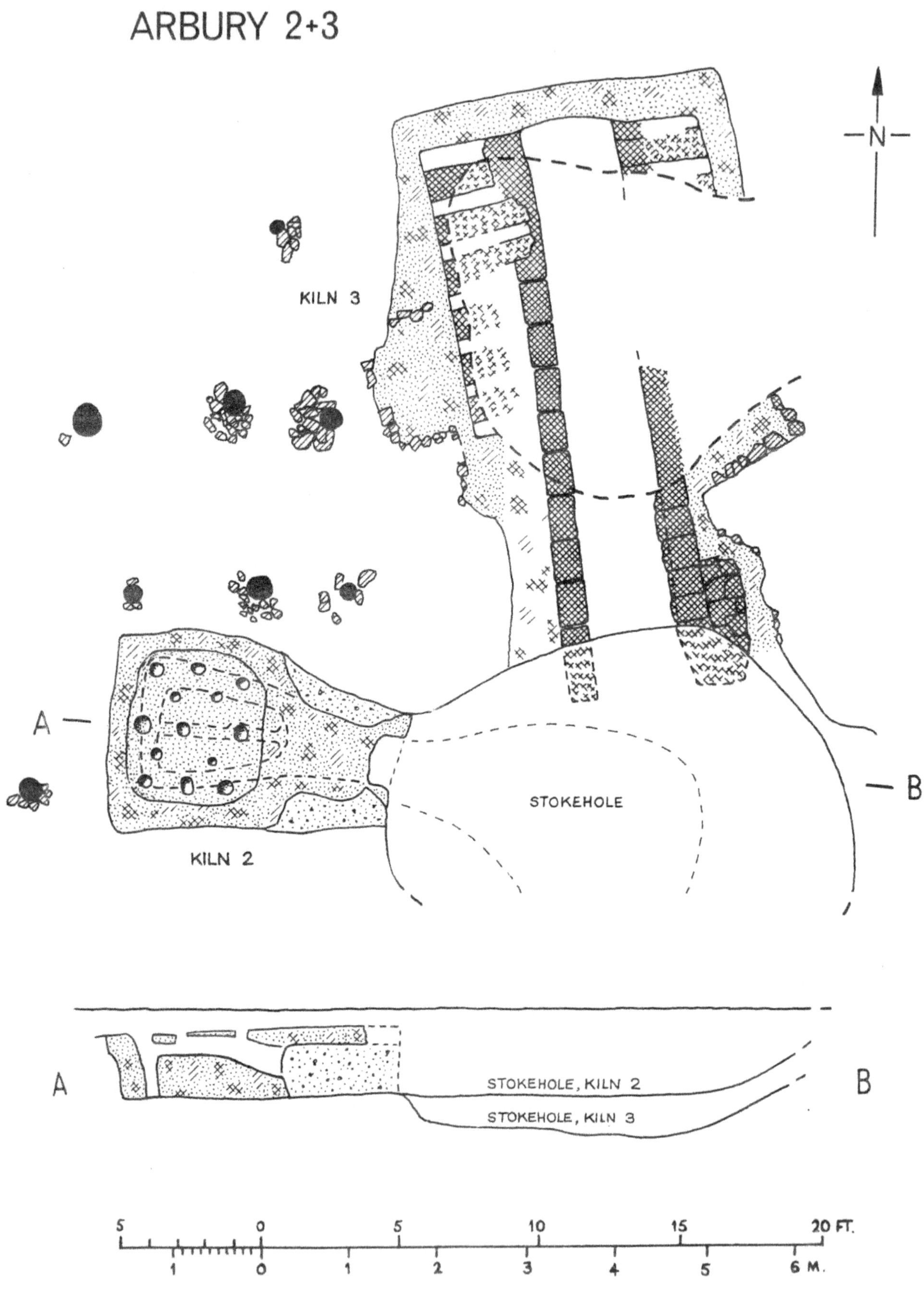
ARBURY 2+3
N
KILN 3
A
B
STOKEHOLE
KILN 2
A
B
STOKEHOLE, KILN 2
STOKEHOLE, KILN 3
5
0
5
10
15
20 FT.
1
0
1
2
3
4
5
6 M.

Fig. 6.30

KILN 2 Fig. 6.30

This was a small, almost square kiln, 1.6 by 1.4 m with an unusual sub-floor. The construction of the stokehole cut through an earlier deeper stokehole for kiln 3. These two kilns lay some 30 m west of kiln 1. The oven floor was seventy-five per cent intact making internal examination difficult; it was only possible to measure rather than dissect the kiln. The oven floor rested upon a central pedestal which itself had a medial flue. There was also a flue around the pedestal, that is, between the pedestal and the outside of the combustion chamber. The flue from the combustion chamber to the stokehole was 1.2 m long. Even though this kiln is more like a traditional pottery kiln the excavator was convinced that it was used for making tile rather than pottery as there was no evidence for the production of pottery from the site. He does offer the suggestion that it might have been a dryer as well as a kiln. Clay and tile were used in its construction and green tiles were identified. This kiln was carefully backfilled and preserved in situ at the request of the land-owner.

KILN 3 Fig. 6.30

This kiln was much disturbed by robbing and its stokehole was damaged by the construction of kiln 2. All that remained were the sides of the main central flue of what appears to be a typical tile-kiln. The outer walls were made of clay, tile and stone and the central flue was lined with tile laid flat. From what remains it is just possible to see that the combustion chamber was c. 3 m square and that the main central flue was one metre wide and projected 2.9 m beyond the combustion chamber to the stokehole.

A series of post holes were found to the west of this kiln and the published plan also shows one to the east. The posts which stood in these holes were 150 mm in diameter. K. Scott, the excavator noted that 'they appear to be aligned on kiln 3 and the building they supported must have been destroyed at the time of the construction of kiln 2..... The building could have been a workshop or drying shed...' The usual range of tile was noted from the site, but only a few box tiles were found.

Date

Pottery from a layer in the stokehole which accumulated after kiln 2 went out of use was dated to the third or early fourth century.

GRIFF HILL FARM Fig. 6.31 SP 3688

Two kilometres north-west of the Arbury complex another kiln was found in 1968 during trenching for a gas main. The kiln was built on a capping of sand and gravel 27 m away from a stream. The combustion chamber measured 2.5 by 1.6 m and was enclosed by walls built of flat tiles bonded with silty clay and laid on a foundation of stone. There were four cross walls (of 6 if the end two are counted) and the side flues were type III construction. The stone-lined flue from the combustion chamber to the stokehole was 1.4 m long, 650 mm wide and corbelled. A trench to the north of the kiln containing much charcoal may have been connected with an earlier kiln. The usual range of tile and brick was found, but it was thought that this kiln was specialising in the production of box-flue tiles.

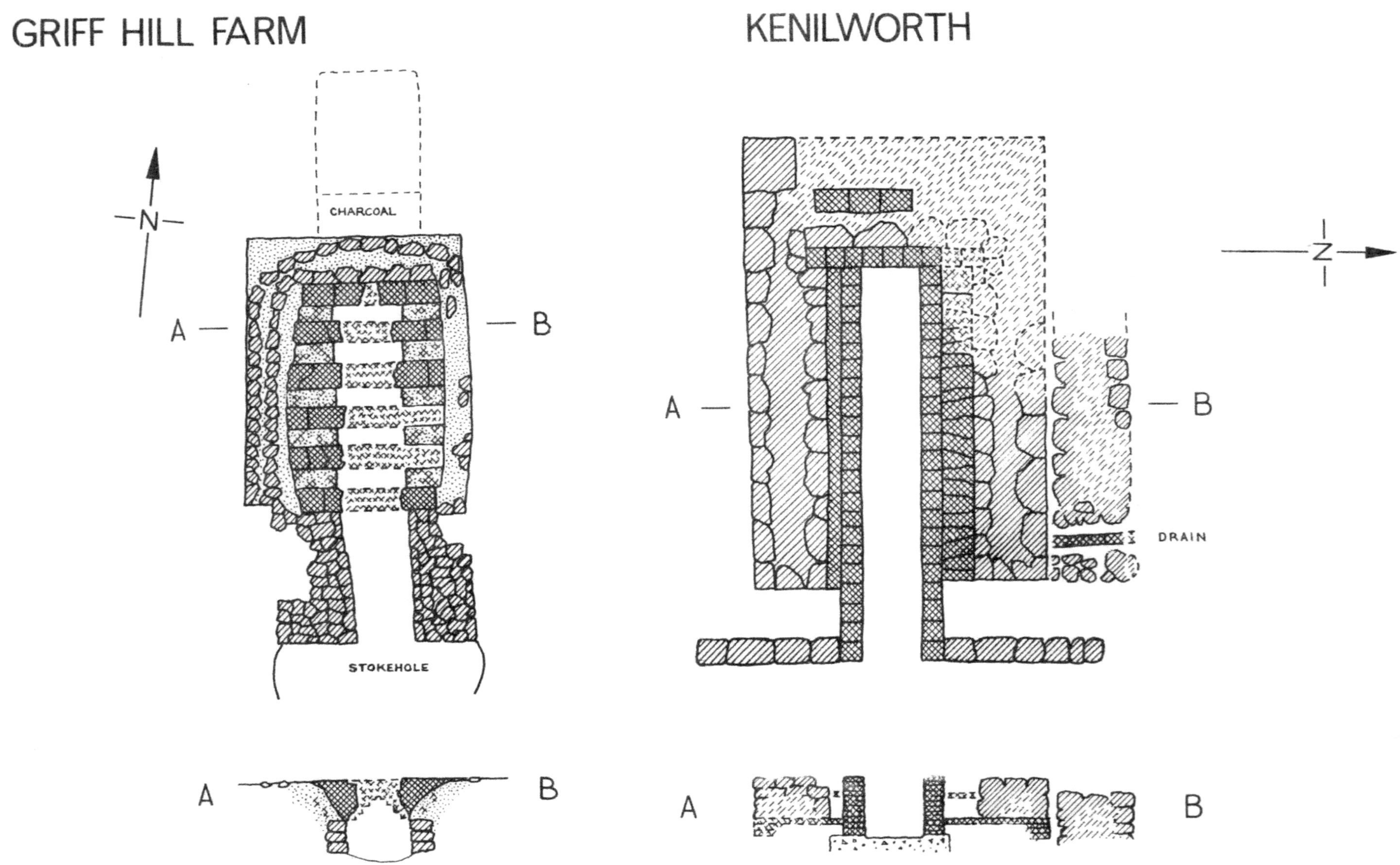

Fig. 6.31

Discussion

These four kilns from the parish of Chilvers Coton are only 5 km south-south-west of the Hartshill potteries and in an area of large-scale clay working. It is possible that more tile-kilns exist on the Arbury site which once belonged to a large tilery for the area. The ultimate destination of the products from these kilns is not known and a research programme to see which sites in the area they reached would be worthwhile. Did, for example, any tile reach Leicester?

References

Scott, K. 'Two Romano-British Tile-Kilns'. Trans. Birmingham and Warks. Arch. Soc. 84 (1971), 7-17.

Scott, K. 'Romano-British Tile-Kilns: The Arbury Tilery'. TBWAS. 87 (1975), 57-67.

KENILWORTH Fig. 6.31 SP 2572

The site, in the vicinity of Chase Woods, was excavated in 1923 and re-excavated in part in 1957. Three possible kiln sites have been identified in the field, but only one has been examined.

The main outer walls of the kiln were built of mortared sandstone blocks, although in the western wall complete roofing tiles were noted 'laid one upon another' to a depth of 610 mm. Within these walls was a central flue 600 mm wide, the sides of which were built of flat tiles bonded with clay. The total length of the flue was 4.2 m long including a projection of 600 mm beyond the east side of the kiln. When the flue reached the stokehole it met a sandstone wall built at right angles to the flue. This wall extended 1.5 m on either side of the flue.

Also on either side of the flue was a 'shelf' upon which, it was argued, the tile was stacked for firing, there being no evidence of cross walls or oven floor. If this is so, then Kenilworth joins a growing list of sites which have produced structures where it was thought that no oven floor existed and that tiles were stacked either side, and over, a flue (page 101). That the structure was used in tile production is indicated by the large number of wasters recorded.

A drain, at right angles to the main flue, is shown on the plan of the excavations and in the drawn section looks like an eighteenth-or nineteenth-century field drain. It could be considered as a badly-drawn imbrex drain, but, as there is no mention of an imbrex drain in the notes of the excavation, it must be assumed to be modern.

Products

Bonding tile or brick and tegulae formed the largest group found on the site, there being only a few imbrices and box-flue tiles. Ninety metres from this kiln is an area of tile debris thought to be the site of another kiln and from among the tile was an imbrex 'stamped with the letters NDVS', presumably the end of a name such as SECVNDVS. It is not clear whether this is a genuine stamp of whether the letters have been written on individually.

LAPWORTH

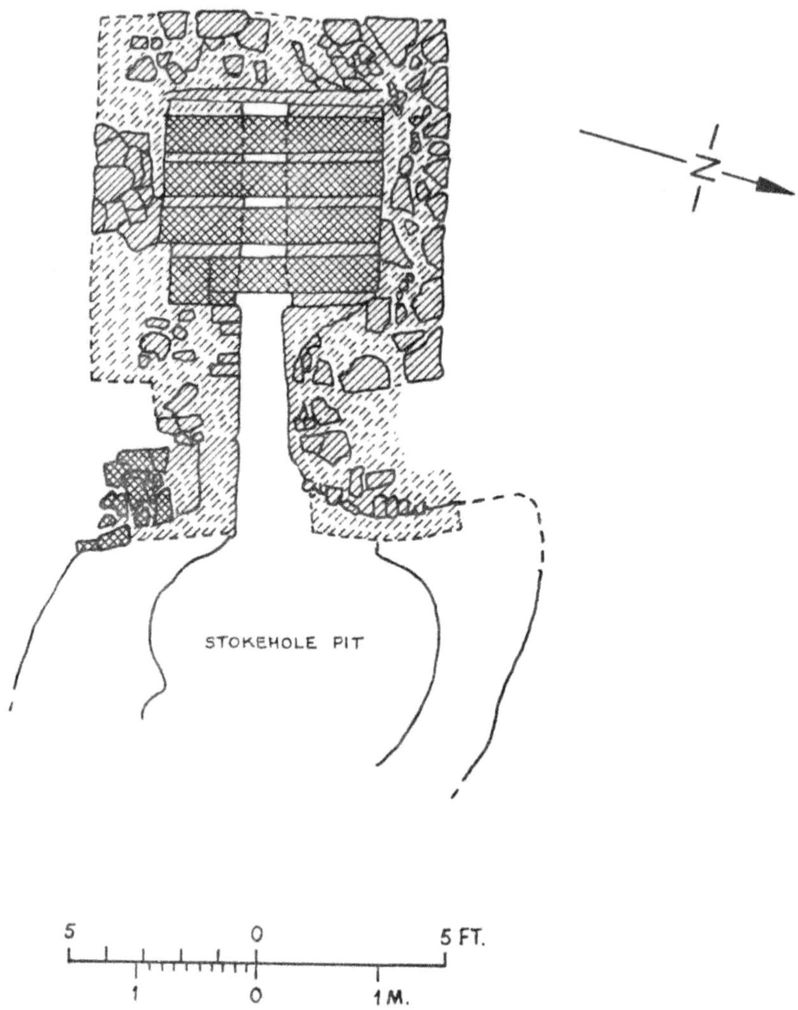

Fig. 6.32

Date

The majority of pottery found is dated to A.D. 300-350 and only one piece of samian need be earlier than A.D. 250.

References

Transactions of the Birmingham Archaeological Society, 50 (1927 for 1924), 54-56.

Unpublished notes from an original typescript written by Brian Stanley.

LAPWORTH Fig. 6.32 SP 1869

In 'Large Far Field', adjacent to Dicks Lane, tile had been recorded on a number of occasions and after ploughing in 1967 much more was noticed, suggesting that further damage was being done to the structure from which the tile came. Between October 1968 and May 1969 excavations were carried out by C. J. Baddeley revealing the presence of two kilns, although only one was examined in detail.

KILN A

The outer walls of this kiln were built of dressed sandstone and they formed a firing chamber 1.8 m square. The oven floor had entirely gone and the sub-floor structure consisted of four cross walls built of tile leaving very narrow cross flues only 80 mm wide. The floors of these flues were sloping and of type III. The fire tunnel projected about 1.4 m, from the combustion chamber to the stokehole and many 'green' tiles were used in its construction. In the stokehole were several layers of ash each sealed by a thin layer of soil, and a radio-carbon test on this ash was carried out giving a date of around A.D. 125, which is similar to that of the pottery found from the site.

KILN B

Fourteen metres to the south of Kiln A a trial trench dug in an area of burning and tile debris located a second kiln which, although not fully examined, showed the same characteristic narrow cross flues.

References

JRS 59 (1969), 216.

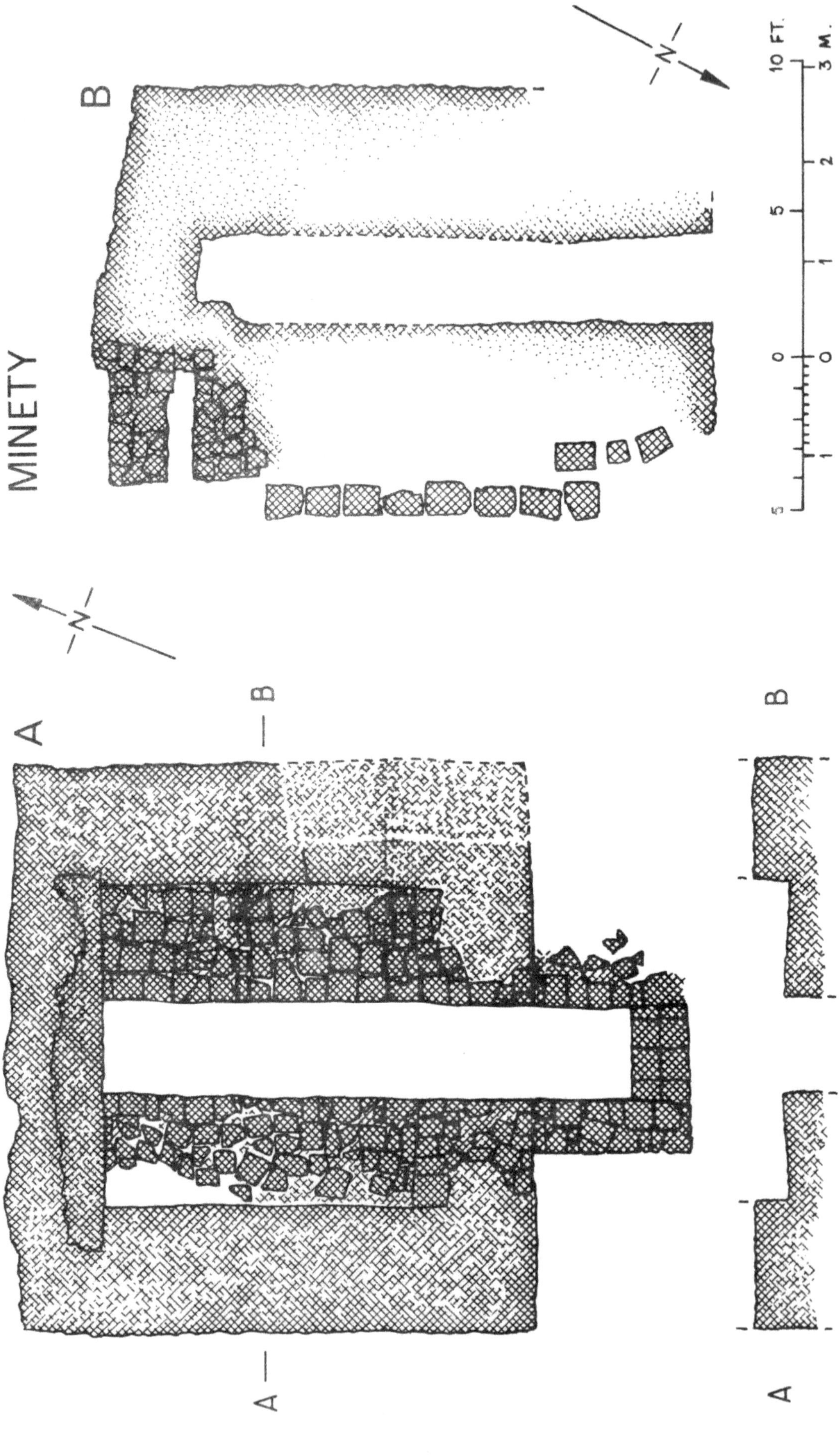

Fig. 6.33

MINETY ST 9992

Field work has now revealed a great deal of tile debris at Park Farm, Oaksey/Minety, suggesting at least ten tile-kilns and two stone buildings, making the site the most extensive Roman brick and tile work noted so far in Britain. Several stamped tiles have been found, including the stamps TPF and LHS.

Excavations were conducted in 1974 by Mr. A. J. Scammell of Bristol but were terminated when it was discovered that the site was scheduled. Mr. Scammell has been most helpful in providing details of his work although the final interpretation of the excavations must await his report. The comments which follow are based on personal observations, comments in letters from Mr. Scammell and on the plans drawn by R. J. Zeepvat who at the time was working for the Cirencester Excavation Committee.

KILN A (No. 5 on sketch plan, fig. 6.1) Fig. 6.33

There seems to be no doubt that these are the remains of a kiln although Mr. Scammell refers to the structure as a fabrication shed, based on the supposition that it was never fired due to the absence of burning and ash. However, the remains compare well with other tile-kilns from Britain and elsewhere, and the posited lack of burning may mean that the structure was intended to be a kiln but was never completed. This would explain the apparent absence of cross walls unless the kiln was of the type in which they never existed (see page 101).

It appears from the plan drawn by Mr. Zeepvat that the combustion chamber was almost square, measuring 3.4 by 3.3 m and formed by a wall of stone 1.3 m thick. The central flue walls were constructed from tile giving a flue 1 m wide, which projected 1.65 m from the front of the kiln or 2.5 m from the combustion chamber. There is no information at present about the stokehole.

KILN B (No. 6 on sketch plan, fig. 6.1) Fig. 6.33

Only part of the kiln was uncovered, namely the main central flue, but at one end attempts were made to examine more of the kiln's structure. The resulting plan is puzzling and far from clear, although there is no doubt that it was a kiln. On available information all that can be said is that there was a main central flue 900 mm wide and 5.2 m long.

Products

There is a wide range of tile and brick from the site and Mr. Scammell is at present analysing the material from his excavations which includes an interesting sequence of comb markings on box-flue tiles, voussoirs and flat bricks which he believes may represent numbers. Also found were what are described as 'kiln props' made from the same material as the tile and brick. Pottery may also have been made in the area.

Date

Pottery from the site has fluctuated in date from fourth to the second century and until there is more information available the value of the pottery for dating purposes cannot be assessed. Mr. Scammell now believes that production was underway possibly in the late first century and certainly by the first half of the second century. Tiles bearing the TPF stamp have been dated to mid-to-late second century at Kingscote, but how long they continued to stamp in unknown.

Discussion

Tiles from Minety reached Cirencester and work reported elsewhere in this volume indicates the distances to which Minety products were distributed (see chapter 18). No site in Britain compares with Minety in the number of kilns, the stone buildings or suggested clay pits indicated. The stimulus for the growth of this brickfield was undoubtedly the town of Cirencester and later the many villas in the region. Much more will be learnt when Mr. Scammell's report is available.

References

Wilts. Arch. Mag. 41 (1921), 424-425.

McWhirr, A. D. and Viner, D. 'The Production and Distribution of Tiles in Roman Britain with particular reference to the Cirencester region'. Britannia, 9 (1978), 359-377.

YORKSHIRE

GRIMESCAR SE 1319

Early discoveries recorded in a sixteenth-century diary described a structure which was probably a kiln and also mentioned a tile stamped COHIIIIBRE. Later, Richmond drew attention to the site and to the occurrence of cohort stamped tiles at the fort at Slack. In 1955/6 'the lower part of a rectangular stone tile-and brick-kiln' was excavated by Mrs. Ann Hallam, but unfortunately the results have not been published and attempts to locate a plan of the kiln have been unsuccessful.

The stone-built kiln measured about 4.8 by 4.0 m and had a series of 'pillars or arch-supports' in the combustion chamber, which was finished with puddled clay to fill cracks between stone. Knuckle imprints were found on the surface of the clay.

Further excavations in Grimescar Wood were carried out in 1964 under the direction of T. G. Manby and J. G. Purdy in order to examine the kiln stokehole and to recover a series of kiln products. Two areas of tile debris were located about 40 m apart.

Products

The usual range of tile and brick seems to have been found apart from a rare circular pila tile 430 mm in diameter. Cohort stamped tiles seems to have been found on all occasions that the site was investigated. They are all of the type stamped COHIIIIBRE. Pottery was also produced on the site.

Discussion

The stamps indicate that the tilery was operated by the Fourth Cohort of the Breuci for the nearby fort at Slack. As with other auxiliary tileries pottery and tile were made side by side at Grimescar.

Date

At Slack, tile first appears in quantity in the Trajanic period and is used extensively in the Hadrianic reconstruction at the fort. The pottery also suggests that the complex was in production c. A.D. 100-120+.

References

Richmond, I. A. Huddersfield in Roman Times. Huddersfield, 1925, 57-9.

Hallam, A. 'The Roman Tilery in Grimescar Wood', Huddersfield and District Archaeological Society Bulletin, 16 (1965), 1-4.

Purdy, J. G. and Manby, T. G. 'Excavations at the Roman Tiler at Grimescar, Huddersfield, 1964', Yorkshire Archaeological Journal 45 (1973), 96-107.

SCOTLAND

MUMRILLS Fig. 6.34 NS 9179

The kiln was found in 1913 some 2 metres south of the Antonine Wall at Mumrills, and it appears to have been built entirely of stone. The side walls of the combustion chamber were 1 metre thick and survived to a height of 1 m. The side walls of the combustion chamber sloped and so the area of the chamber at the bottom was approximately 1.5 m square, but where the oven floor would have been the kiln was about 2 m square. There were two cross walls within the combustion chamber and the floors between these walls were on the same level as the main flue and therefore constitute one of the few examples of the type 1 kiln, although the sloping face of the stone walls would have acted in a similar way to those kilns having sloping floors between the cross walls. The oven floor was made from slabs of stone described as heavy flags and the whole of the kiln showed traces of having been exposed to intense heat. Masses of soot and wood ash were found both inside and outside the kiln. There is no evidence of a substantial stoke hole, although on the published plan a feature which could have been a stoke hole is shown extending northwards beneath the southern curve of the Antonine rampart. The presence of a considerable quantity of brick and tile provided the evidence that this kiln was used for firing them, but it was suggested that it could occasionally have been used for firing pottery, although only two small sherds of pottery were found from the site.

Date

The two small sherds of pottery found were examined by Mr. James Curl who thought them to be first century A.D. in date. However, these cannot be used to date the construction of the kiln.

MUMRILLS

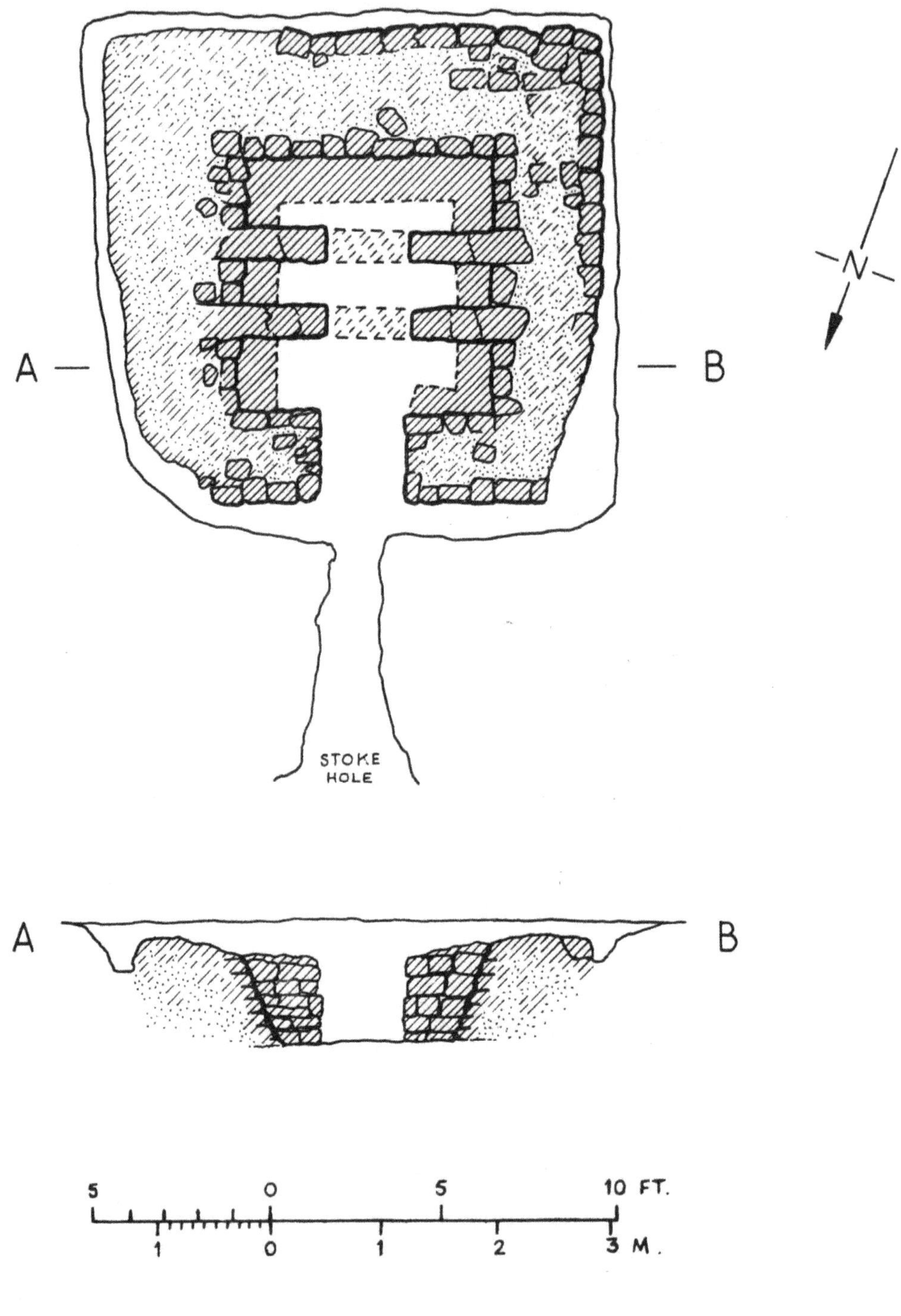

Fig. 6.34

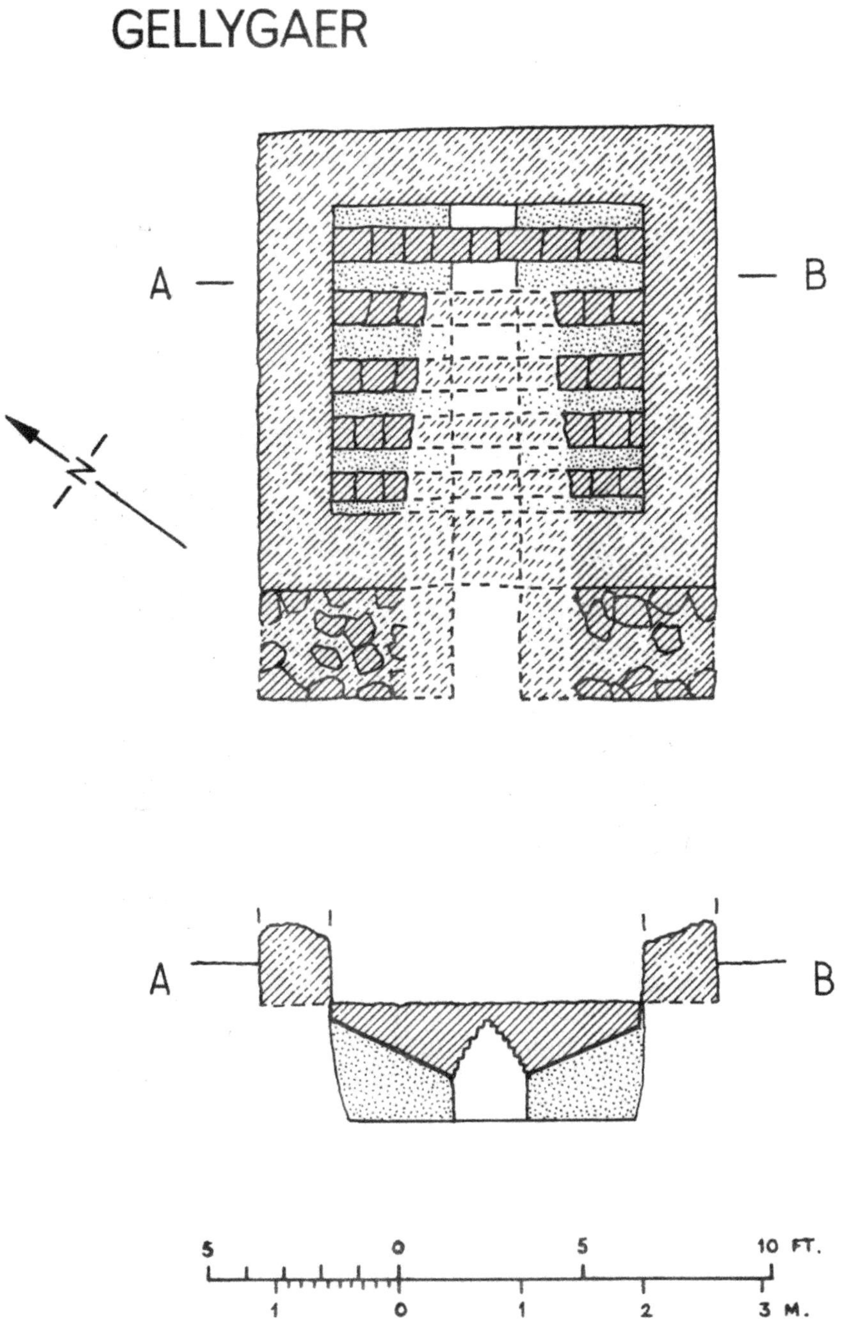

Fig. 6.35

Reference

Proceedings Society Antiquaries Scotland, 49 (1914-15), 123-128.

WALES

CAERNARVON SH 4862

Rescue excavations in 1976 by the Gwynedd Archaeological Trust in an area now thought to be within a fortlet dated to c. A.D. 80-120, revealed part of a tile-kiln or clamp. The remainder of the kiln lay under the driveway to a house which was not removed until 1979 when further observation work was possible.

A large oval pit measuring 6.0 by 3.5 m and 3 m deep was found close to the rampart of the fortlet and from the contents of the pit it was interpreted as a tile-kiln which had only been used once. In the pit was a considerable mass of incompletely-fired tile including fragments of tegulae and imbrices were identified. Beneath this mass of tile was a 0.1 m layer of pure charcoal from the firing of the kiln underneath which, resting on the bed-rock, was a layer of clay mixed with charcoal in what might have been a flue trench. This layer was thought to represent the firing of the kiln when empty in order to harden the clay used in its construction (see page 99).

There was no evidence of any floor to support the kiln load and it is suggested that the tiles were 'self-supporting', i.e. that they were stacked around the flue and mixed with the fuel which fired the kiln or clamp. The tiles nearest the stokehole were well fired whereas those to the rear were soft and crumbly. The presence of the kiln charge still in situ and only one layer of charcoal have lead the excavators to suggest that the kiln was only used once for firing tile. There were no details of the stokehole. Close by was a well and timber building.

Date

Layers above the fill of the pit contained only material dated to A.D. 80-110 which, as there were no datable finds from within the pit itself, must be used to postulate a date for the kiln, along with material found in association.

References

Britannia 8 (1977), 358.

Information from R. B. and S. White.

GELLYGAER Fig. 6.35 ST 1396

During the digging of a grave in the churchyard of Gellygaer Church in 1913, the remains of a Roman kiln were discovered. Subsequent excavation around the graves enabled a complete plan to be reconstructed. Only the sub-floor structure of the kiln was preserved, which consisted of a combustion chamber 2.4 m square with five cross walls and hence six cross flues. These cross flues were at a higher level than the main flue and sloped upwards towards the outside of the kiln. The kiln appears to have been built entirely of stone, which is described as being the rough local pennant stone and during its use the sides of the small flues and part of the main flue became vitrified. Upon this sub-floor structure rested the floor of the oven, which appeared to be of brick, but on closer examination proved not to be paved with whole bricks. It apparently consisted of a mixture of broken bricks and clay daubing which had been baked to a brick-like consistency. This composition was spread over a base of thin stones, firmly cemented together, the total thickness being about 203 mm. From the digging of graves within the churchyard has come much tile, pottery and also coal. Coal was frequently found during the course of excavations, especially in the annex to the fort and the presence of coal within the vicinity of the kiln suggested that it was used to fire the kiln.

Products

Only roofing tiles are specifically mentioned in the brief report. Numerous red fragments are described in the report as "tiles and coarse pottery spoiled in the firing or broken in their removal from the kiln". There is then the possibility that this kiln was used for both tile and pottery. The material recovered suggests in fact that mortaria were being made in the kiln.

Date

There is no evidence in the published report as to the date of the kiln or the mortaria which may have been fired in the kiln.

Reference

Ward, J. 'The Roman Fort at Gellygaer.' Discoveries Made in 1913'. Transactions Cardiff Naturalists Society, 36 (1913), 1-20.

HOLT Fig. 6.36 and 6.37 SJ 4054

This site was excavated between 1907-15 by T. A. Acton and subsequently published by Professor W. F. Grimes in 1930 who took over the task of writing up the excavations soon after Acton's death in 1925. Holt was a works depot for the Twentieth Legion based at Chester and produced pottery and tile including a series of legionary-stamped tiles and antefixes bearing the wild boar emblem of the Legion.

The depot covered an area of at least 20 acres on the banks of the River Dee and the buildings, kilns and finds have been extensively reported upon by Grimes. Consequently little discussion will be included in this Gazetteer, but plans of the double parallel flue kiln and the kiln plant will be included, so that comparison can be made with other kilns. The size of the kilns and the design of the kiln plant clearly illustrate their legionary character.

No precise dates can be given for the individual structures within the depot, but the pottery and occurence of finds in dated structures elsewhere point to it having been established towards the end of the first century A.D. If the legionary tile-stamp ending ANTO represents Antoniniana then it would seem that tile was still being produced in the third century.

Reference

Grimes, W. F. Holt, Denbighshire. The Twentieth Legion at Castle Lyons. Y Cymmrodor 41 (1930).

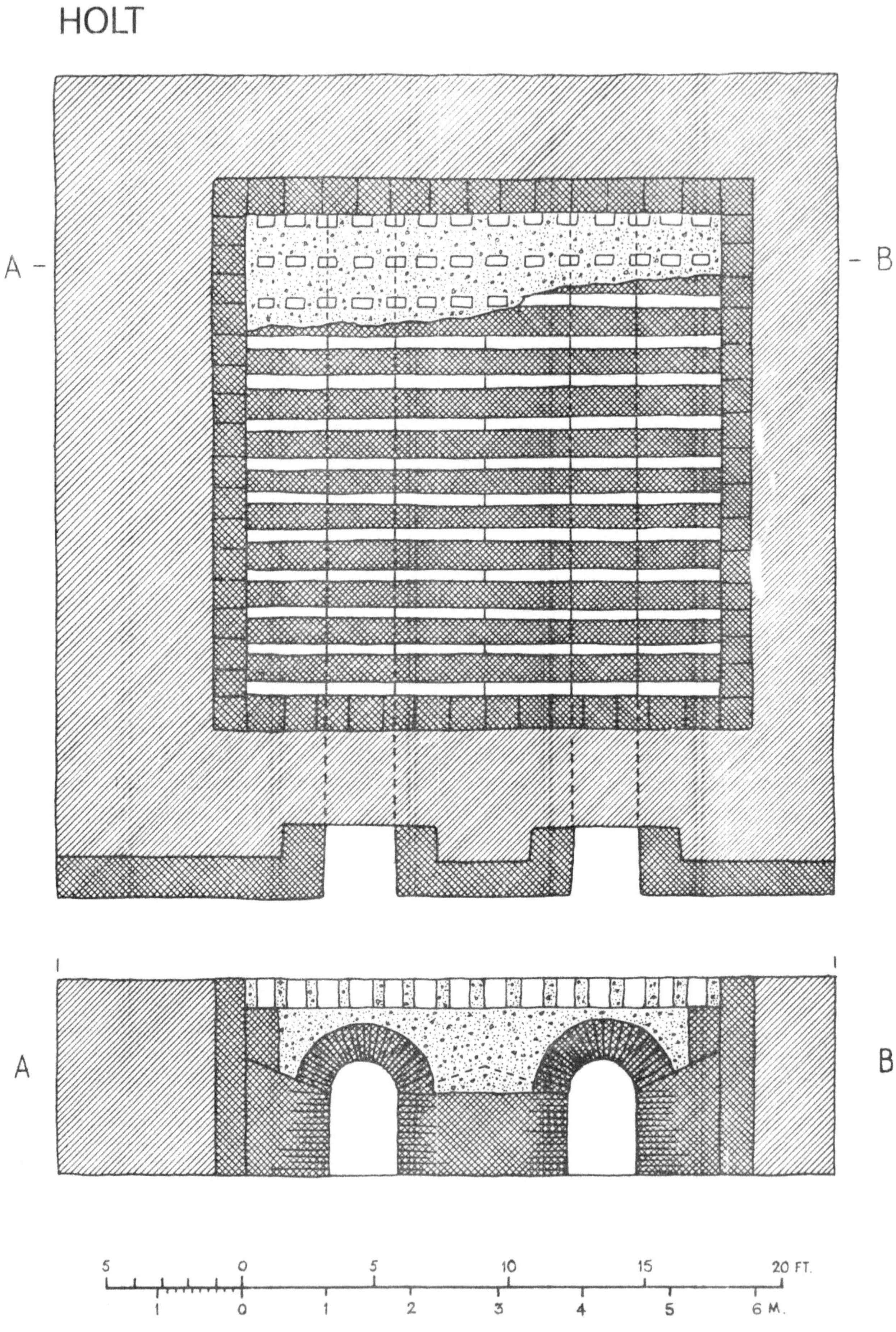

Fig. 6.36

HOLT

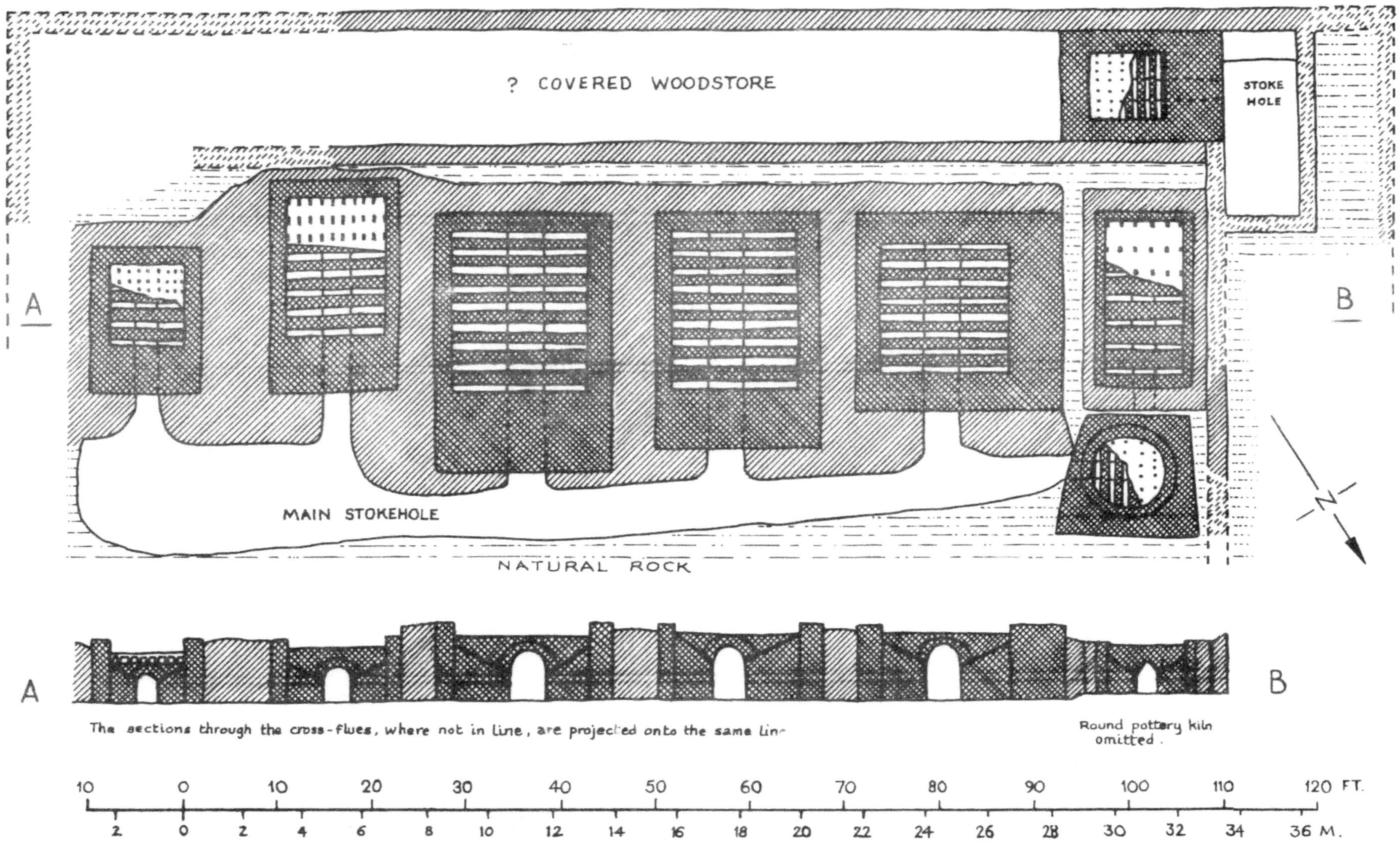
? COVERED WOODSTORE
STOKE HOLE
A
B
MAIN STOKEHOLE
NATURAL ROCK
N
A
B
The sections through the cross-flues, where not in line, are projected onto the same lin-
Round pottery kiln omitted.
10 0 10 20 30 40 50 60 70 80 90 100 110 120 FT.
2 0 2 4 6 8 10 12 14 16 18 20 22 24 26 28 30 32 34 36 M.

Fig. 6.37

A TILEMAKERS' WORKSHOP AT ITCHINGFIELD—A REAPPRAISAL

Talbot Green

Introduction

In the ten years that have elapsed since writing the report on the excavations at Itchingfield, certain ideas have grown in my mind that were not presented therein (Green 1970). They do not alter my belief that, in 1964, a tilemakers' workshop was uncovered and that it was of a purpose-built design: but, on the contrary, the case for this view can be argued more fully and forcefully than was then possible. So I am grateful to the editor for allowing me to contribute this supplementary paper, being stimulated to do so in part by his own recent publication (McWhirr and Viner 1978).

* * * * *

Fig. 7.1 shows a simplified plan of the building excavated at Site 1 at Itchingfield, Sussex (TQ 14092979). The detached wall 5 and stumps of walls 1 and 4 had slid down into the drainage culvert through the action of the newly-deflected stream; that of wall 4 clearly articulated on the end contained in the raft. The south-west extremity of the raft, referred to below as the 'south-west extension', together with the narrowed masonry wall supporting one pedestal, had also slid forward but remained virtually intact. It is, therefore, with reasonable confidence that the restored plan is produced in Fig. 7.2, for it is on this that further arguments are based.

A number of points, mostly made already in the body of the 1970 report or evident in the illustrations, must first be established. They are as follows:

a) The north-east part of the raft had been carefully paved but bore no evidence of a roof or walls (labelled 'paving slabs' on Fig. 7.1).

b) A doorway led from this paving into the building, having a clear stone threshold with herringbone masonry behind for a wooden still. Inside, vertical slabs ran back as if defining a gangway.

c) To the north-west, inside the door, there were rows of small stones set on edge in the clay covering the raft, in contrast to the clear floor of the gangway. They resembled herringbone courses, too.

d) The hard clay floor overlay wall 4 to a depth of between 6 and 11 in. (15 and 28 cm): it did not appear to have carried any internal partitioning in the main period of use, though it may have done so originally (Green 1970, 28).

e) The pedestals in the 'recess' between walls 3 and 4 did not reach to the

---

N.B. Imperial measurements given preference so that comparison with the earlier report is easier. Ed.

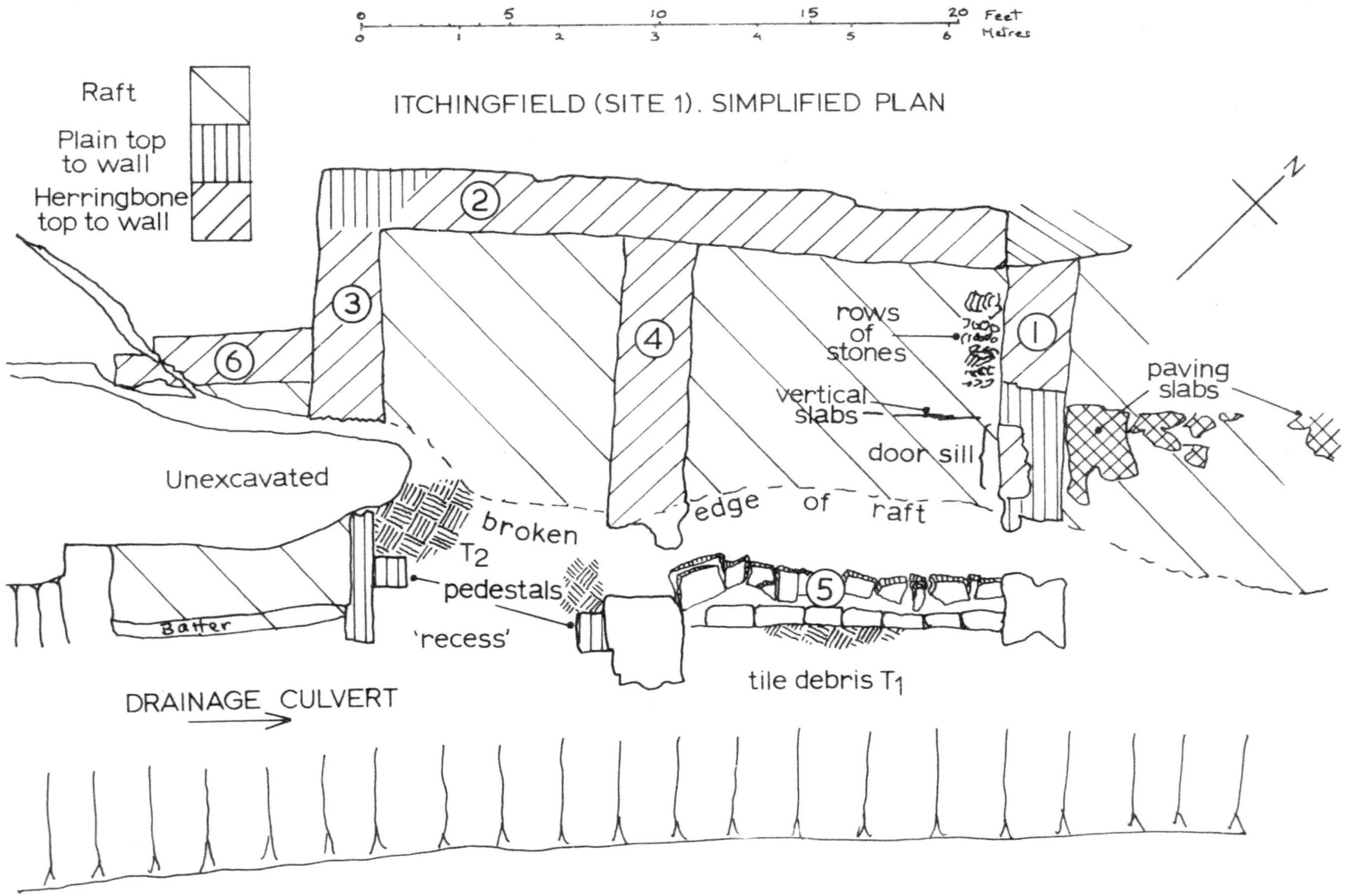
0
5
10
15
20 Feet
0
1
2
3
4
5
6 Metres
ITCHINGFIELD (SITE 1). SIMPLIFIED PLAN
Raft
Plain top to wall
Herringbone top to wall
N
1
2
3
4
5
6
rows of stones
vertical slabs
door sill
paving slabs
edge of raft
Unexcavated
broken T2
pedestals
'recess'
tile debris T1
Batter
DRAINAGE CULVERT

Fig. 7.1

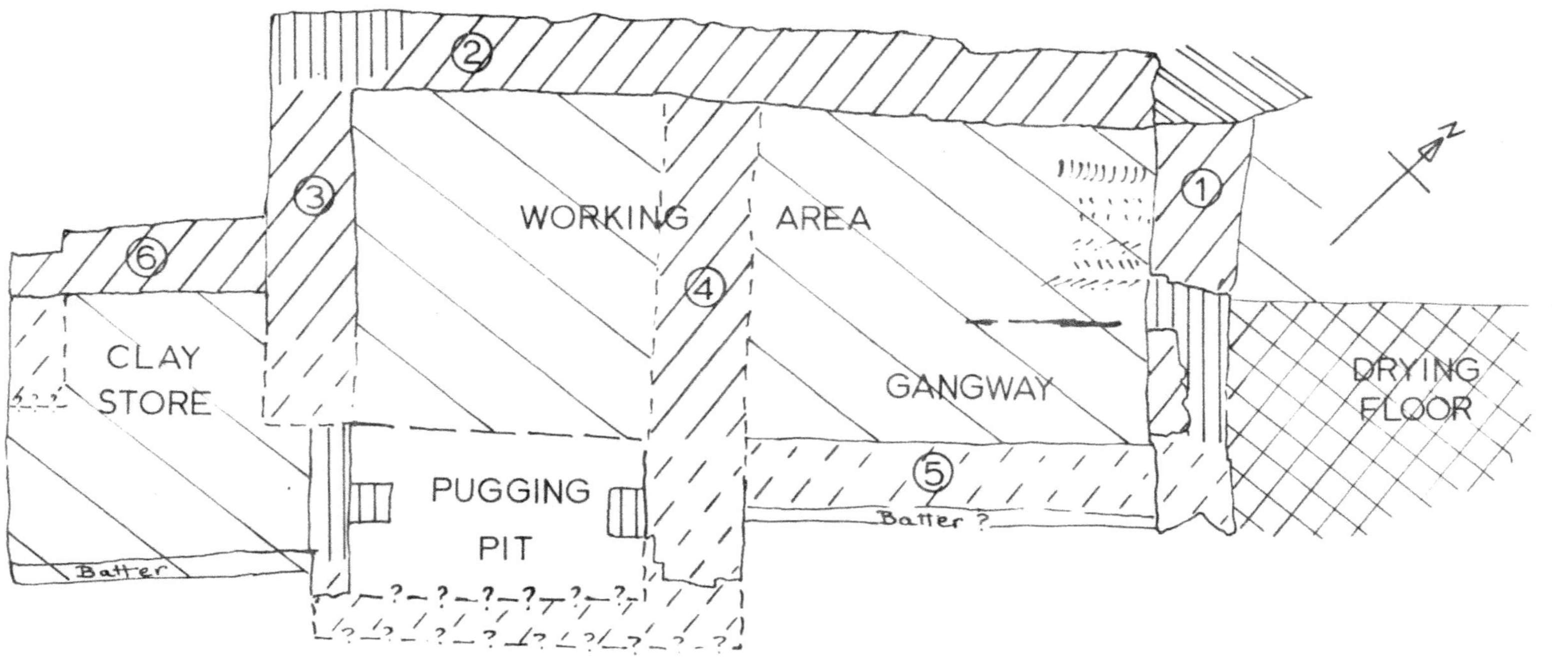

ITCHINGFIELD (SITE 1). FUNCTIONAL INTERPRETATION

Fig. 7.2

top of the raft, let alone the floor level represented by the hard clay. Around them was a mass of clay and behind lay deposits of tile fragments (Fig. 7.1, T2).

f) The face of the raft between the pedestals was left rough, without herring-bone facing, in contrast to the even outside edges along the north-east side and round the south-west extension. The south-east face of this latter, and probably that of wall 5, had a batter (Green 1970, 27).

g) Although mortar was found attached to a few pieces of tile in the various dumps, and therefore was available, none was used either in between the squared masonry of wall 5, in the pedestals, or in the masonry behind that on wall 3. Nor was it used for rendering the walls of the 'recess'.

h) Wall 6 was not continuous around the south-west extension beyond wall 3. It was not possible to clear all this area but observations of the crack faces suggested wall 6 may have turned a right angle to form the south-west side. If so, this wall never reached the south corner, or returned to meet the continuation of wall 3.

i) Wall 3's continuation behind the pedestal and, perhaps, beyond into the area the culvert had destroyed was only half the thickness of the other walls; however, it was built of better masonry. Whether it had any herringbone masonry course on top is uncertain.

From the above observations, it seems reasonable to draw the following conclusions:

1. The indications of a wooden sill, offset in the doorway, show that the north-east paved area was outside as far as the main building was concerned.

2. There is some suggestion of functional diversity in the structure e.g. paved area, gangway, recess, south-west extension.

3. The recess between walls 3 and 5 was not external to the building, since the face of the raft was not faced or battered.

4. This internal recess represents a rectangular pit.

5. The pedestals, set below floor level, were intended to take some downward pressure (with respect to floor level). Only where this sort of thrust was to be exerted was squared masonry used; otherwise, the irregular face of the raft, made of slabs set on end, was considered quite adequate.

6. Assuming that the pedestals were placed symmetrically in the middle of the shorter sides, the pit measured 7 ft. 6 in. (2.30 m) long by 4 ft. 6 in. (1.40 m) wide. No attempt was made to render it waterproof.

7. The south-west extension of the raft, undoubtedly an integral feature, was partially walled round. The walls lay specifically along its north-west and, perhaps, half its south-west sides, thus protecting it from westerly winds and rain. The south-east (downhill) side was open.

At this point it is necessary to consider certain questions relating to the above conclusions. Firstly, was the 'recess' really part of a rectangular pit? Could it have been a stairwell? I think not, for the following reasons.

i) On the arguments set out in (f) above, it was not the site of a flight of external wooden steps.

ii) If it had been the site of a flight of internal wooden steps (which might utilise the pedestals in some way, for support), there was nowhere for them to lead. Had they been leading to a cellar, an angle of 45° seems the most one can expect (cf. Frere 1972, Fig. 28, K-K', 26). This would mean that, to give even a minimum headroom of 6 ft. (1.8 m) such a cellar would have projected far enough out for its walls to have shown up in the opposite bank of the culvert, which was certainly not the case (Green 1970, Fig. 2, F-E). The cellar at Chalk (Johnston 1972, Fig. 4) was at least 7 ft. (2 m) deep; at Shakenoak the excavators believed the same was the case (Brodribb et al. 1971, 20).

iii) The raft was little more than 3 ft. (1 m) thick and wall 1 sca cely went deeper (Green 1970, Fig. 2). Walls 5, although it had slipped out of position, was found resting on unweathered blue clay when sectioned and cannot have had deeper foundations than wall 1. Hence suggestions of any cellar even 6 ft. (1.8 m) deep seems totally untenable.

A second question concerns the function of the pit. An obvious suggestion is that it is a water storage pit, but this seems unlikely for the following reasons:

i) Why have a water pit 7 1/2 ft. (2.3 m) long, 4 1/2 ft. (1.4 m) wide and only 3 ft. (1 m) deep? It would hold over 100 gallons (450 litres) but, in a water storage system, depth not surface area is more important—to allow as much of the water as possible to be collected.

ii) There certainly was a need for water in a tilemaking establishment, but everything suggests it was brought in by hand. There is no evidence that there was any arrangement, such as a leet, to feed water from the spring at TQ 14032970 to the building; such evidence would have been apparent either in the side of the culvert at some point or cutting across Site 2. Despite the local water table under the raft, formed in the hollow in which it sat (ibid., 27), there was no spring.

iii) There was no arrangement to keep the water in this 'water pit' (see (g) above), so any topping-up by hand would involve filling the whole volume of the raft's interstices. Against this, of course, is the fact that, once achieved, one draws on a large reservoir which will allow the level to fall only slightly in the pit for a relatively large withdrawal of water. However, the water level will be no higher than the lowest ground outside, and wall 5, being specially dressed on its outside face and having the tile debris near its base, may have stood largely above ground level.

iv) What function do the pedestals serve? Whence comes the downward thrust they were meant to bear? With a 'water-pit' of such a shape, a well-head seems functionally superfluous, and difficult to envisage from

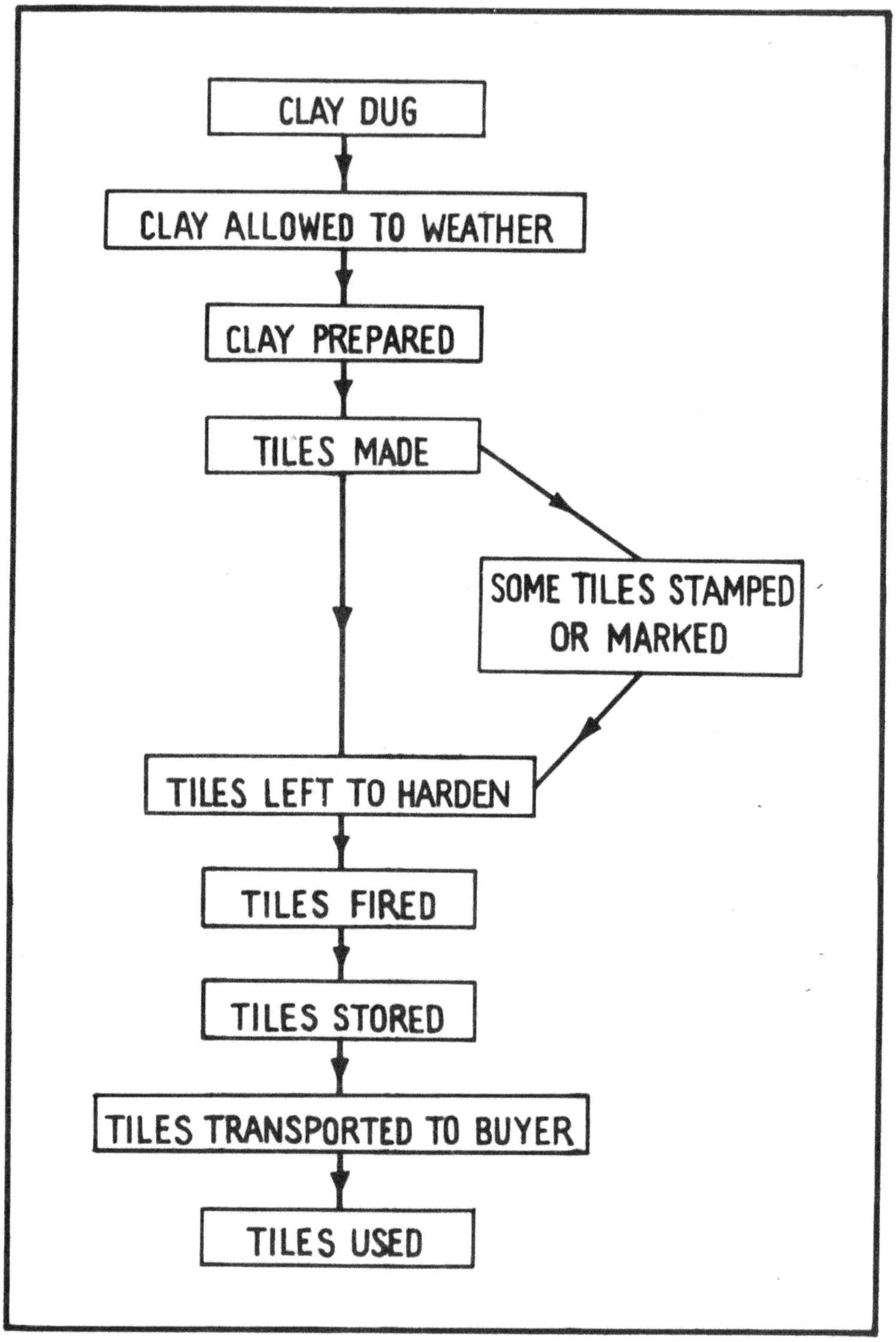
CLAY DUG
CLAY ALLOWED TO WEATHER
CLAY PREPARED
TILES MADE
SOME TILES STAMPED OR MARKED
TILES LEFT TO HARDEN
TILES FIRED
TILES STORED
TILES TRANSPORTED TO BUYER
TILES USED

Fig. 7.3

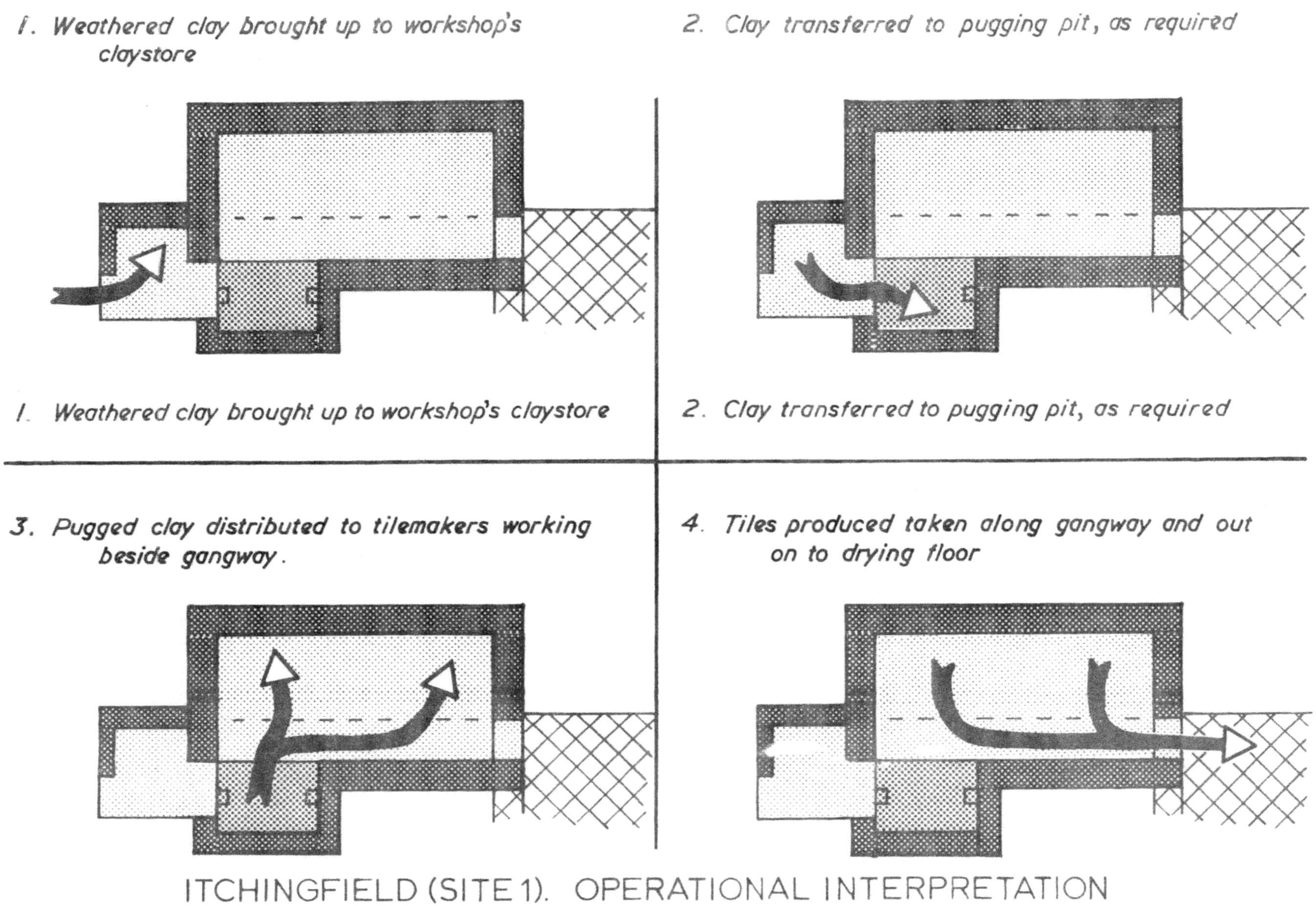
1. Weathered clay brought up to workshop's claystore
2. Clay transferred to pugging pit, as required
1. Weathered clay brought up to workshop's claystore
2. Clay transferred to pugging pit, as required
3. Pugged clay distributed to tilemakers working beside gangway.
4. Tiles produced taken along gangway and out on to drying floor
ITCHINGFIELD (SITE 1). OPERATIONAL INTERPRETATION

Fig. 7.4

the point of view of mechanical construction and operation. The only other suggestion seems to be to support a wooden cover. But such a cover would be highly unstable, having supports only at one point on each of its narrower sides and none along the side nearest the workroom. If trodden on near that edge, it would tilt over.

It seems more likely to have been a pugging pit, on the analogy with practices in use at the modern potteries at Wrecclesham, near Farnham, Surrey, where their pit is supplied by a motor-driven pugging mill mounted above it, and is set in the floor in one corner of the workshop (in which chimney pots, garden urns, flower pots etc. were being made when I visited it in 1970) as is the Itchingfield example. In which case, the thrust that the pedestals were obviously meant to bear could be from a frame supporting the pugger's weight; since the pit had no bottom and the worker would otherwise sink ever deeper into the clay.

On the basis of this 'pugging pit', or even the more unlikely 'water pit', hypothesis, a model of the building's functional purpose can be put forward. The schematic flowchart to illustrate tile production, recently published (McWhirr & Viner 1978, Fig. 1 here reproduced as fig. 7.3), suggests the distinct stages to be catered for. If one equates the 'pugging-pit' with the stage 'clay prepared', and the paved area with 'tiles left to harden', then 'tiles made' may be applied to the main body of the building. This has been shown diagrammatically in Fig. 7.4, which illustrates what a modern production engineer identifies as a 'flow-line' manufacturing process.

Should this flow-line process be applied to the Roman world? The suggestion being made is that the craftsmen analysed the separate activities into which their work fell and planned a building accordingly. The presence of identically-combed flue tile fragments both under and on top of the raft (Green 1970, 32) implies that tile was being made before the building was constructed and consequently the tilemakers would have known their requirements before construction began. The siting of the kilns at Pizzica, at the ends of what must have been the tilemakers' workshops (see above, p. 55), shows such foresight. (Prof. Carter's observations on the lack of pottery sherds in this building parallel our Itchingfield ones.)

As noted above, the flow-line concept holds even on the 'water-pit' hypothesis. Here, the south-west extension serves either as clay-store or pugging floor; in the former case, pugging would have been done in the south-west corner of the main building, tilemaking in the north-east. A doorway beside the 'water-pit' is obviously needed, and is not discounted due to the lack of excavation at this precise point.

Weathered clay was brought to the south-west extension which had partial walls to provide a screen from the rain, but open side(s) to let in light and make access easy for laden men. There is a gap in the main building's south-west wall, conveniently, beside the pit. Inside the main building, movement to and fro along the gangway did not interfere with the tilemaking, and the finished tiles left by a door opposite to that through which clay and water (?) were delivered. The drying/hardening area was handy but out of the way of the manufacturing activities (Green 1970, 33). Where the tiles were taken for

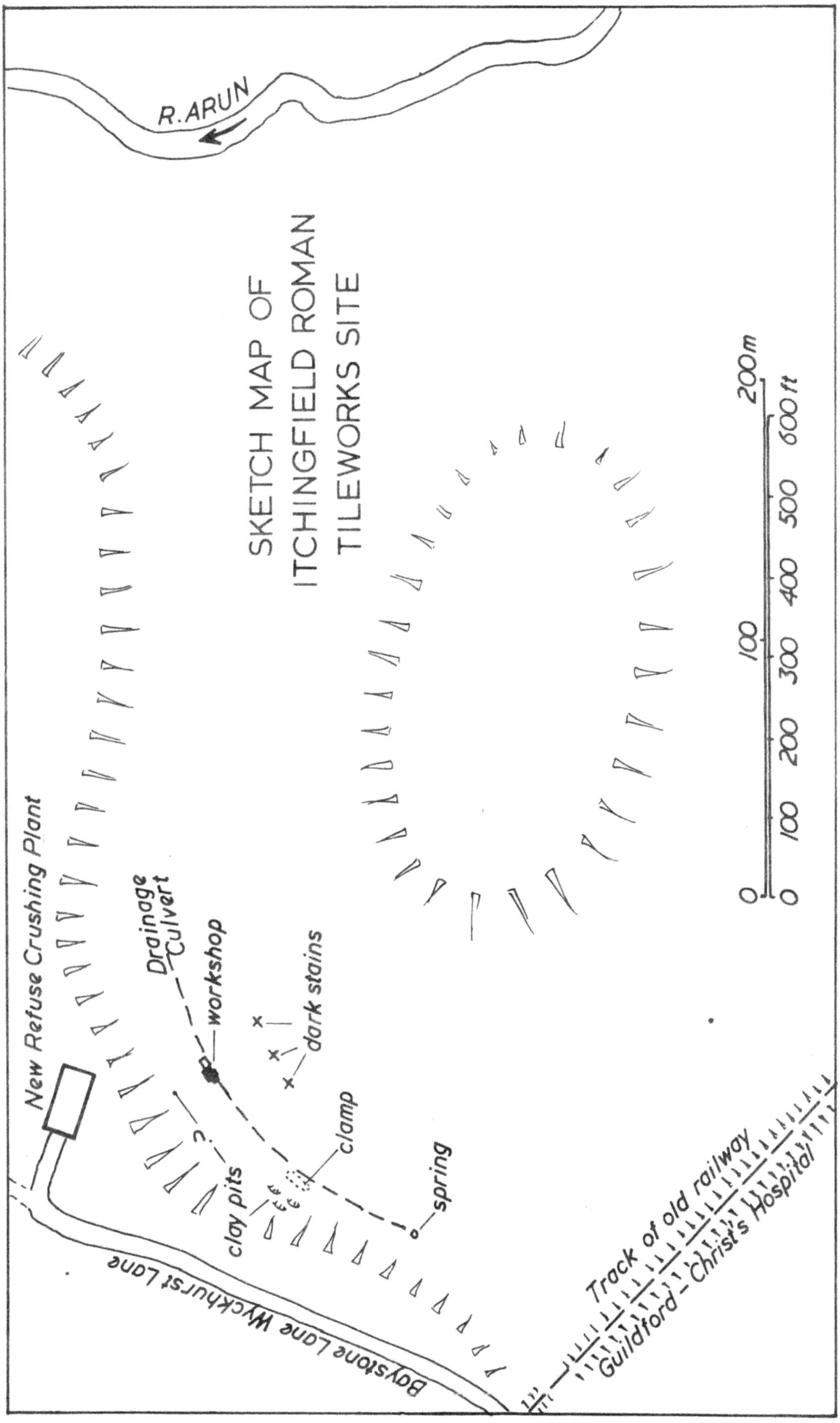
R. ARUN
SKETCH MAP OF
ITCHINGFIELD ROMAN
TILEWORKS SITE
0 100 200m
0 100 200 300 400 500 600 ft
New Refuse Crushing Plant
Drainage Culvert
workshop
dark stains
clamp
clay pits
spring
Baystone Lane Wyckhurst Lane
Track of old railway
Guildford - Christ's Hospital

Fig. 7.5

firing, apart from the probable clamp at Site 2 (ibid., 28), is uncertain: scraping and grading of the whole valley floor south and east of the sites had denied us any chance of answering such questions before we began investigations. Fig. 7.5, apart from illustrating the general topography, shows the location of three dark stains that were found in the valley floor. They were not specifically filled with ashes, however, and need not necessarily mark the bases of kiln stokeholes.

BIBLIOGRAPHY

Brodribb et al. 1971: A. C. C. Brodribb, A. R. Hands and D. R. Walker, Excavations at Shakenoak ii, Oxford 1971.

Frere 1972: S. S. Frere, Verulamium Excavations Vol. 1 (Report Res. Comm. Soc. Antiq. London XXVIII) 1972.

Green 1970: T. K. Green, 'Roman Tileworks at Itchingfield', Sussex A. C., 108 (1970), 23-38.

Johnson 1972: D. E. Johnston, 'A Roman Building at Chalk, near Gravesend', Britannia 3 (1972), 112-48.

McWhirr & Viner 1978: A. D. McWhirr and D. Viner, 'The Production and Distribution of Tiles in Roman Britain with particular reference to the Cirencester region', Britannia 9 (1978), 359-77.

SILCHESTER TILE MAKING—THE FAUNAL ENVIRONMENT

Leslie Cram and Michael Fulford

Introduction

The purpose of this paper is to demonstrate the contribution that a study of foot-impressions on Roman tiles can make to our understanding of both the tile-making process and the animals themselves.

In the first place we should stress that foot-impressed tiles inform us about living animals in the vicinity of the tilery. This information will complement the results of conventional faunal analysis, since the latter is almost entirely limited to food remains or to bones which were particularly suitable for tools. Whereas the foot-impressions do reflect the living animal (whether domesticated or wild) around the tilery, skeletal remains need not have originated from beasts living at or near the site from which they are excavated.

Secondly we should stress that studies need to be made on a regional and provincial basis in order to assess the frequency with which impressed tiles occur in the total tile population as well as the proportion of different animals represented. The aim is not simply to reconstruct the varieties of animals moving amongst the tile-makers, rather the circumstances in which foot-impressions could come about, even if it was to the detriment of the finished product. We may hypothesize that the more the tile-maker was engaged full-time in producing tiles the less likely it is that we will find evidence of animal foot-impressions on his products. An important qualification should be added here: whereas the presence of the impressions of wild animals may simply reflect the degree of security around an individual workshop irrespective of how specialised the tile-maker is, we would not perhaps expect the impression of domesticated animals to be present if the tile-maker was working full-time with no opportunity to farm or keep stock. In areas where demand was high or where year-round tile-making was climatically possible (see below, p. 208) we might expect to find some full-time specialists. Study of the frequency with which tiles are found with foot-impressions is crucial, for it will help to discriminate between stray impressions and those which inevitably result if stock is kept adjacent to the workshop and regularly passes through its yard. Regarding the question of specialisation it is important to consider what differences there are between civil and military workshops. At the latter we might expect greater discipline to the exclusion of foot-impressions (see below, p. 208), but in regions where all-year-round tile-making was not practicable or where demand was slack it is reasonable to expect to find tile-makers farming as well. We suggest therefore that a study of foot-impressed tiles will help to inform us about the very important question of the degree to which the tile-maker was occupied full-time and the degree to which he participated in activiites such as farming.

Clearly the ideal material for study is the waste from tileries themselves. The source of the material is known and, additionally, it may be possible to suggest a close date-range for the workshop. Material from occupation sites may be derived from more than one workshop in the locality as well as tiles imported from some distance. Moreover the very fact that we can examine the tiles for impressions indicates, as is so often the case, that they are no longer mortared together and in their structural context. Thus even for dated and stratified deposits we can only suggest a terminus ante quem for the manufacture of any tile.

Silchester Tiles and Bricks

Our case study is based on the tiles collected from the civitas capital of Calleva Atrebatum (Silchester) which was extensively excavated between 1864 and 1909. The Silchester Collection in Reading Museum contains 314 complete bricks and tiles from Calleva of which 27 (8.6%) have foot impressions (Table 1). We do not know what selection procedures were followed by the excavators but we imagine that complete or otherwise remarkable specimens of tile types were kept. However, it is unlikely that every excavated tile fragment was examined for graffiti or impressions. It is probable, therefore, that the proportion of impressed tiles, selected for their intrinsic interest, is higher than in a random tile-population. Including fragmentary examples there are 92 recorded tiles with foot impressions from Calleva, representing a minimum total of 116 different individual animals or humans. In considering the question of the frequency with which tiles were trodden on by animals we should remember that the tiles used at Calleva were obviously not considered as waste; the proportion of impressed but rejected tiles on kiln sites could well be higher.

Table 1 Roman tiles and bricks in Silchester Collection, Reading Museum, complete examples

Type	No.	Average size (mm)	Foot impressed
Tegula	60		11
Imbrex	33	Length 380-405	1
Box flue	29		-
Flat	8	120 x 90 x 12	-
Flat	42	190 x 190 x 40	3
Flat	6	270 x 270 x 40	2
Flat	5	400 x 150 x 45	2
Flat	25	400 x 300 x 40	7
Round	23	Diameter 270-190	1
Hexagonal/Octagonal	26	Diameter 200-150	-
Brick	80	150 x 80 x 50	-
Total	314		27

Source of the Tiles

Given the nature of the site, it is important to establish the degree of homogeneity of the fabrics of the tiles. If they are likely to have been produced at a number of different workshops or at some distance from Calleva the overall conclusions will be affected. On the basis of macroscopic inspection of hand specimens the tiles divided into three main fabric groups (Table 2). Samples of each group were thin-sectioned and the results of petrological examination appear to support the initial groupings.

Group 1

Tiles of group 1 appeared on examination of the hand specimen to have a light yellow or yellow-brown sandy fabric with rare inclusions of sub-rounded to sub-angular quartz and flint ranging from 5-20 mm in size. Examination of thin-sections revealed rare, etched quartz grains (c. 0.75 mm) in a fine ground silt matrix with haematite crystals.

Group 2

Tiles of group 2 appeared on examination of the hand specimen to have a dark red-brown sandy fabric, often with a grey core with rare sub-rounded to sub-angular quartz and flint of 5-10 mm. The fabric also frequently contained streaks of a lighter, yellow-brown clay. Examination of thin-sections revealed moderately abundant (c. 30% of the matrix) sub-angular to sub-rounded quartz, averaging 0.25-0.3 mm in size with some very rare felspar. There were abundant traces of carbonaceous débris in the matrix.

Group 3

Tiles of group 3 appeared on examination of the hand-specimen as a reddish-brown sandy fabric with extremely rare larger quartz or flint inclusions of 5-10 mm. Examination of thin-sections revealed more abundant (c. 50% of the matrix) sub-angular to sub-rounded quartz, averaging 0.25-0.3 mm in size. Mica (rare) is also present. Carbonaceous material was much less abundant than in group 2.

Group 4

One tile (not thin-sectioned) on examination of the hand specimen appeared to have a fine reddish-brown matrix with an almost conchoidal fracture; added sand tempering appeared to be absent, but grog-temper may be present.

Thus the differences between the larger groups 2 and 3 appear slight; the latter is coarser and with a lower proportion of carbonaceous material in the matrix. Fabric 1 is distinguished by the relative scarcity of the finer quartz tempering and the presence of larger quartz and flint inclusions. Despite these slight differences there is no reason to believe that the fabrics derived from different clay sources and an origin close to Calleva seems highly probable for all fabric groups. The suspected tilery at Little London, 3 km to the south of Calleva, is now in doubt (Boon, 1974, 277-9) and there are no convincing wasters from this site or any other possible workshop to compare with the tile-population from the town.

Date of the Tiles

While the fabric of the tiles does not necessarily help us to define different workshops, it does help to refine overall date ranges. Although the tiles in the Silchester collection have neither precise provenance nor stratified context within the town, subsequent work on structures incorporating tiles in their masonry has helped to refine dating. Fabric 1 is of 1st century date, perhaps belonging to the pre-Flavian period only. It was not found in the masonry of the Forum-Basilica exposed in 1977 and this building was under construction c. A.D. 100 (Goodburn, 1978, 464). Visually it resembles the fabric of the Nero tile (Boon, 1974, 46-7) which thus supports a pre-Flavian date for the group. Tiles of fabric 3 were observed in the primary masonry of the Forum-Basilica and in both the south and south-east gates of the town, dated to the late 2nd century (Goodburn, 1976, 368; Frere, 1977, 418-19). No stratified examples of fabric 2 tiles have been found.

It is interesting to note that limestone rather than tile was preferred for the levelling courses in the town wall of c. A.D. 250-75 and that both limestone and old red sandstone appear to be used for roofing nearby in the Kennet valley in the later Roman period (Cowell, Fulford and Lobb, 1979, 22). Undated examples of identical roofing materials were found at Silchester (Williams, 1971, 178). We may also note that tile is rare in the footings of the Calleva church (Frere, 1975). Stone-built "channelled" hypocausts were also more common in the later Roman period (Williams, 1971, 183) and these are present at Silchester. Although tile is not excluded, flint seems to have been the main material employed. The implication of this evidence is that certainly flat, bonding tiles and perhaps also roofing tiles were more scarce at Silchester from the mid-3rd century onwards. We cannot of course exclude the possibility of their manufacture or their re-use in the later Roman period, but the evidence does suggest that a larger proportion of our tile sample belongs to the early Roman period with fabric 1 belonging to the 1st century and fabric 3 predominantly of late 1st to mid-3rd century date.

Methods of Study

Identifications of the species were obtained using the reference books on modern species (Bang and Dahlstrom, 1974, Lawrence and Brown, 1974). It must be stressed that, in the future, identification work must be based on comparative collections of actual impressions in three dimensions, using archaeological examples or casts and impressions taken from modern animals. The foot impressions on the tiles are often deeply indented so that no two-dimensional representation can show their characteristics, and the material into which the impressions have been made is not similar to the snow or wet earth of the reference book illustrations. As with the identification of bones, so with the identification of foot-impressions, the printed page can only be an indication of identification. This is particularly important for differentiating sheep and goat tracks which proved the hardest to determine and the identifications given here may need to be revised with future work.

For each foot-impressed tile the tile-type and fabric as well as the species were identified. The track was measured wherever possible and identified to left or right side and fore or hind foot. It must be remembered that shrinkage

Table 2 Animal and human foot impressions on tiles from Silchester, minimum number of individuals and fabric type

	Tile and Fabric Type									
	Tegula			Imbrex	Flat				Round	
Species	1	2	3	3	1	2	3	4	3	Total
Horse	-	-	-	-	-	-	1	-	-	1
Cattle	-	-	-	-	3	1	2	-	-	6
Sheep	-	-	-	-	1	4	2	-	-	7
Goat	-	-	3	-	-	1	1	-	-	5
Sheep/goat	-	-	1	-	1	2	6	-	-	10
Dog	1	-	11	2	9	4	7	1	-	35
Cat	1	-	4	-	-	-	3	-	-	8
Dog/cat	1	-	4	-	1	2	4	-	-	12
Bird	-	-	1	1	1	-	4	-	-	7
Human										
barefoot	-	-	-	-	2	2	3	-	-	7
sandalled	1	-	5	-	-	2	9	-	1	18
TOTAL	4	-	29	3	18	18	42	1	1	116

of the clay on drying subsequent to the foot impression will have reduced the measurements, possibly by up to a tenth. The registering of one foot on top of another, where present, and any indication of the gait of the animal (running or walking, turning while its foot was on the tile) were recorded. In a few cases the trail of the animal (the succession of individual foot marks) could be seen and measured.

For tiles with more than one track the minimum number of individuals of the different species was counted. It was often possible to reconstruct the sequence in which different individuals crossed a particular tile or when a foot-impression was made in relation to stamping or decorating or inscribing graffiti.

Results

A future publication will give the detailed archaeozoological results; here only the results relating to tile-making and its faunal environment are given.

Tiles with foot-impressions are overwhelmingly flat, or bonding, tiles, followed by tegulae (Table 2). While some tiles such as imbrices and box-tiles are particularly susceptible to heavy pressure when wet and are likely to have been discarded if walked over, it is odd that small bricks, hexagonal and octagonal tiles in the Silchester collection do not have foot-impressions and only 1 rounded tile is impressed (Table 1-2). This could suggest that the latter were either manufactured in a different environment or by a different procedure to that which yielded foot-impressions.

The undersides of the tiles show that they were laid to dry on a beaten earth or similar floor; fine gravel adhered to the undersides of two (12302,

12382).[1] Unlike the CLBR tiles, impressions of vegetable matter such as straw or cereal were exceptional (12316, 12343), although carbonaceous material was particularly evident in fabric 2 (Peacock, 1977, 242-3). On no tile is there any indication of the obscuring of detail as would surely have happened in some cases if the tiles were put to dry in the open air where rain could fall on them. Where it is possible to tell, the finger-decoration or smearing preceded the foot-impression. This would fit with Boon's interpretation of the smearing marks as a test of dryness before the tiles were removed from the moulds (Boon, 1974, 279).

Thus the foot-impressions were probably made after the tile had been left for the final drying before firing. The depths of tracks vary from deep to hardly discernible, indicating that animals had access to the tiles throughout the drying process after the removal of moulds, etc. However, a light animal such as a vole or mouse, even when crossing a very wet tile might have left no impression. A similar situation might also arise with a cat or dog on a tile dried to leather hardness. Human foot-impressions are usually shallow and suggest that tiles were deliberately avoided when wet.

In considering all tiles first, irrespective of chronological divisions, the commonest animals to cross were domestic dogs and cats, the former more frequently than the latter. Small ruminants were the next most frequent species with sheep possibly more common than goats. Birds (mostly chicken) come next, followed by cattle and a solitary horse. There are no tracks of wild animals or pigs, with the exception of 2 tiles with indistinct tracks which may be fox. Although the sample of impressions on fabric 1 tiles (22) of 1st century or pre-Flavian date is smaller than that of the later fabric 3 (75) (late 1st-mid-3rd century), some differences in the proportions of animals represented may be noted. There are, proportionally, five times as many cattle on fabric 1 tiles as there are on those of fabric 3. Sheep and goat, however, are twice as common on fabric 3 tiles as on fabric 1 and there is a slightly higher proportion of dog and cat on tiles of fabric 1 than those of fabric 3. Obviously a larger sample is necessary before much weight can be attached to these differences.

The small size of the tracks of sheep or goat on some tiles shows that the animal was a recently born lamb or kid and that the tile was therefore made in the spring. Some tiles have an abundance of tracks. There are 26 hoof-tracks of sheep on 12377 from a minimum of 2 individuals. 12313 is covered by the paw-marks of apparently only one dog, together with isolated tracks of cattle and a small ruminant. Two dogs, a sheep and a bird (probably a crow) crossed no. 12318. Sometimes the tracks of the animals are on the front of the foot showing that the animal was running. Some dog-tracks are skewed from the animal turning on the foot that touched the tile.

Of the tiles with human foot-impressions a minimum of 7 individuals were barefoot, one of these being a young child, and 18 wore nail-studded sandals or shoes. Human impressions are 11% more common on tiles of fabric 3 and, whereas the proportion of bare to sandalled feet is twice as great on tiles of fabric 1, the reverse is true of the later group (fabric 3) with five times as many sandalled as bare feet.

Table 3 Measurements of animal tracks from Roman Silchester with comparable measurements of modern adult animals taken from Bang & Dahlstrom, 1974 and Lawrence & Brown, 1974

	Track Length			Track Width		
	Range	Mean	N	Range	Mean	N
Cattle						
Silchester	56-63	58.3	3	50-56	53.3	3
Modern		110			95	
Sheep						
Silchester	24-47	31.5	24	11-30	20.8	29
Modern	50-60			40-50		
Goat						
Silchester	41-55	45.7	9	30-51	37.0	5
Modern	50-60			40-50		
Dog						
Silchester						
Forefoot	31-67	49.1	14	28-69	48.9	21
Hind foot	29-59	47.9	9	26-65	40.1	13
Modern	39-81			31-86		
Cat						
Silchester	23-50	29.1	7	23-44	28.8	12
Modern (wild)	40-60			35-60		

CONCLUSIONS

The Animals Themselves

The foot-impressions can be seen as giving information on the animals themselves in Roman times. Table 3 summarizes the measurements of the tracks of the different species with comparative measurements of modern animals. The track of an animal gives no indication of the age apart from its size and we therefore have to assume that the top of the range of measurements is from adults. No tracks of young carnivores are to be expected on tiles as these animals do not leave their lairs where they are born for the first 6 weeks or more of life. Young ruminants, if they crossed the tiles, would not have been far from their mothers.

The cattle of the Silchester tiles are remarkably small and possibly only young animals are represented. The dog ranges widely in size. The shape of the track of sheep resembles the wild sheep having more pointed tips to the hoof than sheep of today. Bones of sheep from the Roman and prehistoric periods likewise indicate an animal close to the wild sheep. Some of the tracks of cat are large enough to be from the wild cat which was presumably at large in the woods and forests around Silchester in Roman times, but we would expect wild and domesticated cats to have freely interbred. The absence of pigs might be remarked upon but pigs are self-willed, obstreperous and destructive animals and we can well imagine their being kept either in sties or in woodland well away from human settlements. The single horse-track only shows a quarter of the hoof and it is therefore a tentative identification, but it indicates that the animal was not shod. We should note that there are no wild animals (apart from birds and possible fox) ; this suggests that the ground was fenced.

Tile-Making

The evidence of the foot-impression shows a domesticated animal population. We would argue that this indicates the close proximity of a farm or stock yard, suggesting that the tile-makers themselves were farmers as well as artisans. While it could be argued that the impressions of humans are not unexpected and that dogs and cats may have been pets, it should be stressed that 31% of the impressions are of sheep, goat, chicken, cattle and horse. It is difficult to dismiss the animals responsible as casual strays, particularly when wild animals are absent. Sixteen tiles show the presence of more than one animal or of different types of animal on the same tile, and this also supports the case for the workshop(s) being sites close to or within a farmyard. Indeed we may go further and suggest that the tiles were laid out to dry on the ground, but under cover within a barn or shed to which animals could gain access.

Such a picture of tile-making should not surprise us given the semi-specialised or craft character of much of Roman industry. Indeed Boon's study of tiles marked with dates before firing argues for a seasonal activity which would call for other occupations such as farming to support the tile-makers (1974, 279, 365). It is reasonable to expect a similar range of evidence from other tileries in the north-western provinces of the empire. In this respect, it is interesting to compare the data from the civilian tiles of Calleva with the apparently similar range of evidence emerging from the study of the military CLBR tiles of south-east Britain (Peacock, 1977, 242-3).

Future Work

Although there are almost a hundred tiles with foot-impressions from Silchester the sample is not as large as we would wish. In Table 2, for example, we see that more tiles are needed before we can say that the greater number of dog on flat tiles of fabric 1 is by chance or not. We are also aware that, despite similarities in fabric we are probably looking at the product of several Silchester tileries.

The Silchester collection has given us no idea of the frequency of foot-impressed tiles in the whole tile-population. Work on other sites, particularly the tileries themselves, should supply this information and the frequency may vary from site to site.

In considering the tracks themselves, the importance of comparative collections, in three dimensions rather than two, has already been stressed.

We should also warn excavators that some tracks on tiles are so faint that they are hardly discernible. This faintness is to be expected when we consider the lightness of some of the animals crossing the tiles and that the tiles may have been leather-hard. To recognise a faint track it will be necessary for the tile to be well-washed and there to be oblique lighting from a concentrated source. The examination of tiles for foot-impressions and other marks is not something to be done casually but rather by detailed examination of every tile after it has been cleaned.

The study of foot impressions is an unexplored field and promises to supply much information on methods and conditions of manufacture as well as the degree of full-time specialisation in tile-making. Not least should it inform us of the animals themselves. The potential of the subject needs to be realised on both a regional and a provincial level.

Acknowledgements

We are grateful to Dr. Juliet Jewel and Mr. Hugh Carter for examining and discussing the tile-impressions and to Dr. B. Sellwood for his help with the petrological analysis of the tiles.

Note

1. The numbers here and below refer to the Silchester Catalogue in Reading Museum.

Bibliography

Bang & Dahlstrom, 1974: P. Bang and P. Dahlstrom, Animal Tracks and Signs, London, 1974.

Boon, 1974: G. C. Boon, Silchester: The Roman Town of Calleva, Newton Abbot, 1974.

Cowell, Fulford and Lobb, 1979: R. W. Cowell, M. G. Fulford and S. Lobb, 'Excavations of Prehistoric and Roman Settlement at Aldermaston Wharf, 1976-77', Berkshire Archaeol. J. 69 (1979), 22.

Frere, 1975: S. S. Frere, 'The Silchester Church: The Excavation by Sir Ian Richmond in 1961', Archaeologia 105 (1976), 277-302.

Frere, 1977: S. S. Frere, 'Roman Britain in 1976. I. Sites Explored', Britannia 8 (1977), 355-425.

Goodburn, 1976: R. Goodburn, 'Roman Britain in 1975. I. Sites Explored', Britannia 7 (1976), 290-377.

Goodburn, 1978: R. Goodburn, 'Roman Britain in 1977. I. Sites Explored', Britannia 9 (1978), 403-72.

Lawrence & Brown, 1974: M. J. Lawrence and R. W. Brown, Mammals of Britain, Their Tracks, Trails and Signs. London, 1974.

Peacock, 1977: D. P. S. Peacock, 'Bricks and Tiles of the Classis Britannica: petrology and origin', Britannia 7 (1977), 235-48.

Williams, 1971: J. H. Williams, 'Roman Building Materials in South-East England', Britannia 2 (1971), 166-95.

MARKINGS ON TILE AND BRICK

Gerald Brodribb

This paper is concerned with markings of all kinds found on Roman tile and brick which in any way throw light upon the whole business of tile-making. The evidence is based largely upon the tile observed in the course of my present work on a corpus and catalogue of Roman tile throughout Britain. This work is still in progress, but already tile has been examined from over 100 sites, and what has been seen makes it regrettable that more has not been preserved, because tile, so long neglected, can provide a great deal of information.

There are various kinds of marks; stamps, imprints, signatures, tally-marks (cut on the edge of tile), graffiti, and other forms. The most obvious is the stamp. All kinds of authorities, military, naval, civil and private used stamps in the way of officialdom, and perhaps advertisement. I have handled and analysed over 1700 examples of the Classis Britannica stamp (Brodribb, 1969 and in press). It has been discovered that sites in Britain and in France have provided examples of the same stamp, and that French fabric can be found in Britain and vice-versa (Peacock, 1977, 243).

A great variety of stamp was found, especially in the Beauport Park collection, which provides 77 different forms of stamp from 1300 examples, and no fewer than 46 of these were unique, single examples. Though some Classis Britannica sites have stamps confined to that one site, some stamps turn up at several different sites, and there is one stamp (Type 1) which appears on no less than 7 of the 10 Classis Britannica sites in Britain; this ubiquitous stamp may possibly have been the prototype of all British stamps, giving way to more varied and individual designs as time went on.

Type 1 stamp is also of interest because it sometimes shows a hole made by a raised nail-head in the centre of the impression, the nail being used to attach a handle to the die: some legionary stamps also show a similar hole (Fig. 9. 1). On other occasions the nail-head has been driven in so deeply that the resulting hole leaves a blob of clay on the surface. This shows that some stamps had separate handles, though no doubt more often the wooden die had a shaped upper section which would serve as a handle. Other evidence concerning tile-making comes from the Type 20 stamp, for it suggests that the man who applied this stamp also made the cut-aways for the flange.

Personal experience has shown that whereas signatures can be put on when the clay is still soft, stamping and trimming have to wait until the clay is leather-hard. If the clay is too wet the stamp will smudge, and drips of wet clay run down the side, as can be seen on one imbrex. It is possible to establish an order of marking: first comes the signature, followed later by any stamp, for whenever a stamp and signature coincide, the stamp is always

Fig. 9.1 Classis Britannica tile-stamp showing impression of raised nail-head. Scale 1:1

superimposed. At this stage also come the cut-aways of the flange, and the makings of any nail-holes needed. Then comes any tally-mark, and last of all, during the final drying before firing, impressions made by man or beast.

Stamps also suggest that different types of tile were handled by different makers, as shown in Table 1.

Table 1 Stamp types from Beauport Park with over 60 examples.

Type	Total	Tegula	Imbrex	Brick
1	109	94	5	10
2	108	-	108	-
20	211	197	-	14
25	224	194	-	30
27	63	1	62	-
28	80	-	80	-
29	150	2	148	-

It will be seen that apart from a few exceptions a certain type of stamp seems to be kept for use on a particular type of tile.

It is also possible to estimate the frequency with which a stamp occurs. Of some 500 tegulae so far accounted for at Beauport Park, 41 are complete examples, and since all but two of these carry a stamp, it would seem that it was the intended practice to stamp them all. On imbrices a different policy was used. Apart from two examples, all 46 complete imbrices carried either a stamp (80%) or a signature (20%). There are different types and sizes of imbrex, and these have different stamps or signatures. Two examples raise a problem in that they have a signature overprinted by a stamp as though cancelling it. (Brodribb, 1979.)

There are a dozen examples of a stamp having been put on twice, not because the first stamping was faulty, but just as a whim of the stamper: this practice is indulged in mostly by the user of a Type 25 stamp. This stamp is a crude one, and the tegulae it stamps also happen to be roughly made: they carry only one type of signature and never have a tally-mark. The characteristics of certain long-departed tile-makers can be easily identified even now.

On brick, and especially brick for pilae, stamps are less common; they are extremely rare on box-flue tiles and voussoirs. Since all dies have to be cut in reverse, it is not surprising to find errors: sometimes these are of single letters, but there are a dozen types of stamp in which the whole stamp has come out retrograde. The cutter was told to copy a certain stamp, and mistakenly copied just what he saw on the tile before him. Gallic cutters seem to have been more efficient, and have produced few examples of retrograde stamps.

The commonest of all unofficial imprints is the mark left on drying tule

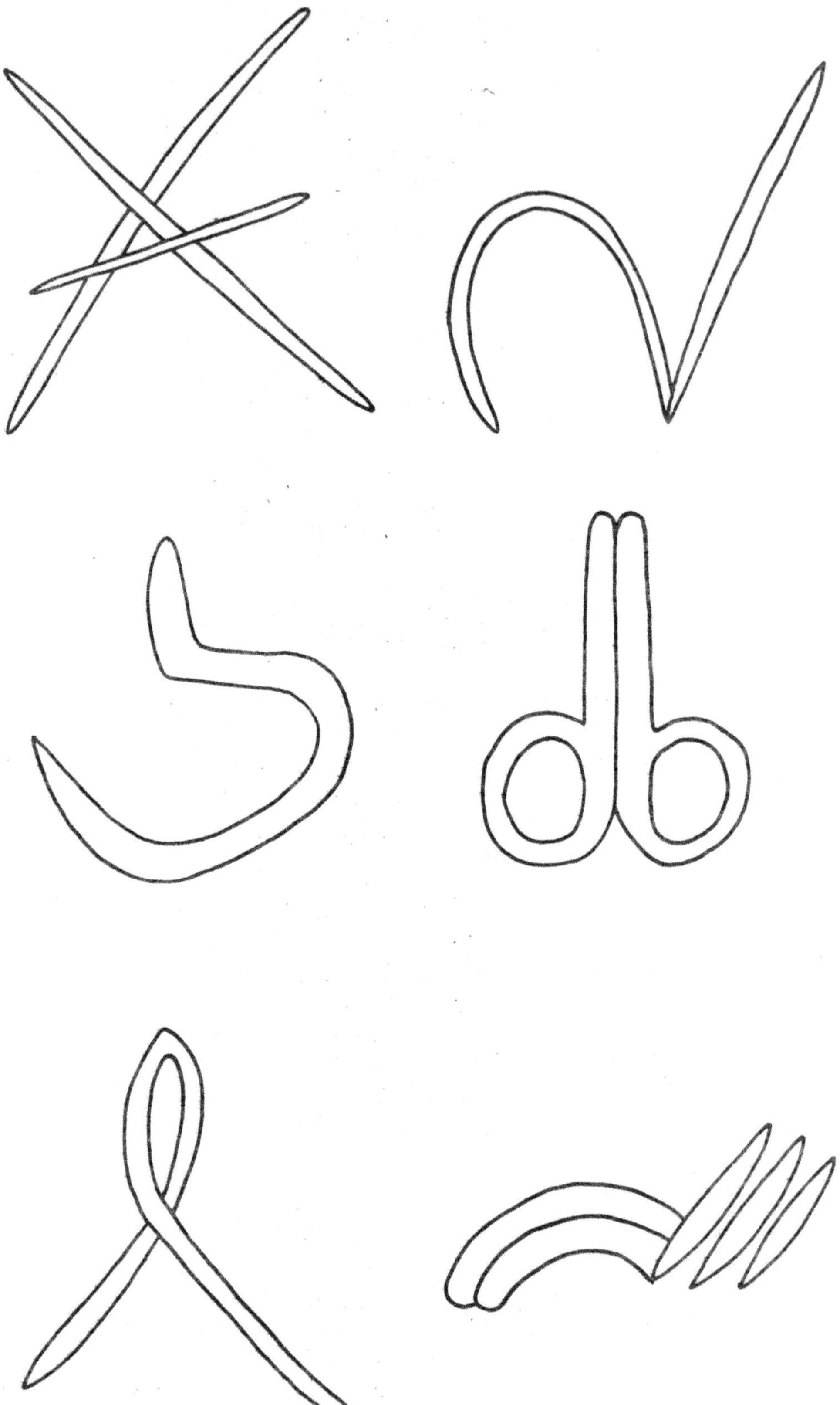

Fig. 9.2 A selection of 'Signatures' found on tile

The commonest of all unofficial imprints is the mark left on drying tile by wandering human or animal (see Chapter 8). That they were discouraged can be seen by a tile fragment from St. Albans which shows a stone just behind a dog print at an angle suggesting that it had been thrown at the dog. Almost all animal prints are found on tegulae or brick, the only three seen on imbrices taking the form of claw-marks on the edge as though a dog had reached up to an imbrex drying on a rack above the ground. There are a few prints which seem to have been made by raindrops, and the rarity of such marks suggests that drying tile was generally under cover, perhaps in a shed with open sides.

Graffiti on tile are not uncommon, and there must always have been a temptation to make use of what was one of the best writing surfaces available at that time (see Chapter 11). The few brief references to tile-production arouse our curiosity without giving us much satisfaction. A tile from London has a drawing of what seems to be a pharos and a large tile from Beauport Park carries the unique impression of a tile-comb (Brodribb, 1973).

Nail-holes are of interest in that of the 237 complete tegulae noted so far, only 60 have nail-holes, divided equally into round and square types. This works out at about 1 in 4, but this figure is influenced by the 29 examples from Silchester. The holes are always placed within the top 30 mm of the tegulae, and one very large tile from Dover has two holes. Tegulae requiring nails were those set lowest on the roof, and since the holes had to be made before firing, a special quantity would have to be readily available before work began on the roof. Other tegulae tended to stay put by sheer weight, and experiment shows that unattached tegulae do not begin to slide until the pitch exceeds 30 degrees (see also page 295 Ed.). Holes are very rare on other kinds of tile, but two 2ft. square bricks, one from Chester (Williams, 1895, 78) and another from Beauport Park, are punctured with frequent holes made with a stick. There are also similar fragments from Purton, Wilts. The purpose of these holes seems to be to assist in the firing and Williams suggests that the holes made in biscuits have the same purpose. But if such holes were really important in the firing of large bricks, why are there so few examples?

Signatures

It has often been noticed that many tegulae have marks on the lower half the face, generally in some form of a semi-circle, and sometimes double or triple. Very little observation has been made of these marks, but in a rare comment by Williams in 1895 on a tile from Chester he referred to a finger-made mark as 'the signum of the maker' with which he 'signed his work'. Ward regarded the marks as having a decorative effect (Ward, 1911, 263) but a label on tile from Fishbourne regards them possibly as a trade-mark. These marks certainly possess a personal identity and are worthy of the name of 'signature'.

From the Beauport Park collection of tegulae there are 27 varieties of 'signum' and of these 8 take the various-sized form of Ward's 'concentric groovings', 4 being single, 3 double and 1 triple. They usually seem to have been made with the finger, but one shows a deep central cut suggesting the use of a stick. The marks seldom show finger-nail and are remarkably symmetrical, with little evidence of any slip of the finger. The concentric type of signature is easily the most common, being found on 41% of the 237 complete tegulae so far noted in Britain: a survey of tile in the Bas-Dauphiné area of France found only 20% made of concentric types (Chauffin, 1956, 83): 80% of the tile had no signature at all, compared with 29% in Britain. There were no elaborate signatures.

The calculations for signatures on tegulae examined so far in Britain fall into three groups: 29% had no signature, 41% had concentric marks of some kind, and the remaining 29% had more elaborate types of signature (Fig. 9. 2). This last 29% is made up of 69 examples, covering as many as 32 varieties which are usually finger-made, but at times a tool or a comb has been used. I have also noted partial tegulae as well as complete ones, and if the signatures on these are included the total variety of signatures comes to 58. It seems hardly believable that so many different markings can be devised, especially when such obvious possibilities as the cross and the circle are very seldom used.

A study of roof-tile shows that of all the many thousands of tegulae once used in Britain only a very few have survived complete, for I am sure that for all their readiness to reject tile and throw it away on the spoil-heap, excavators in the past have seldom failed to keep examples of complete tile. Seeing that tile is such tough material, it is puzzling to know where all the rest have gone to. I presume it has been mostly used as foundations for other later buildings and is now deeply buried and lost for ever. Not long ago it was observed that rejected tile from the site at Great Witcombe was being used as foundation for a new barn.

It was hoped that the wide range of signatures might help in solving the problem of whether tiles were made at some large centre and widely distributed from the centre, or whether tile-makers went from site to site and made tiles as required. Here a distinction might be made between official tile, carrying a legionary or naval stamp, and tile made for private use. The official tile is more likely to come from a central depôt (e.g. Holt), whereas that for private building may have come from a local tilery, as, for example, that found in the complex at Eccles. Apart from the ubiquitous semicircle markings, the more distinctive forms of signature tend to match only between nearby sites, with little evidence of being found further afield, as is the case with some examples of relief-patterned box-flue tiles. Lowther suggested that the makers of box-flues travelled about taking their special rollers with them or at least allowing them to be used (Lowther, 1929, 10). Some finger-made signatures are very elaborate and no doubt handwriting experts could say something about the characters of the tile-makers from their signatures, especially from the more flamboyant efforts. One interesting example of signature is that of a possibly left-handed tile-maker who produced tegulae found at Langstone, near Havant. The signature is lop-sided and he had such difficulty in shaping the cut-aways at the bottom of the flange, that he made one in way and the other in another way, in a unique combination. One wonders if his fellows found him in any way 'sinister' ! All we know is that he seems once to have lost control, and broken out into demented squiggles over the face of one tegula towards it right-hand flange.

So far signatures on tegulae only have been considered, but they occur elsewhere. The total number of signatures found on complete brick of all kinds was added together in order to see whether signatures on brick are less common than on tegulae and to see if the same signatures occur on both tegulae and brick. Of 389 examples of complete brick of all varieties from tiles for pilae to the large bipedalis, as many as 293 (or 75%) have no signature (as against 29% with tegulae). On the others there are 35 different signatures, of which 9 occur also on tegulae, so there are as many as 26 signatures confined to brick, which supports the idea that workers making brick were different from those concerned with tegulae.

It would be interesting to know why some tegulae and even more brick remained unsigned; perhaps a tile-maker had to qualify to be allowed to sign his work and certainly the high proportion of unsigned brick suggests that the simpler job of making flat brick was left to less skilled workers while the more experienced men dealt with the tegulae.

The approach to imbrices seems to be different. Those from Beauport Park that do not carry stamps have signatures of six different types. Elsewhere such signatures are almost unknown. There is a partial imbrex from Park Street, St. Albans, which carried what Lowther called a 'decoration' (Lowther, 1945, 99); there are some truly decorated imbrex fragments from Corinium (McWhirr and Viner, 1978, 362); there is a fragment of imbrex from Northchurch combed across the end 'to make the mortar adhere between two overlapping imbrices' (Neal, 1976, 27); but there is little other marking from elsewhere apart from the occasional thin lines across the 'gable' or finger-marks running along the ridge, which are not like signatures. Perhaps imbrex-making was not regarded as work worthy of bearing a personal mark.

The matter of signatures has been looked at more carefully by writers from the Continent. Jahn suggests that they may be some form of traditional craft-mark, denoting procedure at the kiln, possibly connected with shift-work. He finds the marks are limited to basic forms, and that these forms are in accordance with the practice of the guild, and not left to the whim of the individual (Jahn, 1909).

On box flue-tiles and hollow voussoirs there is little sign of signatures of the kind already discussed, but there is a considerable variety of combings and scorings put on to assist the attachment of mortar, and there are also the roller-made relief patterns. Most sites seem to produce a wide range of combings, suggesting the use of different combs and plenty of ideas of how to employ them. Sometimes on the same face two different combs have been used and the pattern on one face may be quite different from that on another. Occasionally however, there are certain characteristics which seem to be particular, such as the examples from Bignor where the gaps between the lines are filled in with little half-moons, or that from Itchingfield where the gaps are filled with marks jabbed on with the end of the comb (see Plate 20. I).

Payne in his report on material from Darenth (Payne, 1897,70) found four different designs on 50 examples of box flue-tiles and believed that they belonged to different 'gangs' or 'stools' of tile-makers at a factory, each gang having its own scoring design to identify the makers of what was always a skilled piece of work. An elaborate form of combing, when some simple

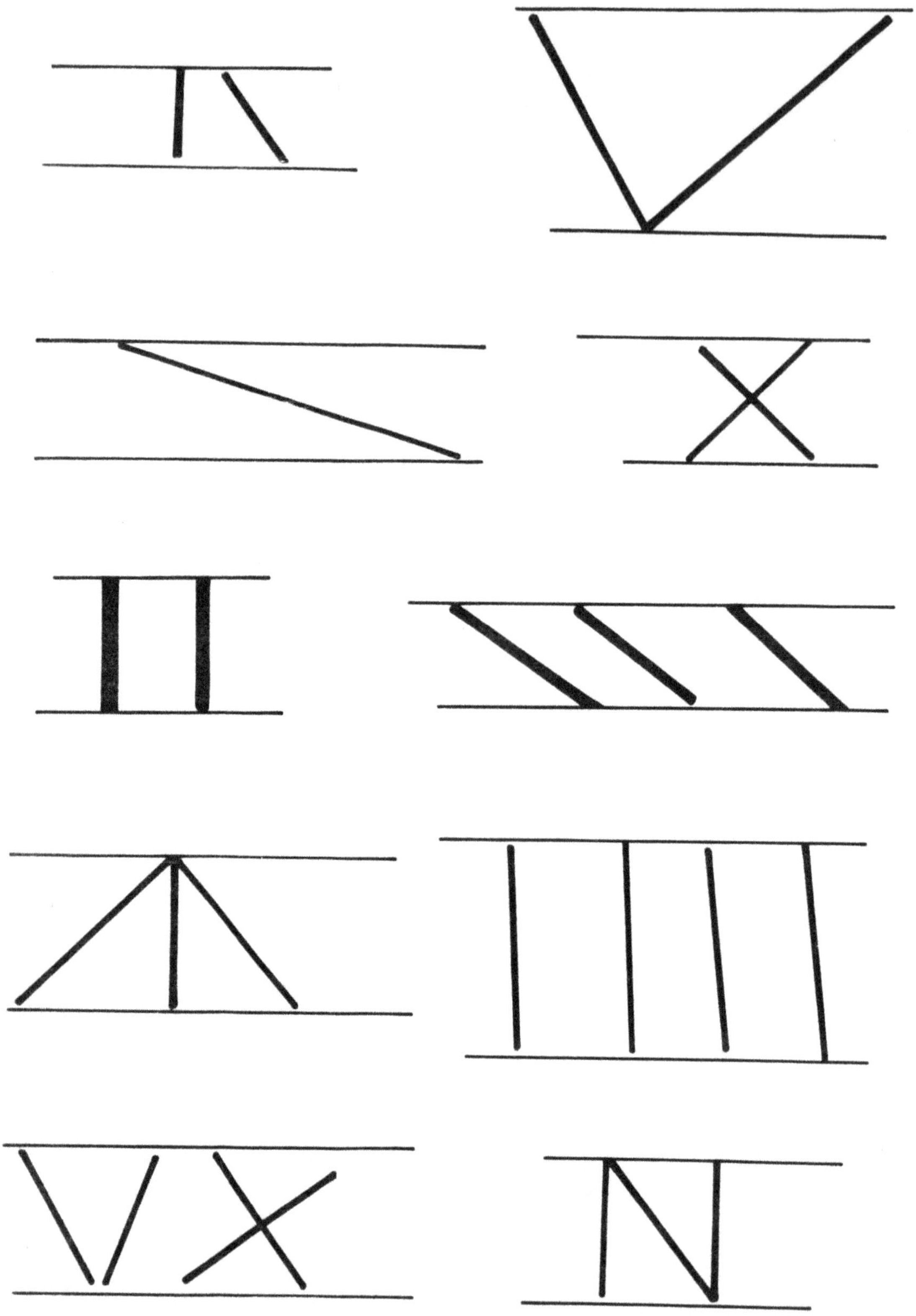

Fig. 9.3 Tally-marks found on edge of tiles. Scale 1:1

scoring would have been quite adequate, can only be the result of craftsmen taking some special pride in his work. Hollow voussoir tiles have the same treatment as box flue-tiles and are much commoner than reports would suggest. It has been observed that the fabric of voussoir tiles sometimes seems different from that of box flue-tiles.

Tally-Marks

Finally, we come to a mark that seems seldom to have been observed. I refer to a mark cut on the fore-edge of tegulae and brick, placed usually below any signature and which often seems to represent a number. Since first noticing these marks on tile from Beauport Park, I have found at least one example from most of the sites producing Classis Britannica stamps, and it seems that it might be a feature confined to Fleet sites, until recently when several other sites, many of them with legionary connections, have produced examples. I am sure there are many more examples to be found now that their existence has been established. So far the total of tally-marks stands at under 300, and 249 of these come from Beauport Park, of which 151 have been found on tegulae and 98 on brick. The frequency with which they occur on tegulae seem to be about 1 in 4. They fall into a number of varieties, of which there is a selection in Fig. 9.3 and since they mostly seem to represent numbers, I have for this reason called them tally-marks. The meaning and use of them is not yet clear. Sometimes the lines are fine and superficial, sometimes they are deeply cut in a V-shaped groove. Sometimes the lines are vertical, sometimes at an angle and they vary in size according to the thickness of the tile. Though most of the marks look like numbers, some do not (unless they are an unusual form of the number) and there is an opinion that they are not really numbers at all.

However, the apparently unique reference in an English report refers clearly to 'a numeral IV' on the edge of a tile from Dover (Amos, 1921), and Jahn in his report on tile from Vindonissa (Jahn, 1909, 121) finds on the edge of the tile combinations of the marks I, V and X which can be read in upright or inverted position as a 'number sign'. The highest figure is XXI. There is another brief reference to such marks concerning tile from the Bas-Dauphiné area of France. These are the marks I or X and the writer regards them as 'marques de stockage', which have no connection with the numbers of the Legions (Chauffin, 1956).

Some tiles from Mèdes (France) which were once part of a ship's cargo have a feature of an unusual kind. Near the lowest part of the face of the tegulae, but not cut on the fore-edge like tally-marks and within the semicircle of the signature, there are various small finger-made marks taking the form of a V, a triangle, or two large dots: Carraze regards them as a 'marque de tacheron', a worker's mark concerned with manufacture, possibly of the same purpose as a tally-mark, but not of a numerical kind. There is not enough evidence at present to link the signature with these marks and identify the worker (Carraze, 1976, 29).

It is evident that there is a wide range of marking features to be found on tile: add to this the various differences of style shown by the tile-maker in the types of cut-away, the shape of the flanges, the groovings beside them and the general finish and the nature of the fabric, and it will be clear that every tile is an individual handmade object, often with a message to tell. It is to be hoped that further research in this hitherto untrodden path may eventually enable tile to take its place alongside pottery as a valuable source of dating material.

BIBLIOGRAPHY

Amos, 1921: G. Amos, note in Journal of Roman Studies, II (1921), 239.

Brodribb, 1969: G. Brodribb, 'Stamped Tiles of the Classis Britannica', Sussex Arch. Coll., 107 (1969), 102-125. and revised version in Sussex Arch. Coll. in press.

Brodribb, 1973: G. Brodribb, note in Britannia, 4 (1973), 333 and pl. XLI a.

Brodribb, 1979: G. Brodribb, 'Tile from Beauport Park', Britannia 10 (1979), forthcoming.

Carraze, 1976: F. Carraze, 'Centre de coordination et de documentation Archéologique de Provence', Cahier 4, 26-32.

Chauffin, 1956: J. Chauffin, 'Tuiles du Bas-Dauphiné', Gallia 14 (1956), 83 and 85.

Jahn, 1909: V. Jahn, 'Die römischen dachziegel von windisch,' Anzeiger Für Schneizerische Altertumskunde, 11 (1909), 111-129.

Lowther, 1929: A. W. G. Lowther, 'A study of patterns on flue-tiles', Surrey Research Paper I (1929).

Lowther, 1945: A. W. G. Lowther, 'The Roman villa at Park Street, St. Alban's', Arch. Journal, 102 (1945), 21-110.

Neal, 1976: D. S. Neal, 'Three Roman buildings in the Bulbourne Valley', Herts. Arch. 4 (1974-6), 1-131.

Payne, 1897: G. Payne, 'The Roman villa at Darenth', Arch. Cant. 22 (1897), 49-84.

Peacock, 1977: D. P. S. Peacock, 'Bricks of the Classis Britannica', Britannia 8 (1977), 235-247.

McWhirr and Viner, 1978: A. McWhirr and D. Viner, 'Production and distribution of tiles', Britannia 9 (1978), 359-377.

Ward, 1911: J. Ward, Romano-British Buildings and Earthworks, London.

Williams, 1895: F. Williams, 'Deva', J. B. A. A. I. NS. (1895) 77-8.

TILE-STAMPS AND ROMAN NOMENCLATURE

T. P. Wiseman

This paper originated in a request by Alan McWhirr a year or two ago to comment on the tiles from the Cirencester area—stamped TPF. etc.—on which he and David Viner have written in volume IX of _Britannia_ (McWhirr and Viner, 1978). The results of his request were, in fact, incorporated into their article (_ibid._ p. 367); but since, quite apart from the misprints, my comments there now seem rather jejune, I am glad of the opportunity to expand them. What follows is essentially a recapitulation of Roman habits of nomenclature, but I hope that some of the comparative evidence adduced for abbreviated names will justify the repetition of familiar material—and perhaps even be helpful, for those who are more at home with pottery than _praenomina_.

In his recent book on the brick-stamps of the city of Rome (Helen, 1975), Tapio Helen has demonstrated—convincingly, I think—that the names on the brick-stamps of the first and second centuries A.D. represent (a) the landowner in whose property the claypits were situated, and (b) the businessman (or woman) who ran the factory that made the bricks. But that was a large-scale and comparatively complex industry; in less metropolitan circumstances one would imagine a simpler structure, consisting of just the owner and the workman—slave or free—whom he employed to turn his clay into bricks. That too is demonstrable from brick-stamps; I cite two examples out of many, _CIL_ XV 2440-1;

C. FIDICLANI CLEMEN. PUDENS SER. FECIT.
C. FIDICLANI PRIMI L. POMPONIUS CRESCES FEC.

C. Fidiclanius Clemens and C. Fidiclanius Primus are the landowners; the men who made the bricks are respectively Pudens, a slave, and L. Pomponius Crescens, a citizen.

This preamble is to establish the obvious—that, other things being equal, one would expect brick-stamp inscriptions, however obscurely abbreviated, to represent names—of land-owner, entrepreneur and/or brick-maker. What we must now do is to see whether Roman habits of nomenclature can help us to decipher the abbreviations, or at least narrow down the areas of uncertainty.

A free-born Roman should have at least two names, one for himself and one for his _gens_, or extended family. The personal name is the _praenomen_, the family name the _nomen gentilicium_, as in "Marcus Antonius", "Titus Labienus", "Lucius Vitellius". After these two elements, a citizen's formal nomenclature also includes his filiation (the father's _praenomen_ plus "f." for _filius_, the grandfather's plus "n." for _nepos_) and his _tribus_, the voting district to which he is assigned by place of birth. But he may also have a _cognomen_, as in "C.Iulius Caesar", "M.Tullius Cicero", "L.Cornelius Sulla"; if so, it appears in the full nomenclature _after_ the filiation and _tribus_.

Praenomen, *nomen gentilicium* and *cognomen* are the *tria nomina*, the "three names" of the Roman citizen: for the sake of clarity, let us call them 1, 2 and 3. The original form of name was presumably just 1+2, and this survived in some families well into the first century A.D.: several of the legionaries buried at Wroxeter and Lincoln, for instance, have no *cognomina* (*RIB* 254, 255, 257, 292, 294 etc.). But already that was unusual: Suetonius, writing under Hadrian, makes a point of mentioning that the four sons of L. Vitellius—one of whom became emperor in 69—were distinguished from each other only by their *praenomina*, A.Q.P. and L. (Suet. *Vit*. 2.2). None of them had a *cognomen*, and by Suetonius' time that needed explaining.

The form 1 + 2 + 3, however, was already very old, at least among the Roman nobility: the sarcophagus of a P. Cornelius Scapula is known from at least the third century B.C. (*ILLRP* 1274a). Originally, no doubt, *cognomina* were nicknames, but in many families they became hereditary, to the extent that the combination 2 + 3 represented practically a single unit, such as Cornelius Scipio, Calpurnius Piso, Iunius Silanus, and so on. In such cases the *cognomen* served to distinguish not an individual but the branch of a *gens*; the middle name becomes less important—all Scipiones were Cornelii—and so the 1 + 3 form is frequently found in aristocratic nomenclature (on this, see Syme, 1958, pp. 172-5).

In other families, however, *cognomina* were still used as individual names: one thinks of the P. Rutilii Calvus, Rufus, Lupus and Nudus in the second and first centuries B.C.—Rutilius the Bald, Rutilius the Red, Rutilius the Wolf, Rutilius the Naked, all with the same *praenomen*, Publius. In such cases it is the *praenomen* that becomes less important, and a 2 + 3 name-form would be enough to identify an individual.

These are aristocratic examples, because in the Republic that is what the evidence consists of. But before I come to discuss brick stamps under the Empire, I want to draw attention to the way these aristocratic names could be abbreviated. In Appendix I are listed a selection of moneyer's names on Roman coins, from their first appearance down to the time when practically the whole name was stamped on the coin. We can see examples of 1 + 2 names (nos. 138, 154 etc), 1 + 2 + 3 names (nos. 216, 217, 223, 227) and filiations in each category (nos. 210, 228); there are many examples of 1 + 3 names (nos. 74, 173, 189, 202 etc.), and one important example of a 2 + 3 name (no. 221), probably to be identified as T. Annius Rufus, son of T. Annius Luscus—Red Annius the son of One-Eyed Annius, identification being by *cognomen* rather than *praenomen*.

We must notice too that though many of these names are abbreviated as we might naturally do it, either by a simple initial (nos. 51, 63) or by a whole syllable (nos. 65, 113, 146 etc), there are also several examples of two-letter abbreviations—CO for Cornelius, ME for Metellus, and so on (nos. 81, 211 etc.)—which is a convention we must bear in mind when considering abbreviations elsewhere.

Perhaps the most interesting name-formulae in the list are nos. 125 and 144, where the moneyers are identified by the initials of the *tria nomina*. No. 134 is presumably another example of this, with the middle name abbreviated to two letters, while nos. 64, 93, 103 and 111 are probably initials of 1 + 2-

type names without cognomina. The identification of moneyers with magistrates known from the literary evidence provides a valuable external check for the interpretation of "QLC", "LPF" etc., as it does for the most elaborate example the coinage provides of a name in initial form (Crawford, 1974. no. 374)—QCMPI for "Quintus Caecilius Metellus Pius imperator".

Three-letter formulae representing the tria nomina are quite common on dedicatory inscriptions (e.g. ILS 3032, 3225, 3604, 4900, 9295, 9336), perhaps because the dedicator modestly avoided obtruding his name in competition with that of the god. They are also common on seals, for the same reason as on the coinage—lack of space. On large circular seals the full name is written round the edge and the three initials in the middle (CIL XV 7944, 7945, 7947, 7950); on smaller rectangular ones only the initials appear. In four of the fourteen examples of the latter in CIL XV 7996, two initials instead of three (DM, CR, PR, TN) evidently represent 1 + 2-type names, without cognomina. (Or they could be women, who do not have praenomina: see CIL XV 548-557, in Appendix III.)

Three-initial formulae also turn up more unexpectedly, in places where identification would seem to matter more than brevity—even on funerary inscriptions (e.g. ILS 3586: "dis manibus C.H.C."). And Sir Ronald Syme has recently drawn attention to an inscription from Dacia, not at all laconic in other respects, honouring the imperial legate P. Furius Saturninus merely as P.F.S. (ILS 7155). As always, Sir Ronald knows more than he says: "the rubric might be expanded, not without instruction and perplexities" (Syme, 1978, p. 14).

Before we leave the basic name-formations (1 + 2 and 1 + 2 + 3), we should consider the question of brothers. Since in these cases it is still the praenomen which distinguishes the individual, that is the only element that varies between the nomenclature of brothers. I offer three examples, two from brick-stamps (1 + 2, 1 + 2 with filiation) and one from an amphora (1 + 2 + 3):

P. C. Q. Septicior(um)	(CIL XV 2385)
C. L. Tossieis C.f.	(ibid. 2501)
Q. L. Corneli Placidi	(Callender, 1965, no. 1470).

If the three Septicii, the two Tossii and the two Cornelii Placidi had chosen to be more economical with their stamps we might have had PCQS, CLTCF and QLCP—which is worth bearing in mind when we consider four- or five-letter formulae.

Now we must turn to the 2 + 3 formation, in evidence as far back as the second century B.C. with the senatorial Rutilii and Annii. Here it is the cognomen, not the praenomen, which identifies the individual. With the spread of the Roman citizenship by manumission and enfranchisement, this formation became the norm, for reasons which will be clearly apparent.

Take Pudens, the brick-making slave of C. Fidiclanius Clemens mentioned above (CIL XV 2440). If he were freed, he would take his master's first two names, and become C. Fidiclanius Pudens. (By the tria nomina alone he would be indistinguishable from a freeborn citizen; but his full nomenclature would include, instead of a filiation, "C.l." for "Gaii libertus".) Similarly with the enfranchisement of foreigners: the Gallic Caburus, whose son Caesar sent as an envoy to Ariovistus, was given the Roman citizenship by C. Valerius Flaccus the consul of 93 B.C.; he too took his benefactor's first two names, and became C. Valerius Caburus (Caes. B.G. I 47.4).

Under the principate, the recipients of the Roman citizenship take the emperor's family names—C. Iulius for Augustus and Caligula; Ti. Claudius for Tiberius, Claudius and Nero; T. Flavius for Vespasian, Titus and Domitian; M. Ulpius for Trajan; and so on. The only example I shall cite, honoris causa, is Novantico the son of Adcobrovatus, the first known citizen of Leicester: granted the citizenship along with the rest of his unit after service in Trajan's Dacian wars, he becomes M. Ulpius Novantico (CIL XVI 160). Such men and their descendants, along with all the imperial freedmen and their descendants, cause a vast proliferation of citizens with "imperial" praenomina and nomina gentilicia. Only the cognomen—that is, in the first generation at least, the original foreign or slave name—serves to identify the individual.

This is presumably the reason for the steady increase in importance of the cognomen at the expense of the praenomen in the imperial period. Let me quote an illuminating example from Britain (RIB 324, 326: Caerleon). P. Sallienus P. f. Mae. Thalamus, of Hadria in north Italy, an officer in the II Augusta Legion, has his name written out, tribus and all, in the formal old-fashioned way. But his sons are identified not by praenomen—as, for instance, Gaius and Lucius Sallienus Thalamus—but by cognomen, as just "Ampeianus" and "Lucilianus", with the first two elements of the name ignored. Another indicator is the identification of centuries in military inscriptions; the centurion's name is given in the genitive, and though RIB does show one or two examples which include the praenomen (e.g. RIB 346, 472, 474), the vast majority give it in the 2 + 3 form—"centuria Valeri Flacci"—or with the cognomen alone.

Freedmen's names, and the growing importance of the third element in free-born nomenclature as well, have their effect on abbreviated brick-stamps. If our landowner C. Fidiclanius Clemens had manumitted not just Pudens but a dozen other slaves as well, from Amatus to Zosimus, then a series of 3-letter formulae from CFA to CFZ would result. And even without manumission the same effect would apply: the landowner C. Fidiclanius Primus of CIL XV 2441 could easily be Clemens' brother, the cognomen now being the distinguishing mark. Suppose three brothers—Primus, Secundus, Tertius—as well as Clemens, and again you have 3-letter formulae differing only in the final element.

In Appendix II I have selected some sequences of amphora-stamps which seem to illustrate this phenomenon. The four-, five- and even six-letter formulae are not easy to interpret, but the partially-resolved names show that the three-letter ones represent the tria nomina. (The fact that the sequences all begin with praenomen initials—for G. at 324-5 is an alternative to C. for Gaius—strongly suggests it in any case). In the second, fourth and fifth group in Appendix III, we have proof of it from unabbreviated brick-stamps.

Appendix III also shows, in the third and fourth groups, other elements which might confuse the issue in an abbreviated stamp: S. or SER. for "servus", FE. or FEC. for "fecit". At this point we must remember the tile with "Primus fecit X" in the Jewry Wall Museum (what would we make of PFX?)—though that is not a stamped inscription. We must also remember the use of F. for *fecit* (e.g. *CIL* XV 1167. cf. 1226) as well as for filius. Indeed, there is one stamp in *CIL* XV which has both at once: 1471, P. Tettius L.f.f.

The F's in the Cirencester series (TPF, TPFA, TPFB, TPFC, TPFP, TPLF) are, with one exception (TPLF), in the wrong position for a filiation; and also, with the same exception, in the wrong position for *fecit*. Moreover, it is conceivable that the apparent exception is not really an exception at all. Even for the coinage of the Roman people, a die-stamp could be made with the letters in the wrong order: the denarii of Sextus Pompeius (Crawford, 1974, no. 235: 137 B.C.) show not only SEX. POM. but also SEX. PMO. The final group of brick-stamps in Appendix III, of abbreviated names from the Praeneste area, shows two possible examples of the same phenomenon: CST/SCT (nos. 2366-7) and QST/QTS (2368-9). They *could* be separate names, since all four begin with *praenomen* initials (S. may stand for Sextus: see *CIL* XV 7944, and Appendix I, no. 151), but the double coincidence would be very remarkable. So perhaps TPLF should be TPFL, another in the TPF series: TPF looks like the initials of the *tria nomina*, but what the fourth element represents we cannot tell.

Of the other Cirencester stamps, ARVERI is evidently the genitive of an unabbreviated name (though one otherwise unknown to the *Thesaurus Linguae Latinae*); the first element of IVC.DIGNI must be an abbreviation of Iucundus, which would give us two *cognomina*, probably best resolved as "Iucundus (the slave) of Dignus"; LHS obviously represents *tria nomina*, beginning with the *praenomen* Lucius; so do LLH and LLQ, and the LLS from Colchester, the varying third element representing either three freed slaves of the same master, of three free-born relatives in a family that used the *cognomen* to distinguish individuals; TCM could be taken with the Alcester TCD as another example of the same phenomenon, or else with the SCM from London as two relatives distinguished by their *praenomina* (Titus and Sextus respectively); as for VLA, that could only be an abbreviation of the *tria nomina* if we allow the archaic *praenomen* Vibius, which just survives into the early Empire (*ILS* 1901, 6757), but as Professor Rivet points out (McWhirr and Viner, 1978, p. 367), it may be better explained as an abbreviated Celtic name.

Finally, a reminder that the Cirencester TPF series is not unique in being obscure. At *CIL* XV 1352 we have a brick stamped EX F. (i.e. *ex figlinis*, from the claypits of...) S.P.S.G.C.C.P.S., while at 6972-3 there are two glass bottles stamped S.P.S.C.P.D. and S.P.S.G.A.D. The common element is S.P.S.—perhaps something like "Sex. Pomponius Secundus"—but what the rest of the formulae represent is anybody's guess. Even at the time, initials like that must have been singularly unhelpful.

There is a nice story from the town of Tarracina about 100 B.C., which illustrates how variously initial formulae could be resolved. Several of the buildings of the town bore the mark LLLMM, which presumably represented the names of the *duoviri*, the local magistrates of the colony—as it might be, L. Licinius Largus and M. Memmius. In fact, there evidently were two citizens of the town called Largus and Memmius; the Roman orator and wit L.

Crassus alleged that they had had a fight over a girl, during which Memmius had bitten Largus in the arm. "So that's what it stands for", said Crassus, "lacerat lacertum Largi mordax Memmius"—mordacious Memmius lacerates Largus' limb.

Bibliography

Callender, M. H. Roman Amphorae, with an index of stamps, Oxford 1965 (University of Durham Publications).

Crawford, M. H. Roman Republican Coinage, Cambridge 1974.

Helen, T. The Organization of Roman Brick Production in the First and Second Centuries A.D.: an Interpretation of the Roman Brick-Stamps. Helsinki 1975 (Annales Academiae Sci. Fenn., Dissertationes Humanarum Litterarum, no. 5).

McWhirr, A. and Viner, D. The Production and Distribution of Tiles in Roman Britain with particular reference to the Cirencester region, Britannia 9 (1978), 359-377.

Syme, R. Imperator Caesar: a Study in Nomenclature, Historia 7 (1958), 172-188.

Syme, R. Antonius Saturninus, JRS 68 (1978), 12-21.

APPENDICES

I. Denarii 210-140 B.C.
M. H. Crawford, Roman Republican Coinage (Cambridge 1974).

51.	M.	
63.	C.	
64.	MA.	
65.	AVR.	C. Aurunculeius?
74.	C.VAR.	C. Terentius Varro
75.	C.AL.	
81.	CN.CO.	Cn. Cornelius Dolabella?
93.	MP.	
95.	VB.	
103.	MT.	
111.	AL.	
125.	QLC.	Q. Lutatius Catulus?
132.	ME.	(-.) Caecilius Metellus?
133.	TAMP.	(-.) Baebius Tampilus
134.	LPLH.	L. Plautius Hypsaeus?
138.	P.MAE.	P. Maenius
141.	TOD.	
144.	LFP.	L. Furius Philus?
146.	AVTR.	(-.) Autronius
147.	CN.DOM.	Cn. Domitius Ahenobarbus
151.	S.FV.	
152.	SX.Q.	
153.	CN.CALP.	Cn. Calpurnius Piso?
154.	L.COIL.	L. Coelius
155.	PVR.	(-.) Furius Purpurio
161.	TAL.	(-.) Iuventius Talna
162.	MAT.	(-.) Matienus
173.	C.SAX.	C. Cluvius Saxula
174.	A.CAE.	
175.	C.SAE.	
176.	PAE.	
178.	CINA	(-.) Cornelius Cinna
179.	BAL.	
189.	P.BLAS.	P. Cornelius Blasio
190.	OPEI.	(-.) Opimius
191.	VAL.	(-.) Valerius
193.	TVRD.	(-.) Papirius Turdus?
199.	SAR.	(-.) Atilius Saranus
200.	NAT.	(-.) Pinarius Natta
201.	C.SCR.	C. Scribonius
202.	C.TAL.	C. Iuventius Talna
204.	L. SAVF.	L. Saufeius

205.	P. SVLA	P. Cornelius Sulla
206.	SAFRA.	
207.	FLAVS.	(-.) Decimius Flavus?
209.	L.ITI.	
210.	C.IVNI. C.F.	
211.	Q.ME.	Q. Caecilius Metellus
216.	L.SEMPR. PITIO	L. Sempronius Pitio
217.	C.TER. LVC.	C. Terentius Lucanus
221.	AN. RVF.	(-.) Annius Rufus
223.	C. CVR. TRIGE.	C. Curiatius Trigeminus
227.	M.AVF. RVS.	
228.	C.VAL.C.F.FLAC.	C. Valerius Flaccus

II. Amphorae.

M. H. Callender, Roman Amphorae (Oxford 1965).

5.	AAF
6.	AAG
9.	AARV
96.	APC
97.	APCO
98.	APE
100.	APF
100b.	APFS
101.	APH
104.	APLA
105.	APM
121.	AR
122.	ARA
124.	ARD
127.	ARN
131.	AS
132.	ASA
134.	ASCL
271a.	CCF
271b.	CCFP
276.	CCM
277.	CCNS
281.	CDA
284.	CDN
286.	CDV
307.	CERC
309.	CERH
324.	GGE
325.	GGR
333.	C.I.ALB.
334.	CIB
335.	CIE
336.	CIF
340.	CIM
343.	CIPE
344.	CIR
348.	CIS
349.	CIT
350.	CIV
352.	CIVK
353.	C.IVL.HILA.
354.	C.IVLI MARCELLI
355.	C.IVLI NICI
356.	C.IVLI ZOS.
359.	C.IVL. REBVR.
360.	C.IVL.SVRI
385.	CM
386.	CMA
387.	C.MARI. SILVANI.
388.	C.M.CALAIS
389.	CME
392.	CMM
395.	CMR
396.	CMS
397.	CMT
398.	CMV
399.	C.MUSSIDI.NEPOTIS.
468.	CSA
478.	CSH
479.	CSL
481.	CSP
483.	CST

519. DC
520. DCD
521. D.C.IAN.

817. LCA
818. L.CAESI.AMAT.

823. LCF
824. L.C.FIRMIN.
825a. LCH
825b. LCHE
827. LCM
828. LCO
830. LCP
831. LCQ
834. LCRS
836. LCT

1039. MC
1040. MCA
1041. MCC
1042. MCFRIG
1043. MCLF
MCLSH
MCLSP
MCLSR
1048. MCR
1049. MCS
1051. MCV

1158. MPF
1159. MPM
1162. MPR
1163. MPS

1743. T.VAL.RVF.
1745. TVG
1746. TVP
1747. TVS

III. Tiles and other <u>instrumentum domesticum</u>, Rome and district. <u>CIL</u>.XV.

18-27. MRL; M.R.LVPI.
84-86. TRP; T.RAVSI PAMPHILI.
97-100. CCC; C.C.COSANI; C.CVRIATI COSANI.
288-298. TSM; T.STAT. MAXIM; T.STATILI MAXIMI.
548-557. DD; DOM.DOM.; DOMITIAE DOMITIANAE.
887-8. LBS; L.BELLICI SOLLERTIS.

1149. T.FLAVI. AVG.L.CLONI.
1150. T.FLAVI. EVCHRI.
1152. T.FLAVI. HERMETIS.
1153. T.FLAVI. IANVAR.
1155. T.FLAVI. RHODONIS.

1179. EX FIG. M.HERENNI POLLIONIS...
1181a. COSMI M.HER.POL.SER.
1181b. COSMI M.H.P.S.
1182. SUCCESSI M.H.P.S.

1247. L.LVRI.APRIONIS.
1248. L.LVRI.BLANDI.
1249. L.LVRI.MARTIALIS.
1250. L. LVRI. MARTIALIS FE.
1252. L.LVRI.MYRNI.
1253. L.LVRIVS PROCVLVS FEC.
1254. DECEMBER L.LVRI.PROCULI FECIT.

1341. C.OPPI.FELICIS.
1343. Q.OPPI.IVSTI.
1345. Q.OPPI.NATALIS.
1347. Q.OPPI.PRISCI.
1348. Q.OPPI.VERECVNDI.
1349. C.OPPI.VEIENTI.

1429-1441. QSP; Q.SERV.PVD.; Q.SER.PVDENT.; PVDENS.

2359. LLS ET ACFS
2360. SCPR
2361. CA
2362. AM
2363. CLV
2364. CSN
2365. CSR
2366. CST
2367. SCT
2368. QST
2369. QTS
2370. LVF

GRAFFITI ON ROMAN BRICKS AND TILES FOUND IN BRITAIN

Roger Tomlin

1. Preface

By 'graffiti' I mean letters, whether capitals or 'cursive' (the two categories overlap), inscribed on bricks and tiles with a variety of instruments—the finger, a stick, a knife, a writing-stylus, etc. Almost all graffiti are made before firing, as far as I know, though this has not always been noted when they were published. I have excluded tile-stamps (though I quote a couple for comparison), brickmaker's 'signatures' of a non-literate nature like the 'S'-shaped swirls found on CLBR tiles and the combed patterns noticed by A. W. G. Lowther (except for a tile [68] with lettering made with a comb), and 'tally' marks cut into the end of tiles like those from Colchester (JRS 1944, 17) and Verulamium (JRS 1943, 11). I have collected just over a hundred published and unpublished examples in a check-list at the end of this paper, reference being made to them by site and number. I have seen nearly twenty for myself, and photographs are available of a few others, but it must be emphasised that graffiti are often hard to decipher, let alone understand, and are often fragmentary. Thus the reading is sometimes doubtful.

I will use the generic term 'tile' to describe both 'bricks' and 'tiles' (the distinction is obscure), that is, not only the flanged roofing tile (tegula) and its accompanying ridge-tile (imbrex), but the hollow box-tiles whether 'wedge-shaped' (cuneatus) or straight (tubulus) used for water (e.g. 71) as well as hot gases, and even building tiles like the 'foot-square' pedalis commonly used for making hypocaust pillars (hence pilaris). The type of tile is often specified in the original publication, but sometimes the graffito is too fragmentary for this to be determined, and the Latin text of CIL and EE uses tegula generically ('tile'). It can be said, however, that graffiti on an imbrex (e.g. 17, 27) are most unusual, a tegula being the commonest medium. This is what one would have expected, since a moist flat tile would present an attractive writing surface, rather like a big waxed tablet. A skilled penman could produce a long text (e.g. 14, published at the end of this paper) within such a compass.

One of the tiles from Witcombe (102), it has been noted, was inscribed over 'a number of roughly parallel scratches', presumably the marks caused by levelling off the tegula in its mould. (A graffito from Aquileia [CIL V 8110. 176] refers to the process of 'scraping' bricks, 'cave malum si non raseris lateres DC'.) This is generally the case: graffiti were made after the tile had been completed. On the other hand, the last two letters on the Harpsden tile (56) were trodden on by a dog, which suggests that graffiti might be made before the tiles were laid out to dry. This is likely to be true of 'signed' tiles, or tiles recording some stage in the manufacturing process.

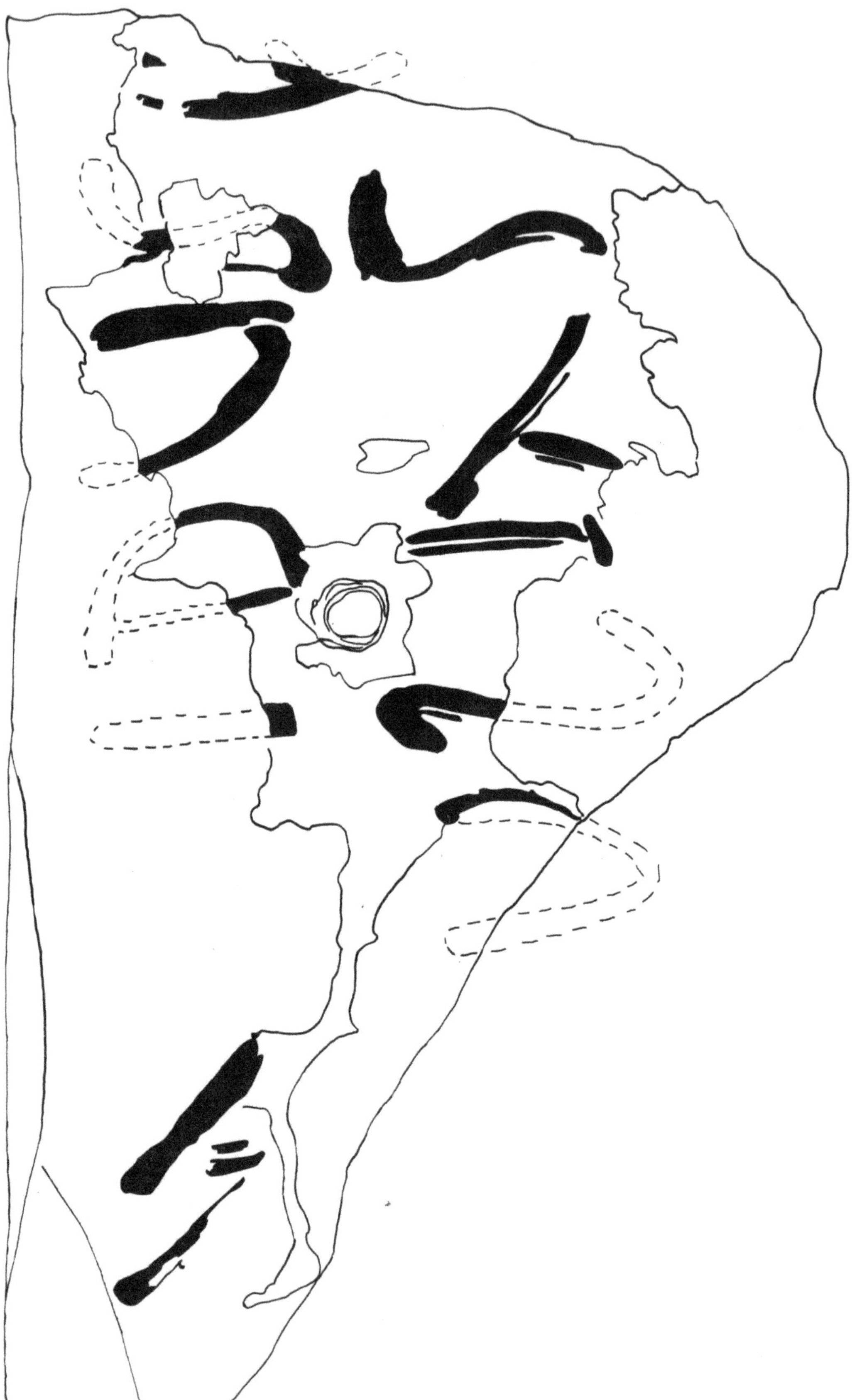

Fig. 11.1 Graffito from Cirencester. Scale 1:1 (Drawn by Mark Hassall)

2. Graffiti in Roman Tile-Making

There were three good reasons for inscribing graffiti on tiles: to date a batch, to record a total, to note (perhaps unofficially) the maker's name. In passing they preserve some of the technical terms for tiles current in the brickyards of Roman Britain.

It was worth knowing when a batch of 'green' tiles started drying, to judge when they were ready to be fired, and perhaps to check that they did not dry too quickly—which is why Vitruvius (II. 3. 2) confined Italian tile-making to spring and autumn. This precaution was not necessary in the summers of northern Europe. Dated tiles are common from the Rhine and Danube (see Saalburg Jahrbuch 25 [1968], 65f.), like one of a series from Siscia: 'V Kal. Aug. Severus et Fortunatus later. CCCCXXXX' (CIL III 11383), two workmen made 440 bricks on 28 July, year unstated as usual. Such graffiti are surprisingly rare from Britain. There is in fact only one certain example, one from Silchester (84) of 26 September, a date later in the year than its Continental parallels. A fragmentary numeral, also from Silchester (86), may be part of a date; and a graffito from Westhampnett (97) may be a date towards the end of May or June. A damaged tile from Cirencester (38) looks to me as if it may have borne a date in early July, but another reading may be preferred (Fig. 11. 1).

Batch-totals are more common. The most informative is the pilaris from Wigginholt (100) which itemises at least 20 pillar-tiles, four voussoir-tiles (spelt cuniati), and 560 box-tiles, perhaps the total needed to build the villa bath-house. A puzzling tile from Dover (47) notes a batch of 550 box-tiles (on a likely restoration of the text); then someone else, apparently, has written that he 'broke 51' (unless quassiavi is some technical term hitherto unattested). Some tiles are simply marked with numerals like 'tally' marks, Chester (28) with '400' Woodchester (104) with '44, 46'., Holwood Hill (65) with '10'. There are fragmentary numerals of this kind on the tiles from Aylesford (3, 4), another from Chester (27), one from York (108), and perhaps one from Farningham (52). The tile from Tomen-y-Mur (94) baldly says 'I have made (more than 20)', but the box-tile from Leicester (68) is a work of art which ranks in its way with the Cirencester panther (?) exhibited at the Conference, the charging bull of Dover (48), and Silchester's 'shaggy-coated ox' (Boon, Silchester [1974], 63). 'Primus has made 10 (of these)' has been deftly inscribed with a comb, and framed within an elegant combed pattern of wavy lines. Two tiles from Cirencester present an interesting palaeographic problem. A tegula (35) exhibited at the Conference (Museum No. B1404) has been variously read as CLLTXVIIII, CETVS IIII, CCCXXIIII, BLXVIIII, ARTVCI, CCCXVIIII and ALTVM (see Antiquity 26 [1952], 88f.), an indication of how difficult these short snatches of 'cursive' can be. The graffito was inscribed rapidly if with assurance (see Fig. 11. 2), and I wonder whether the problem can be resolved by reading the initial 'letters' as an incomplete CXↃ, the usual epigraphic symbol for mille, sometimes looking like a modern '8' on its side. It appears on two of the distance-slabs from the Antonine Wall (RIB 2200, 2205) and on a tile from Carnuntum (CIL III 11452), and is common in the sense of milliaria (of cohorts). A 'C' (100) was ligatured to it, followed by an 'X' (10) whose bottom left-top right diagonal was cut first, like that of

Fig. 11.2 Graffito from Cirencester (B 1404). Scale 1:1
(Drawn by Nick Griffiths)

Fig. 11.3 Graffito from Cirencester. Scale 1:1

Primus (68) or the 'XIII' from Silchester (87). Thus the Cirencester tegula reads MCXVIIII, '1,119 (tiles)'. The other tile (36) I have not seen (Fig. 11.3), but I wonder whether the difficult reading AVCXIEX[... might not be resolved in the same way, as MCXI ... (?). Batch-totals of this kind find many parallels in the Rhine and Danube provinces, for example CIL III 11381, one of the series from Siscia: 'Kal. Iulis, Severus CCXX, Fortis CCXX, Candidus CCXX, Felicio CCXX, in uno DCCCLXXX', a note that on 1 July four workmen made 220 each, total 880. Roman Britain has produced the most laconic of batch-totals, perhaps the most eloquent: at Silchester (81) someone scrawled SATIS with his finger in the wet clay. 'Enough.'

This comment is anonymous, but sometimes the workman 'signed' his tile. Like Primus (68), Clementinus of Silchester cut 'Clementinus made [this] box-tile' (80) in capital letters of somewhat 'cursive' form. The dated tile from Silchester (84) concludes with 'from the hand of Icc[... '. Another (82) was explicitly made (fecit) by someone whose name ended in ...]rionus. 'Silur made' one of the tiles from Witcombe (103). A few soldiers signed their tiles. A stamped tegula of the Twentieth Legion from Holt (60) bears the well-cut but mis-spelt graffito of Iulius Aventinus, a soldier of the First Cohort of Sunici—a Roman citizen serving in an auxiliary unit, apparently seconded to the legionary brickyard, perhaps from Caernarvon (cf. RIB 430). An unusual tile-stamp found at Chester bears a graffito-like text: 'tegula A. Vidu. Vero III cos. Leg. XX' (Brit. 1978, 16), a legionary tile 'signed' by Aulus Vidu(... ?) in the year 167. Compare the stamp from Maryport: 'Coh. I Hispa. Indutius fec.', First Cohort of Spaniards, made by Indutius (CIL VII 1232). To return to graffiti: a tile from Chesters (34) carries the name of Copa (?) of the century of Urfus; one of the group of almost indecipherable tile fragments from South Shields (90) may read in part OFFL, perhaps of[ficina] Fl[aviani?], 'workshop of ...'.

Roller-stamped tiles from Plaxtol, Kent, also bear a graffito-like text: 'parietalem (tegulam) Cabriabanu. farbicavi', a wall-tile by Cabriabanus (Brit. 1971, pp. 297-8). Like them a few graffiti, as I have said, name tiles by type. Wigginholt (100) has been mentioned already: pilares, cuneati and tubuli. Tubuli recur at Barnsley Park (5) and Silchester (80), and can be restored at Dover (47). Teg(u)la could well be restored on the tile from Willowford (101), but is only doubtfully read on one of the South Shields fragments (90, see note). It is also doubtful whether some case of pedalis should be restored on the tegulae from Colchester (43) and Farningham (51). They are both the wrong sort of tile; Colchester has been inscribed after firing (and thus the graffito need have no connection with the manufacturing process), and the upright which survives of the fourth letter is more appropriate to 'E' or 'I' than 'A'. Perhaps, if more than three letters (PED[...) had survived in either case, the obscene verb p(a)edicare would have been indecently exposed. The graffito ...]pedalis[... which survives on a samian sherd from Caistor St. Edmund (Brit. 1971, 66) is also a problem: it is presumably whole or part of some unattested personal name, and irrelevant to our purpose. The generic term laterc(u)lus, 'brick', which is current in the Rhine area (e.g. G. Brambach, CIR 114, 1397) and elsewhere, does not occur in Britain.

3. Tile-graffiti in general

One of the tiles from Silchester exhibited at the Conference had been successively trodden on by a small dog, a goat, and someone wearing a studded sandal. This suggests that access to the brickyard was not restricted. Thus, unless there is explicit reason for believing so, we cannot assume that a graffito was made by a tile-maker, though it is more likely than not. The graffiti which follow, therefore, should be seen as evidence of daily life in Roman Britain, rather than as documents of the tile-making 'industry'. Most of them, like the majority of graffiti in general, are little more than personal names. It may be convenient simply to tabulate them:

Roman citizens

(73) Quintus Avidius Amabilis, Aulus Antistius Felix (restored)
(8) L. C. C. (abbreviated tria nomina?)
(62) ...]enus Sedatus
(60) Iulius Aventinus
(7)? Iulius
(54)? Iunius
(19) Messius Secundus (restored)
(6) Lucius Septimius
(45) Ti(...) Primus

Cognomina only

(51) Atti[...
(72) Austalis
(18) Bellicianus
(89, 96) Calvus
(83) Campanus, Campester
(106) Car[...
(91) Cerialis (restored)
(80) Clementinus
(56) Datius (for Dati[v]us)
(30) Fidel[...
(83) Lucilianus
(55)? Pius
(107) Pollio
(68, 39) Primus
(31) ...]rinus
(29) Secundus
(76) Seneca
(16) Sirica
(31) ...]talis
(48)? Taurus
(70) ...]vanus

Greek

(88) Philodespotus

'Celtic'

(78) Birginus (for Briginus, cf. JRS 1969, 24)
(84) Icc[... (Iccius?)
(32) Litegenus (for Litugenus)
(32) Maccus
(85)? ...]marus
(103)? Silur (but cf. CIL II 5923)
(15) Tarcarus

Uncertain

(12) Armea
(34) Copa, Urfus (not Ursus)
(83) Pertacus (perhaps coined from per and taceo on the analogy of the next word, 'perfidus', from per and fido?)
(82) ...]rionus

This list does not include the twenty-five names from the recently-discovered brick from Binchester (14) which is published below. The list contains few pleasures or surprises for the student of Roman (and Romano-British) nomenclature. The high proportion of non-'Celtic' names among the inhabitants of Roman Britain literate enough to write their own (or someone else's) name on a damp tile may be noted; but all except one of the 'Celtic' names (Maccus, presumably) are autographs.

A few graffiti refer to what may be called third parties, by conveying gratuitous (and hopelessly allusive) information about someone else. A 'girl' (puellam) at both Caerleon (16) and Silchester (77) was the object of some verb left to the understanding, or more likely, now lost. The obscene verb paedicare, I have suggested, may lurk on the Farningham (51) and Colchester (43) fragments; perhaps the fragment from Silchester (79) with the letters ...]DICA should be added. These wretched shards are outclassed by the garrulous but obscure tile from Budapest (CIL III 10716) which details what Gratus did with the slave girl of an optio of the Second Legion. At York (107) Pollio gave an unspecified social club his good wishes: 'Pollio collegio feliciter'. The word feliciter recurs in the undeciphered tile from Chesterford (32). A broken tegula from Caves Inn (24) makes what may be a tantalising reference to the Civitas Coriosolitum (Corseul, in north-eastern Brittany), perhaps the writer's place of origin. If so, does this mean that the tile 'industry' in Britain attracted (or needed to attract) skilled immigrants?

A famous graffito, from London (72), informs us that Austalis has been wandering off on his own for thirteen days. To the ear of faith at least, this jingle (with its pair of end-line half-rhymes tredecim and cotidim) sounds like two accentual iambic dimeters. If 'Austalis' really is metrical, then he can take a humble place beside Binchester (12)'s 'Armea me docuit recte benedicere cunctis'. This hexameter line of unknown authorship employs a commonplace of amatory verse, the notion that falling in love teaches one (me docuit) to see things in a new light. Its author, or at least the copyist who inscribed it in such neat small capitals, blundering benedicere in the process, was no less sophisticated than the man who inscribed the Virgilian tag 'conticuere omnes' (Aen. II. 1) on the tubulus from Silchester (83). The words were a favourite with the graffiti-writers of Pompeii (CIL IV 3889, 4036, 6707). Someone at Sirmium was more original: by inscribing the hexameter line 'Protinus Aeneas procedere longius iras' (CIL III 10717) on his tile, he conflated two passages from Aeneid V (461, 485), without a finite verb to hold them together. Hexameter lines of unknown authorship like 'Armea' also come from Pannonia, written on tiles, for example a couplet from Savaria (III 11411):

> 'Credere vix dubito, sed amicum amittere nolim;
> si tibi credidero, non te tam saepe videbo'.

The Roman equivalent of neither a borrower nor a lender be; for if you lend money to a friend in need, you will be seeing rather less of him.

G. C. Boon has made the attractive suggestion (in Silchester [1974], 64) that graffiti like the one ending with the Aeneid tag (83) were used to give tile-workers a grounding in literacy for them to mark their batches correctly. It is the regular counsel of despair to call difficult Roman inscriptions 'writing practice'. The first two lines of a fragmentary graffito from Corbridge (44) make no obvious sense, but the third line can be restored as 'lege fel[iciter]', perhaps a lesson followed by the necessary exhortation, 'Read, and good luck to you'. There are half a dozen attempts at an alphabet among tile-graffiti (23, 37, 50, 67, 105), one of them blundered (21). There are two attempts to practise a single letter, 'S' (47) and 'M' (53); and what may be attempts to practise pairs of letters, consonant-'I' (61), 'NO' (66), 'QQ' and 'TT' with 'VI' intervening (103, cf. 61). Not that this is peculiar to tile-makers: at Wroxeter (JRS 1960, 24) someone was practising single letters and pairs of letters on a small rectangle of mudstone. At Slack (88) Philodespotus seems to have had three goes at writing his own name, though we must remember that when Cabriabanus cut one of his roller-stamps for marking tiles, he thought his name was worth repeating (Brit. 1971, 53). At Caerwent (18) someone wrote the name 'Bellicianus' on a tile, which three people copied in turn, more or less successfully. Alphabets seem to have had a value both educational and decorative. Thus, to take some random examples found in the course of searching for tile-graffiti, they have been incised round the neck of a flagon from Littlechester fort, Derby (JRS 1967, 45), and underneath a pot from Richborough (Richborough IV, 253, no. 1), each time with errors; inside the rim of a bowl from Chedworth (JRS 1955, 27) and, perhaps 'for magical protection', on the leaden plate of a kind of brooch from Holt (JRS 1966, 22). Were the users of these alphabets gaining literacy thereby—or demonstrating its possession to others?

Tile-makers, some of them at least, as well as people who gained access to drying tiles, could clearly write, or practise writing. But few of these graffiti were a necessary part of the manufacturing process. Dated tiles come from only two or three sites, batch-totals from a few more; but if we exclude simple signatures, these two categories amount to only eighteen graffiti, about one-sixth of my total. These hundred-odd inscribed tiles are themselves only a small proportion of the thousands of tiles and fragments which must have been handled during the last hundred years. It looks as if tile-batches in Roman Britain were not, as a rule, dated or numbered (at least on the tiles themselves), however useful we might think this would have been. Most tile-graffiti may not have been much different from the general run of graffiti, except in the lack of ownership inscriptions: 'signatures' in a way way fall into this category, and I wonder whether the drawing of a bull from Dover (48) labelled TAVRI is in fact a rebus of the artist's name ('Taurus, his tile') rather than 'bulls' or 'of the bull'. The quality of tile-graffiti ranges widely, from the illegible or painfully scratchy (TARCARVS [15], for example) to accomplished pieces like 'Armea' (12), whose authors, like Iulius Aventinus (60), had left the lementary skills far behind. They were writing for the fun of it, perhaps in default of more convenient materials, like the great wood-engraver Thomas Bewick (1753-1828), a self-taught artist, who began by drawing in chalk on gravestones and stone-flagged floors, until someone took pity on him and gave him paper.

Five graffiti remain to be mentioned, which fall into no previous category. A floor-tile from Abridge (1) carries a bold drawing of a cantharus (tankard) done with the finger, labelled CISS: the artist seems to have intended a cissybium, a Greek loan-word for 'drinking cup' mentioned uniquely by Macrobius in the fifth century A. D. There are three Christian symbols, which unfortunately cannot be dated, Chi-Rhos from Leicester (69) and York (110), and an alpha straddled by an omega from Wickford (99). It has been noticed that both the latter and the Chi-Rho from York were cut on a tegula near the flange, so that they would have been covered (should we say concealed?) by the imbrex. What is in some sense the predecessor of these potent symbols, the phallus, popular as a graffito in Britain as elsewhere, does not seem to occur on tiles. Lastly, at Lincoln (71), some optimist cut LIBER ESTO before firing on a flue-tile which was later buried as part of a pipe-line. 'Be free.'

The spelling of tile-graffiti, as one would expect, sometimes derives from the spoken language. Most instances have been collected already by J. C. Mann in Britannia 1971 (218f.), and the list which follows is based on his tabulation:

VOWELS

-e- for -i-	32 feleciter
-i- for -e-	60 milis
	100 cuniati
-i- inserted	47 quassiavi (see R. P. Wright's note ad loc.)
-o- for -u-	24 Corielsoliliorom

CONSONANTS

-c- for -qu-	74 ecum
-v- omitted	56 Datius

CONTRACTIONS

5 tubl
12 bidicere (in error?)
60 cortis
72 Austalis, dibus, cotidim
101 tegla
107 Polio colegio felictr

TRANSPOSITION

78 Birginus (cf. Brit. 1971, p. 297, farbicavi)

4. An inscribed brick from Binchester

The longest tile-graffito yet known from Roman Britain, indeed one of the longest 'cursive' texts yet recovered, was found as recently as 1978 during excavations at the fort of Binchester (Vinovia) in County Durham. (I am grateful to the excavators, Mr. I. M. Ferris and Mr. R. F. J. Jones, and to Mr. D. Coggins on behalf of the Bowes Museum, for information, and the opportunity of studying and publishing this interesting item. It will also appear in Brit. 1979.) This graffito was inscribed before firing with a stylus or similar instrument on a square brick of which about half remains, measuring 195 mm wide by a maximum height of 115 mm, 50 mm thick. The writing is 'cursive' of late-third or fourth century date, which suits the circumstances of the find in what seemed to be an undisturbed build-up of debris over the floor of a room added to the fourth-century bath-house inside the fort. The surviving text (see Fig. 11.4) is ordered in four headed columns, the first two now incomplete; a line was scored down the first three columns while the clay was still moist. The text reads:

Column 1	Column 2	Column 3	Column 4
CIIII	V	CII CO[N]STANTIS	CIII
MA[T]VTINI	VINCENTI	TITIANI	IVVENA[LIS]
COGITATI	CVNOVENDI	KARINIVS	KARINII
VIRILIS	CATVGNAVI		VITALIS
FLORENTINI	CARINI		
CLEMENTIS	MARINI		
MARINI	LVCILLIANI		
BELLICI	SENICIANII		
IVVENTIANVS	VITALIN[...		
CVPITI	M[...		
SOLINI	[...]		
SENI[C]IANI			
[...]			

The reading of a few letters is doubtful (they have been dotted in the Brit. text), but the writing as a whole is comparatively straightforward. There is, however, a word of uncertain reading between CVPITI (Col. 1) and the bottom of Col. 2.

The text may be more or less clear, what survives of it, but the meaning is obscure. It is obviously a list of twenty-five or more personal names, all but two (Iuventianus and Karinius) in the genitive case. All but three of them are common Latin cognomina, virtually all of them (if one includes variant spellings and cognate forms) already attested in Roman Britain. Karinius, strictly speaking, is a nomen formed from a cognomen (Carus), but this distinction was irrelevant by the time the text was written. There are only two 'Celtic' names, Cunovendus (already attested from the northern frontier zone as 'Cunovindus' [Brit. 1971, 14]) and Catugnavus (a plausible formation akin

Fig. 11.4 Graffito from Binchester

to the well-attested 'Catugnatus'). In passing it is worth noting that the proportion of 'Celtic' names is even lower than in the random sample of names gathered from other tile-graffiti. Nor is this a 'random' sample: whatever its purpose, it is likely to be a list of members of a military unit stationed in the hinterland of the fourth-century frontier. By this date, such a unit of limitanei would have been locally recruited for generations. These names are a cross-section of local nomenclature in the late-Roman period, after two centuries or more of Roman rule.

It would be easier if we knew the meaning of the four column-headings. (That in Col. 3 has been inadvertently combined with the first name on the list, one of some celebrity in fourth-century Britain, as it happens.) 'C' could be a numeral (100), but the resulting 'totals' (104, 5, 102, 103) fall into no plausible pattern. 'C' is more likely to be an abbreviation, perhaps a well-known one, perhaps one private to the writer. A numeral follows (4, ?, 2 and 3); we might suppose that the writer had intended to add a horizontal stroke to the 'V' of Col. 2, when it would become 'CI' (C. 1), thus completing the sequence. (There is no sign that this stroke was ever made, but this may also be true of the vertical stroke of the first 'T' in 'Matutini'). If the abbreciation is a well-known one, then C(ohors) is the strongest candidate, especially as legionary cohorts were numbered. But the names which follow cannot be those of centuries as one might expect (centuria understood, with the centurion's name in the genitive as usual), since there are too many of them in Col. 1 at least. But if the names are those of legionaries (ignoring for the moment the question of why they are in the genitive case), one would have expected them to be identified not only by cohort, but by century as well. A further difficulty is that the stamped bricks from this bath-house are not legionary, but were made, presumably locally, by a unit which seems to have taken its name, in the late-Roman manner, from the neighbouring fort of Chester-le-Street (Brit. 1978, 25, n(umerus) Con(cangiensium)).

Two lead sheets were found at Leintwardine fort (JRS 1969, 31) inscribed with an inconsequential list of names whose purpose is unknown. Someone may find a closer parallel than this to the Binchester list (clearly purposive, judging by its headings nand the way columns have been checked off), by which it can be explained. Although it has been written on a brick, it is quite unlike any other tile-graffito from Roman Britain. One genitive case is just plausible (e.g. TAVRI [48]), but twenty-three strain belief. The solution may lie in this direction. The names of witnesses to a Roman legal document (a Will for instance, or a military discharge diploma), usually seven in number, are listed in the genitive case—'So-and-so (his seal)'. Is the Binchester brick a memorandum of signatures copied, and subsequently checked, from an original document or documents, for example the receipts there must have been for rations issued to individual soldiers? This clerical activity presumably occurred at the brickyard and related to soldiers seconded to it—tempting though it is to posit an eccentric late-Roman unit in the North-East which kept its records in cuneiform, perhaps the Tigris Bargemen of South Shields (Not. Dig., Oc. 40.22) reverting to ancestral practice.

5. A Check-List of Roman Tile Graffiti from Britain

I have collected published graffiti from the following sources:

Corpus Inscriptionum Latinarum VII (1873) (abbreviated as C)

Ephemeris Epigraphica III (1877), IV (1881), VII (1892), IX (1913) (abbreviated as E, with volume number)

F. J. Haverfield, Roman Britain in 1913 (1914) and Roman Britain in 1914 (1915) (abbreviated as RB 1913 and RB 1914)

Journal of Roman Studies, Roman Britain survey (from 1921 to 1968) (J)

Britannia, Roman Britain Survey (from 1969) (B)

All items are cited by number, not page. The two journals are cited by year only. I have not made a systematic search for unpublished graffiti, but I have included a few made known to me by courtesy of the following museums: Chester (Grosvenor), Cirencester, Colchester, Glasgow (Hunterian), Gloucester, Reading, South Shields. There must be quite a number more. I have excluded tile-stamps; 'signatures' consisting of combed patterns or 'S' (and similar) made with the finger; 'tally' marks cut into the ends of tiles.

The text given is usually that of the published source, though I have sometimes been able to verify it by autopsy or from photographs. Where I have adopted a new reading I have marked it with an asterisk. Epigraphic conventions are adapted from RIB, in particular by not dotting doubtful letters, and by representing fragmentary letters by a stop (e.g. no. 9, Bignor). Square brackets enclose letters which can be restored with fair confidence; round brackets enclose letters omitted in the orignal, by contraction, abbreviation, etc. Three stops bounded by a square bracket represent an unknown number of missing letters. Line division is as here. Texts are transcribed in capitals, whether or not they were originally written in capitals or in 'cursive'. The open 'A' (like a 'V' upside down) is transcribed as 'A'; but the two-stroke 'E' (II) remains as 'II' (e.g. no. 32, Chesterford).

The check-list is intended to simplify references in the body of the paper, and as a conveniently brief survey of what one may expect to find in a tile graffito from Britain. It does not supersede the original publication (though it may occasionally comment upon it); nor does it in any way anticipate the publication of these graffiti in RIB II and RIB III.

No.	Site	Reference	Text	Remarks
1	Abridge	B 1975, 9	CISS(YBIVM?)	drawing of cantharus
2	Aldington	B 1970, 33	TA[. . .	
3	Aylesford	J 1963, 39a	. . .]I	
4	Aylesford	J 1963, 39b	. . .]XIII	
5	Barnsley Park	Cirencester Mus.	TVBL	for tub(u)l(us)?
6	Beauport Park	B 1971, 81	LVCIVS [SEP]TIMIVS	
7	Beauport Park	B 1973, 33	. . .]IVLIVS	
			. . .]IVC. OAANIVSVS	
8	Bignor	E VII 1142a	L(stop)C(stop)C(stop)	
9	Bignor	E VII 1142b	. . .]. RSVE. [. . .	
10	Binchester	E VII 1146	ARMEA ME DOCVIT[. . .	complete example now known (12)
11	Binchester	B 1978, 26	. . .]NA[. . .	
			. . .]ATO]. . .	
12	Binchester	B 1978, 27	ARMEA ME DOCVIT RECTE	
		(+ 1979 add.)	BIDICERE CVNCTIS	for b(ene)dicere? Or [ti]bi dicere?
13	Binchester	B 1978, 28	BOVM[. . .	or BOVA[. . .
14	Binchester	B 1979	25+ names	for text see this paper
15	Cadder	Hunterian Mus.	TARCARVS	
16	Caerleon	J 1964, 30	. . .]SIRICA VIDVA. [. . .	
			. . .]PVELLAM[. . .	
17	Caernarvon	B 1976, 65	. . .]ABSO[. . .	imbrex frag.
18	Caerwent	C 1255	BELLICIANVS	in four hands
			BELLICIANVS	
			BELLICIANVS	
			BELLICIANV	
19	Camelon	E IX 1295	MESSIVS[. . .	
			SE[. . .	traces suggest Secundus
			. . . CI[. . .	

No.	Site	Reference		Text	Remarks
20	Canterbury	J 1947, 16		. . .]GII[. . .	
				. . .]AVA[. . .	
21	Canterbury	J 1955, 19		ILKIM[. . .	
22	Carpow	J 1962, 42		. . .]TIT[. . .	
23	Caves Inn	J 1966, 31		ABCDEFGHIKLMNOPQ[. . .	
				ABCDEF. HIKL[. . .	
24	Caves Inn	J 1966, 33		[CIV]ITATIS CORIELSOLILIOROM[. . .	
				. . .]NIOM	
				. . .]M	
				. . .]CESOM	
25	Caves Inn	B 1973, 35	(a)	. . .]BRA[. . .	Four fragments of tegulae
			(b)	. . .]MLA[. . .	
			(c)	. . .]BVREIS[. . .	
			(d)	A[. . .	
				BII[. . .	
				CAV[. . .	
				ICA[. . .	
26	Chelmsford	B 1979		. . .]AG[. . .	
27	Chester	B 1971, 21		. . .]X[. . .	imbrex frag.
28	Chester	B 1971, 22		CCCC	
29	Chester	B 1977, 47		. . .]DAI[. . .	piece of pottery?
				SIICVN[DVS?]	
30	Chester	Grosvenor Mus.		FIDEL[. . .	
31	Chester	Grosvenor Mus.		. . .]RINVS[. . .	
				. . .]TALIS	
32	Chesterford	C 1256		LITIIGIINVS MACCVS. . .	only a drawing survives; not yet deciphered
				. . .	
				FIILIICITIIR	
				. . .	

No.	Site	Reference		Text	Remarks
33	Chesters	E VII 1145		. . .]IGEIT. [. . .	
34	Chesters	E IX 1291		COPA 7 VRFI	7 = centuria
35	Cirencester	J 1291, 14		* MCXVIIII	for this reading, see text
36	Cirencester	J 1961, 33		AVCXIEX[. . .	another numeral (MC. . .)?
37	Cirencester	B 1977, 71		. ABBDDEE	
38	Cirencester	B 1977, 72		* . . .]IDVS	a date? But B reads Candidus f.
				. . .] LIAS	tegulas
39	Colchester	C 1257		PRIMV[. . .	
40	Colchester	J 1944, 15		IM. [. . .	after firing
41	Colchester	J 1944, 16		L. DERE. [. . .	after firing
42	Colchester	B 1978, 40		. . .]A	
43	Colchester	Colchester Mus.		PED. [. . .	ped[i], rather than peda. . . ?
				CLAM[. . .	
44	Corbridge	RB 1914, 6		OMQ. [. . .	CMO. . . ?
				LIIND. [. . .	
				LEGEFEL[. . .	lege fel[iciter]?
45	Cramlington	C 1258		TI PRIMVS	
46	Cranleigh	J 1958, 32	(a)	EP	inscribed in four different hands
			(b)	mason's mark	
			(c)	FRAB	
			(d)	VMAN	
47	Dover	B 1973, 26		[TVBVL]	
				OS X DL F	barred D (=500) (1.3) cursive, in
				QVASSIAVI LI	second hand?
				SSSSSSSSS	
				SSSSSSS	
48	Dover	B 1978, 44		TAVRI	drawing of bull
49	Dover	B 1978, 45		. . .]NOSPII I VS	
50	Exeter	B 1976, 20		IABCDIIFF	

No.	Site	Reference	Text	Remarks
51	Farningham	J 1965, 28a	ATTI[. . . PED[. . . [. . .]	 ped[ales] ? (on a tegula)
52	Farningham	J 1965, 28b	M[. . . D[. . .	numerals ?
53	Froxfield	J 1939, 16	M (repeated at least 7 times)	
54	Godmanchester	J 1968, 52	. . .]IVNIV[S. . .	
55	Hamstead Marshall	J 1963, 38	PII	
56	Harpsden	J 1953, 14	DATIVS	dog's paw-mark obscures
57	Hartfield	B 1973, 34	OST retrograde	perhaps TRO recto
58	Hartfield	B 1978, 67	IV	
59	Havant	J 1927, 28	. . .]ORIS	
60	Holt	RB 1913, 5	IVLIVS AVENTINVS MILIS CORTIS PRIMA SVNICOR	Leg. XX VV stamp
61	Holt	RB 1914, 9	. . .]LIVITILI[. . . . . .]IT TAL[. . . [. . .]	
62	Holt	B 1971, 92	. . .]IINVS [S]EDATVS	
63	Holt	B 1972, 59	. . .]. TS. [. . . . . .]AR[. . .	
64	Holt	unpublished	COSSEVFVERR (?) AGMIIM[. . . (?)	undeciphered; this reading conjectural
65	Holwood Hill	C 1254	X	
66	Hucclecote	Gloucester Mus.	. . .]NONON[. . .	
67	Lancaster	B 1974, 28	AB[. . .	
68	Leicester	E VII 1143	PRIMVS FECIT X	written with comb; with combed decoration
69	Leicester	B 1976, 41	A XP	Chi-Rho (barred Rho)

No.	Site	Reference	Text	Remarks
70	Lincoln	J 1925, 17a	|VANVS	[Sil]vanus (or sim.);
71	Lincoln	B 1976, 42	LIBIIR IISTO	
72	London	E VII 1141	AVSTALIS DIBVS XIII VAGATVR SIB COTIDIM	
73	Oakenholt	B 1978, 74	QVINTVS AVI. [. . . AMABILIS. [. . . . VS ANTI[. . . . . .]V[. . .	
74	Park Street	J 1945, 8	ECVM[. . .	after firing; for equum?
75	Reculver	C 1261	[. . .] [. . .]VS [. . .]	only a drawing survives; undeciphered
76	Richborough	J 1925, 16	. . . ?]SENECA	Richborough II, pl. xii, 2
77	Silchester	C 1259	[. . .]. PVELLAM	
78	Silchester	E III 126	* BIRGIN VS	for Briginus? (E reads Birgaius)
79	Silchester	E IX 1292a	. . .]VDII . . .]DICA . . .]VS	
80	Silchester	E IX 1292b	FECIT TVBVL CLEMENTI NVS	tubul(um)
81	Silchester	E IX 1292c	SATIS	
82	Silchester	E IX 1292d	. . .]S FECIT . . .]RIONVS	
83	Silchester	E IX 1293	PERTACVS PERFIDVS CAMPESTER LVCILIANVS CAMPANVS CONTICVERE OMNES	cf. Aen. II. 1

No.	Site	Reference	Text	Remarks
84	Silchester	E IX 1294	VI K OCTO[BRES] MANV ICC[. . .	26 September
85	Silchester	Archaeologia 56, 123	. . .]MARI	
86	Silchester	B 1972, 27	QQI[. . . CNL[. . . IM[. . . ITVC[. . . V	
87	Silchester	Reading Mus.	. . .]XIII[. . .	
88	Slack	J 1963, 37	PHILOD[. . . TVS. PH[. . . V[. . . PHIL[. . .	Philodespotus ?
	89-93 (South Shields) are illustrated in Arch. Ael. n. s. 10 (1884-5), 250-1, and part-published as E VII 1144.			
89	South Shields	E VII 1144 (a)	CALVI * IIIIGRIS(?)	 E reads filia X
90	South Shields	E VII 1144 (b)	. . . OFEL . . . SE[. . . QC FECLA	 Of(ficina) Fl(. . .)? E reads teg(u)la only
91	South Shields	E VII 1144 (c)	•[C]ERIALIS (?)	E reads .erianus
92	South Shields	unpublished	CAE. [. . . (?) S PVG	conjectural
93	South Shields	unpublished	. . .]MS. [. . . . . .]QOX[. . .	conjectural
94	Tomen-y-Mur	J 1948, 18	. . .]XX FIICI	

No.	Site	Reference		Text	Remarks
95	Twyford	J 1925, 18		...]VoEIIIC[...	
96	Westhampnett	J 1936, 7	(a)	CALVI	
97	Westhampnett	J 1936, 7	(b)	VKI	K reversed; for K(al.)?
98	White Notley	J 1966, 34		...	
				...].IDV[...	
				...]IND[...	
99	Wickford	B 1972, 26		A inside Omega	
100	Wigginholt	J 1940, 20		P[ILARES ...]XX	
				CVNIATI IIII	
				TVBV(LI) N DLX	barred N (numero) and D (500)
101	Willowford	J 1947, 15		...].[...	
				...]GLA[...	retrograde; [te]g(u)la?
102	Witcombe	J 1940, 21		QQVITTVITTVV (?)	
103	Witcombe	J 1951, 18		SILVR F	f(ecit)
104	Woodchester	Lysons, Woodchester, 13		XXXXIIII	
		(with pl. xxvi.7)		XXXXVI	
105	Wroxeter	J 1943, 16		ABCDEF	
106	Wroxeter	J 1966, 32		CAR[...	
				N[...	
107	York	E VII 1139		POLIO	see RCHM Eburacum, inscr. 24
				COLEGIO	
				FELICTR	
108	York	J 1963, 40		...]XXXV	
109	York	B 1972, 56		...]MSA[...	
				...]M[...	
110	York	B 1974, 58		XP	three-stroke Chi Rho

Note: I have excluded one of the South Shields fragments (Arch. Ael. n.s. 10, 251) as being indecipherable and possible falsum (the 'E' (?) is un-Roman); and C 1260 (Winchester, IKLM) as being medieval, since it is an 'encaustic tile' found with four others which certainly sound medieval (JBAA 1 [1846], 312).

Addendum: Since compiling and checking this list, I have received a photograph of a tegula fragment recently found during excavation at Binchester. It appears to read ...].TOVERS. It will be published in Brit. 1980.

ORIGINS OF LEGIONARY TILE-STAMPING IN BRITAIN

Alan McWhirr

There is no recent study on the beginnings of legionary tile-stamping and in considering the origins of this practice in Britain it is necessary to examine briefly the evidence from the continent, in particular from areas where the legions involved in the Conquest originated. Latest views on this subject have been communicated to me by Professor H. V. Petrikovits (2. 11. 78) and Dr. C. Ruger (28. 11. 78) and the comments which follow are based on opinions contained in letters along with a study of the published literature. Further evidence should be forthcoming on the origins of tile-stamping on the continent when Dr. Gustav Millar, the excavator at Neuss, has completed his investigation of the earliest stratified stamps from that site.

The views still held by scholars on the continent conform to those put forward in 1933 (CIL 13 part 6 page vii) by Ritterling, that is, the earliest stamps are those made by Legio IV Macedonica and found at Mainz. Consequently, the earliest possible date for these tile-stamps hinges on the date at which the legion moved from Spain to the Rhine to occupy part of the double fortress at Mainz which is currently put at around A. D. 45. Dr. Ruger, whilst accepting a date of 43/45, is a little more cautious and feels that this date may be a shade too late. If the date of 43/45 is accepted then one would not expect to find any tile-stamps produced before A. D. 43 on the continent of the legions which took part in the conquest. No such stamps have been found and it now seems certain that the conquest legions adopted the practice of stamping during their stay in Britain, that is, those legions which stayed long enough to do so.

THE EVIDENCE FROM BRITAIN

i. Legio II Augusta

The earliest military building to be found so far in Britain which used tile and brick is the legionary bathhouse at Exeter constructed by Legio II Augusta. Tile was in production before the construction of the bathhouse as tile fragments were found in the metalling of the fortress streets which were laid between A. D. 55-60. The bathhouse, which was built between A. D. 60-65 incorporated the usual range of tile and brick including *pilae* tiles, roofing tiles, antefixes and box tiles which P. Bidwell suggests were made in an area immediately to the north-east of the fortress where tile-making debris has been found (Bidwell, 1979, 13). No stamps of Legio II have been found at Exeter which indicates that it was not the practice for this legion to stamp tiles in the late Neronian period. Boon and Bidwell have pointed out that no stamps of Legio II have been found at Strasbourg, where the legion was stationed before coming to Britain, the only significant find being a tile bearing a handwritten Leg II (Bidwell and Boon, 1976, 280).

Frere suggests that Legio II was moved from Exeter to Gloucester before being stationed in South Wales (Frere, 1967, 75), but Bidwell now believes that the fortress remained in full occupation down to c. A. D. 75 when it then moved to Caerleon (Bidwell, 1979, 1). There is a LEG II tile-stamp in Gloucester Museum, but it is unprovenanced and there is no indication as to how it arrived in the collections (information from J. Rhodes). The tile is 'unwashed of mud' which might indicate that it was found locally, but this is far from certain. The stamps from Caerleon are considered by Boon to date from the second century; there are no tile-stamps from the first century baths at Caerleon. Boon suggests that the tile-stamps of Legio II from Seaton and Sea Mills are second or third century in date and indicate maintenance work on outlying military posts (Bidwell and Boon, 1976, 279).

There are four tiles stamped by Legio II from Usk, all occurring in areas of paving or in hearths overlying early levels and Dr. Manning states that the most probable date for them is second or early third century—'certainly not before the mid-second century' and that they 'must be contemporary with the Caerleon examples'. The fact that the Usk examples were found in hearths and paving might suggest that they were reused and therefore earlier in date, but until we see full details of the Usk excavations we must accept the excavator's opinion (letter dated 7. 11. 78). Dr. Manning suggests that in the second century Usk was, in fact, being used as a works depot supplying Caerleon.

The early use of tile at Usk is shown by a tegula mammata from a fortress pit and almost certainly from the fortress baths, indicating that tile was being made in the pre-Flavian period by troops at Usk (Manning and Webster, 1978, 381), which would not be surprising in view of the fact that there was a legionary pottery in operation at this time and often the two activities were carried out together. Dr. Manning, however, believes that it was the Twentieth legion who was in residence at this time.

The evidence from Legio II indicates that although legionary tile-makers were producing the usual range of tile and brick in the first century A. D. as well as sophisticated antefixes, they did not stamp tiles with the legion's title until about the mid-second century.

ii. Legio IX Hispana

There is little doubt that before the legion was stationed at Lincoln in c. A. D. 60, it was on campaign in the east of England either as a complete legion, or divided into vexillations. The legion remained at Lincoln until A. D. 71 when it was moved to York. Little of the internal arrangements of the fortress were known until in 1978/79 when part of the principia was excavated and preliminary indications are that although a quantity of tile and brick was found in military levels, none were stamped. There is one tile stamped LEG IX HISP from Lincoln, but Mr. Andrew White, Keeper of Archaeology at Lincoln Museum, believes that this tile originally came from York. The tile was bought at the Stamford Institution sale in 1910 and the Lincoln Museums accession register states, 'Tile with stamp of IX Legion (found) Lincoln about 1850 ex Stamford Institution'. Mr. White, however, draws attention to the sale catalogue issued by Cade and Son on 1st June 1910 which only records 'Lot 184 Portions of Roman tiles from York'. The origin therefore, of the

Lincoln provenance is not clear. Mr. White cannot find any evidence for stamped tiles being found at Lincoln in c. 1850 and in view of the details contained in the sale catalogue a York provenance is more likely.

Two other tile stamps of Legio IX have been noted in the East Midlands, one from Old Winteringham, Lincs., and the other from Hilly Wood, Northants. The Old Winteringham find only contains two letters LE .. and R. P. Wright points out that 'in so far as two letters are enough this matches one stamp of LEG IX HISP found at Lincoln and York'. (JRS 59 (1969), 242, f 49 Stead, 1976, 190). But, as argued above, there is no good reason for supposing that the Lincoln example was produced in the area and so the only parallel for the Old Winteringham stamp is from York and cannot be associated with the legion when based at Lincoln. The application of analytical techniques in the future may help to unravel this particular problem.

The Hilly Wood stamp found in 1867 in the parish of Bainton has recently been described by A. Challands (Durobrivae 3 (1975), 21 and Fig. 9). He describes the tile as a rib voussoir. The circumstances of discovery are a little confused, but perhaps the comment in the unpublished papers of J. T. Irving in 1886 that it was 'ploughed up' is the most apt; it may or may not have been associated with a burial. Dr. Webster has claimed that the stamp 'is identical in shape, die stamp and material with one found in York' (Webster, 1958, 51). In fact, J. T. Irving sketched his interpretation of how the tile could be used in a tile grave—presumably based on a tile-burial found at York which contained a stamped tile and which had been published in Archaeologia in 1773. Indeed, the tile itself may have come from York. On present evidence it cannot be used to show that the ninth legion was stamping tiles in the first century A.D.

The legion was moved to York by Cerialis where it stayed until the first quarter of the second century A.D. and there it certainly stamped tiles (RCHM, 1962, 114). The discovery of a tile-stamp and a mortar-stamp of Legion IX at Nijmegan in Holland had led Professor Bogaers to suggest that the legion replaced Legio X Gemina at Nijmegen in about A.D. 121. An inscription shows that the legion was still at York in A.D. 107/8 (RIB, 665), but how long it remained after 107/8 can only be inferred from the evidence in Holland. Whatever the case may be the legion was stamping tiles before it left Britain, perhaps in A.D. 120, which is earlier than the date postulated above for the Legio II. Other stamps of this legion have been found at Carlisle and Scalesceugh (Wright, 1978).

iii. Legio XIV Gemina

Legio XIV was part of the conquest army, but only stayed in Britain until A.D. 67/69 and consequently the discovery of any tile-stamps of this legion from Britain are crucial in the study of the beginnings of the practice of tile-stamping in this country.

No stamps of this legion have been listed from Britain, but a tile stamped LEG XIV is in the possession of Mr. Perry the Curator of Horsham Museum, Sussex. Mr. Perry purchased the tile in the Portobello Road in about 1965 from a Mr. James Keggie, Hildreths Arcade who had purchased it earlier in a mixed lot at Sotheby's. Mr. Perry was unable to find out from Sotheby's the vendor, but he is under the impression that the lot originally belonged to a collector and if so, the tile need not have been found in Britain. If this is the case then one should look perhaps to the continent for parallels to this stamp. CIL which lists (in volume 13 part 6) stamps from Gaul and Germany found before 1933 has in all some 290 stamps from Gaul and Germany. None of these stamps uses the XIV form for the legion, they are all XIIII. There is, therefore, no exact parallel for this stamp from Gaul or Germany and doubt must be cast on its authenticity. If it is genuine and came from Britain it must be pre- A.D. 69 and be the earliest dated tile-stamp to have been found here.

The tile eas examined by Dr. D. Williams at Southampton University who was unable to determine where it was made. No tests for date have been carried out.

iv. Legio II Adiutix

This legion was brought to Britain by Cerialis and replaced the IXth at Lincoln probably in A.D. 71. It moved to Chester in c. A.D. 77 where it stayed until late 86 or 87 when it was withdrawn and sent to Moesia. No stamped tiles of this legion have been found in Britain.

v. Legio XX

The legion remained at Colchester for several years after Claudius' departure before moving, perhaps to Gloucester. Frere has suggested that Legio XX replaced Legio XIV at Wroxeter in A.D. 67 (Frere, 1967, 93). Later the legion was involved in campaigns in Scotland and stationed at Inchtuthill before moving to Chester around A.D. 87.

The bulk of tile-stamps of this legion come from Chester and the legion's works' depot at Holt. Grimes was not able to attribute dates to the stamps from Holt, but those found at Chester during excavations were known to have been in use in the period ranging from the early second century to the third. Grimes suggests those stamped with an A as the last letter, interpreted at Antoniniana, should be dated to A.D. 212-22. A first century date is attributed to one of the antefixes found in Chester in 1929 by Professor Newstead. This carries the legionary number and emblem, the boar.

The majority of stamps so far recorded have been found in an area of the country in which one would expect to find Legio XX active. They have been found at Carlisle, Manchester, Prestatyn, Heronbridge, Caernarvon, Ffrith, Caerhun, Caersws, Wilderspool as well as at Chester and Holt. Two stray finds have been made at Silchester and Whittlebury in Northants. The latter was examined by Dr. Webster who made use of X-ray fluorescent spectroscopy and concluded that it was of Holt manufacture and consisted of two separate tiles (Webster, 1958, 51).

There are indeed two separate tiles with stamps in Northampton Museum, but the stamps appear to be identical and to be the same as examples illustrated by Grimes in the Holt report (Grimes, 1930, Fig. 59 no. 1). The circumstances of the discovery of these stamped tiles suggest that they may have been part of a collection. They were found in 1822 along with a number of Greek, consular and imperial Roman coins, when trees were felled.

The Silchester stamp was found by chance in 1961 within the walls of the town and read LEGXXVV (Britannia, VIII (1977), 441). The circumstances of this discovery are not clear.

vi. Legio VI Victrix

An inscription from Halton Chesters (RIB, 1427) records the presence of this legion working under Aulus Platorius Nepos who was governor from A.D. 122 to 126 and it seems likely that the legion came over with Hadrian and Nepos in A.D. 122. Eventually it was based at York. Before coming to Britain it was based at Vetera in Lower Germany where tile-stamps are recorded (CIL, 12160).

R. P. Wright has recently listed all the known stamps of this legion in Britain (Britannia, VII (1976), 224-235) and there seems to be a fairly wide distribution and many different types of stamp. Wright identifies some 79 dies. The bulk of stamps come from York, but others have been found at Carlisle, Carrawburgh, Corbridge, Gayton Thorpe, Carpow, Catterick, Ebchester, Wallsend, Chesterholm, Aldborough, Slack, High Rochester, Chester and Halifax. Wright's study analyses the stamps into types, but offers no help with the dating of these different dies, although this may become possible in due course. A detailed study of the distribution of these stamps is not relevant to this discussion on the origins of military tile-stamping, as they cannot have stamped tiles in Britain earlier than the ninth legion.

Miscellaneous Legionary Tile-Stamps from Britain

Legio XXII

W. St. C. Baddeley, in his book on the history of Hailes, entitled 'A Cotteswold Shrine' published in 1908, illustrates a tile stamped LEG XXIIPRPFE. The tile is said to have come from 'above Kemerton and Westmancote on Bredon' and is listed under Stanway by the Royal Commission (RCHM, 1976, 111). CIL lists some 299 different stamps of Legio XXII Primigenia, but not one matches the design illustrated by Baddeley although one style has similar motifs. There is no way of knowing whether this tile is genuine or not, or whether it was in fact found in the ground at the place stated above. Whatever the case may be the fact that the stamp contains the legion's title Primigenia dates it to the Hadriannic period or later, as it was Hadrian who conferred the title upon the legion, and is of no help in tracing the origins of legionary tile-stamping in Britain.

Legio VIII

A tile discovered in Leicester in 1855 bears the stamp LVIII retrograde and is usually considered to stand for Legio VIII. A close examination of the stamp shows it to be complete; there is no break before the first I so it cannot have originally been VIIII. There are several indications for the presence of this legion, or detachments, in Britain. It is recorded at the time of the conquest and may have accompanied Claudius to Britain. However that may be it seems too early for the legion to be stamping tiles and hardly the occassion for it. RIB 782 suggests that men from the legion were in Britain, probably sent as reinforcements, during the first half of the second century. There is also a shield-boss from theTyne bearing a punched inscription LEGVIIIAVG, and although there is no evidence of date, it is tempting to link this shield-boss with the landing of troops on the Tyne in the second century. It is difficult to see, therefore, any context for a stamped tile of the eighth legion, if indeed the LVIIII does represent that legion and is not some form of civilian stamp.

Discussion

The evidence from the four legions which took part in the conquest indicates that although they were capable of producing tile and brick quite early in Britain, certainly by the late 50s, there is no example of a stamped tile positively dated to the first century A.D. There are no stamped tiles from the legionary bathhouse at Exeter or Caerleon which are both dated to the first century, and there is no tile from Britain stamped by the XIVth Legion which was withdrawn in A.D. 69. The IXth Legion was not stamping tiles at Lincoln between c. A.D. 60 and A.D. 71, but did so before it left Britain in perhaps A.D. 120. Legio II Adiutrix was in Britain from A.D. 71 until 86/7 and in this period may not have been involved in tile and brick making; certainly there are no tile-stamps of this legion in Britain. It seems quite likely that the legions stationed in Britain adopted the practice of stamping tiles at different times and on the evidence at present available, this does not seem to have occured until the second century.

Acknowledgements

Many people have provided information and discussed various problems with me for which I am most grateful. They include Paul Bidwell, Bill Manning, Robert Moore, Even Perry, Professor H. V. Petrikovits, John Rhodes, Dr. Chris Ruger, Malcolm Watkins and Andrew White.

Bibliography

Bidwell, P. T., 1979. The Legionary Bathhouse and Basilica and Forum at Exeter, Exter, 1979.

Bidwell, P. T. and Boon, G. C., 1976. 'An Antefix Type of the Second Augustan Legion from Exeter', Britannia, 7 (1976), 278-80.

CIL Corpus Inscriptionum Latinarum, Berlin, 1863-.

Frere, S. S., 1967. Britannia - A History of Roman Britain. London, 1967.

Grimes, W. F., 1930. Holt, Denbighshire. The Works Depot of the Twentieth Legion at Castle Lyons being volume 41 of Y Cymmrodor, 1930.

Manning, and Webster, 1978. 'Romano-British Tile from Usk, Gwent'. Antiq. Journ. 58 (1978), 381-3.

RCHM, 1962. Royal Commission on Historical Monuments, An Inventory of the Historical Monuments in the City of York, I: Eburacum, Roman York. London, 1962.

RCHM, 1976. Royal Commission on Historical Monuments, Ancient and Historical Monuments in the County of Gloucestershire I: Iron Age and Romano-British Monuments in the Gloucestershire Cotswolds. London, 1976.

RIB Collingwood, R. G. and Wright, R. P., The Roman Inscriptions of Britain: I, Inscriptions on Stone, Oxford, 1965.

Stead, I. M., 1976. Excavations at Winterton Roman Villa, London, 1976.

Webster, G. A., 1958. 'The Roman Military Advance Under Ostorius Scapula', Arch. Journ. 115 (1958), 49-98.

Wright, R. P., 1978. 'Tile-Stamps of the Ninth legion', Britannia, 9 (1978), 379-382.

MILITARY TILE-STAMPS FROM BRITAIN

Mark Hassall

As has been shown in the previous chapter, with the possible exception of legion XI, (Wright, 1978), none of the legions stationed in Britain in the first century A.D. stamped the tiles they were producing in their tile-works: the same is almost certainly true for the auxiliary units and the *Classis Britannica*. However, from some point early in the second century, all three legions then stationed in the province are found stamping tiles and their example is followed by the British fleet and at least two *alae* and ten cohorts of Britain's auxiliary garrison. It will be convenient to examine the evidence for all three branches of the military in Britain separately, starting with the 'senior service', the legions.

Legionary Tile-Stamps

From the time of Hadrian, the legionary dispositions in Britain were made permanent both as regards unit and location, with legion II Augusta based at Caerleon, legion VI Victrix at York and legion XX Valeria Victrix at Chester. The evidence for the practice of stamping tiles by the first of these has, in effect, been reviewed by George Boon in his comprehensive study of the legion's base (Boon, 1972). At Caerleon no stamped tiles have been found in contexts securely dated to the first century and indeed the earliest context that Boon actually cites for a stamped tile of this legion is in the auxiliary fort at Aberyscir, not constructed in stone (the time at which the buildings will have received tiled roofs) until after *c*. A.D. 140. The presumption is that the legion was already stamping tiles for its own use at this date. Aberyscir, incidentally was just one of a number of forts in South Wales which received tiles from Caerleon (others being Usk, Abergavenny and Llandovery in the second century, and Pennal in the Dyfi esturary in the third).

In the course of the third century the legion received various honorific titles derived from the dynastic or personal names of the current emperor, and these are included in abbreviated form on its stamps. Thus we find the title *Antoniniana* under Caracalla in *c*. A.D. 212 and this was retained under Elagabalus but dropped during the reign of Severus Alexander (222-235) in favour of *Severiana*. A single stamp of the legion carries the title *Vi(ctoriniana)* after the "Gallic" emperor Victorinus 269-271, and is in fact the latest dateable evidence for the existence of the legion at Caerleon. It might be thought that tiles bearing the common *Antoniniana* title would be equally valuable as a source for dating the buildings in which they are found. Unfortunately this is not the case since tiles seem to have been often re-used or even stockpiled at time of manufacture for use at a later date. Thus for example numbers of *Antoniniana* tiles have been found from the barracks of the legion's seventh cohort, but the restoration to which they belong probably dates to A.D. 255-60 (thirty years after the tiles were originally stamped) as we know from an inscription which records the rebuilding (RIB 334).

Legion VI, like legion II, produced tiles not only for its own use but for attached auxiliary forts, in the case of legion VI as far north as Hadrian's Wall and including the great base at Corbridge (Wright, 1976). Tiles stamped by the legion mostly carry the inscription leg(io) VI v(ictrix) and are examples of Wright's types 42-5. Besides the titles victrix and pia fidelis, which the legion already bore before its transfer to Britain in 122, it was, like legion II, given the further title Antoniniana and Severiana by Caracalla and Severus Alexander respectively, as well as Gordiana by Gordian II. 238-244. The legion was also, uniquely, awarded the title Britannica which occurs on stamps from Carpow (occupied 208-212).

The tile-works of legion XX have been excavated at Holt and the stamps found there have been discussed by Grimes (1930), who published his report of the excavations in exemplary fashion from the notes left by the original excavators, though it should be noted that not all the known stamps of the legion have actually been attested at Holt. Legion XX like the other two British legions was apparently given the title Antoniniana while other stamps (JRS 45 (1955), 176) carry the letters DE perhaps for De(ciana) derived from the name of the emperor Trajan Decius (249-251). One stamp seems to give the name of an officer in abbreviated form and reads SUB LOGO PR variously expanded as sub L(...) O(...) G(...) o(ptione) pr(incipis) or sub Logo pr(aefecto), depending on whether a subordinate officer or the actual commander of the legion is being indicated. An interesting graffito shows that a soldier in the first cohort of Sunuci (stationed at Caernarvon under Severus, RIB 430), Aventinus by name, was working at Holt. This suggests that although the legion might supply auxiliary units in the vicinity with tiles, the auxiliary units might, on a reciprocal basis, send unskilled labour to the works depot. The question of how labour was organised in military tile works however is one about which we know all too little, though from military tile works on the continent we have records of magistri figlinarum, in one case in charge of a squad of sixty men (Bogaers, 1969, 31 and note 24). Finally two tiles have recently been published from Chester (Fig. 13.1a) which uniquely among all tile-stamps from Britain conclude with a consular date, A.D. 167. They will have read tegula(m) Auulus) Viduc(ius) f(ecit) Vero III co(n)s(ule) Leg(ioni) XX V(aleriae) V(ictrici), 'Aulus Viducius (?) made this roof tile for the Twentieth legion Valeria Victrix in the third consulship of Verus' (Britannia 9 (1978), 476, no. 16).

Tile Stamps of the Fleet

The tile-stamps of the Classis Britannica have been admirably reviewed by Gerald Brodribb (1969) though his work will need revision once the very extensive material from Dover has been published. The Dover material will not however effect Brodribb's observations on the extremely restricted distribution of finds of stamped material. Apart from examples from the major British and continental bases at Dover and Boulogne they are confined to a relatively short stretch of the south coast running from Pevensey to Richborough and iron-working sites in its immediate hinterland. Lympne has also produced stamped tiles of the fleet, but the tiles themselves both here, and at

a

b

Fig. 13.1 a. Composite drawing of legionary tile-stamp from Chester. Scale 1:1

b. Auxiliary tile-stamp from Brancaster. Scale 1:1

Dover, Pevensey and Richborough, appear to belong to structures which antedate the construction of the shore forts which were later built at all four sites. The absence of stamped tiles of the fleet from the shore forts does not necessarily mean that the Classis Britannica had been disbanded at the time of their construction, perhaps only that the tile-works operated by the fleet were not called upon to supply material (for auxiliary tile-stamps from two of the shore forts, Reculver and Brancaster see below).

Finally, just as we do not know when the fleet ceased to exist, let alone ceased to stamp tiles, so the date at which it first started stamping tiles is not known though new evidence from Dover may be expected to throw light on this particular problem.

Auxiliary Tile Stamps

In reviewing the evidence for tiles stamped by auxiliary units in Britain it will be convenient to divide the country up into a series of zones.

(a) Hadrian's Wall. Tile-stamps of cohors V Gallorum have been found at South Shields and the unit is known to have been in garrison there in the early third century (RIB 1060). Similarly stamped tiles of ala I Asturum come from Benwell where an inscription (RIB 1334) attests the unit as the third-century garrison (a tile stamped with the same die is also known from Wallsend). Great Chesters has produced tiles stamped cohors II Asturum and this too is known as the third-century garrison (RIB 1738). From Hare Hill comes a stray tile of cohors I Tungrorum known from numerous inscriptions to have been the garrison at Housesteads (e.g. RIB 1584-6); this is slightly odd since one would have expected cohors II Tungrorum, the third-century garrison at Castlesteads (RIB 1981-3), which is closer to Hare Hill.

(b) The Wall Hinterland. Maryport has produced tiles of cohors I Hispanorum. There certainly was a unit of this name stationed at Maryport under Hadrian (RIB 823 and note). But the Notitia Dignitatum (xl 49), whose information for the garrisons of the Wall and its hinterland has been shown to be valid for the third and fourth centuries, places a cohors I Hispanorum (not necessarily the same unit) at a site which may or may not be identical with Maryport, so that the tile-stamp could possibly be third or fourth century in date, rather than second.

From Lancaster come tiles of the ala Sebosiana and an inscription originally set up to theGallic emperor Postumus, 262-6, records the presence of the unit there (RIB 605). Tiles of cohors III Bracaraugustanorum have been found at both Manchester and Melandra Castle. From the fort at Slack come tiles stamped by cohors IIII Breucorum and the tile-works of this unit have been located at nearby Grimscar (see page 182), while a detachment probably of this unit (though possibly cohors III Bracaraugustanorum, the reading of the stamp is not completely certain) was outposted at the fortlet of Castleshaw (Bogaers, 1969, 40). The dating at these sites is interesintg: Slack was occupied from c. A.D. 80-140, being rebuilt in stone some time after 90, while Castleshaw was only occupied from 90-125 (Bogaers, 1969 loc.cit.). The tiles from these two sites are some of the earliest stamped by auxiliary units in Britain. Cohors IIII Breucorum was transfered at an uncertain date to Ebchester where it was certainly in garrison in the third century (RIB 1101) and where it continued to stamp tiles. Stamped tiles also come from two other third or fourth century sites in the general vicinity of Ebchester: the N CON. stamps from Binchester, probably the numerus Concangiensium, 'unit of men from Concangium' (= Chester-le-Street) and the ΛBOΛCI stamps from Chester-le-Street itself. The latter could conceivably be intended for Abulci, the numerus Abulcorum being one of the garrisons attested by the Notitia Dignitatum (xxviii 20) in Britain—at Pevensey to which it could have been trasfered from the north (Hassall, 1977, 9).

(c) Wales: Only one fort in Wales, Caersws has produced auxiliary tile-stamps (Nash-Williams, 1954, 107-8). They read C·I·F·/S·P·P, of which the first line is to be expanded c(ohors) I F(risiavonum) and C·IC·F perhaps for c(hortes) I C(eltiberorum * et I) F(risiavonum) though I can find no parallels for two auxiliary units jointly stamping tiles in this way. However the alternative, c(ohors) I C(eltiberoum) f(ecit) seems even less likely since the word f(ecit), though usually to be understood is not normally included.

(d) South Eastern Britain (The Saxon Shore): Two of the earliest Shore Forts, Reculver, guarding the approaches to the Thames estuary and Brancaster sited in a similar position in relation to the Wash, have both produced auxiliary tile-stamps (Hassall, 1977, 8-9). From Reculver come tiles stamped by Cohors I Baetasiorum, the garrison recorded in the Notitia Dignitatum (xxviii 18). From Brancaster two tiles are now known stamped C(hors) I Aq(uitanorum) (Fig. 13.1b) the Notitia garrison being the Equites Dalmatae (xxviii 16). This is the first time that a Shore Fort has yielded evidence for a garrison other than that given in the Notitia. The probable explanation is that the cavalry regiment (one known to have been raised after the middle of the third century) replaced the infantry one.

Conclusion

Legionary tile-stamps are clearly basically a phenomenon of the second and third centuries, though a few of the legion IX stamps may date from the late first century. The Classis Britannica were probably stamping tiles more or less contemporaneously but the published dating evidence is meagre. Of the twelve auxiliary units that stamped tiles, six did so during the third century, one cohors IIII Breucorum did so in the early second and also in the early third while two, cohors III Bracaraugustanorum and cohors I Hispanorum may have stamped tiles in the early second century but the evidence in both cases is ambiguous. For the remainder there is no evidence. The conclusion to be drawn is that in Britain auxiliary units normally stamped tiles only in the third century. This is in direct contrast to the Rhineland where the stamping of tiles by auxiliary units was usually a century earlier, during the late first and early second century (Bogaers, 1969, 32).

Bibliography and Abbreviations

The following bibliography is not intended to be comprehensive, but gives only those works cited in the text. Full documentation of finds of military tile stamps made up to 1956 will be found in RIB volume 2 (forthcoming).

---

* or C(ugernorum), if the tile stamps are late first or very early second century, or C(ornoviorum) if they are third century.

Bogaers, J. E., 1969. 'Cohortes Breucorum', B.R.O.B. 19 (1969), 27-50.

Boon, G. C., 1972. Isca the Roman Legionary Fortress, Cardiff, 1972.

B.R.O.B. Berichten van de Rijksdienst voor het Oudheidkundig Bodemonderzoek.

Brodribb, G., 1969. 'Stamped Tiles of the Classis Britannica', Sussex Archaeological Collections, 107 (1969), 102-125.

Grimes, W. F., 1930. Holt, Denbighshire: the Works Depot of the Twentieth Legion at Castle Lyons. Y Cymmrodor, 41 (1930).

Hassall, M. W. C., 1977. 'The Historical Background and Military Units of the Saxon Shore', in D. E. Johnson (Ed.), The Saxon Shore. CBA Research Report No. 18, 1977, 7-10.

JRS Journal of Roman Studies.

Nash-Williams, V. E., 1954. The Roman Frontier in Wales.

Notitia Dignitatum, Notitia Dignitatum Occidentalis, O. Seeck (Berlin 1876, reissue Frankfurt, 1962).

RIB The Roman Inscriptions of Britain, eds. R. G. Collingwood and R. P. Wright, Oxford, 1965.

Wright, R. P., 1976. 'Tile-Stamps of the Sixth Legion found in Britain'. Britannia 7 (1976), 224-235.

Wright, R. P., 1978. 'Tile-Stamps of the Ninth Legion found in Britain'. Britannia 9 (1978), 379-382.

THE USE OF TERRA-COTTA FOR ARCHITECTURAL ORNAMENT IN ITALY AND THE WESTERN PROVINCES

Thomas Blagg

There are three general questions underlying this review of a rather wide range of material. First, what part did terracotta play in Roman architectural decoration? Secondly, and closely connected with the first, in what way was this related to developments in Roman building construction? Thirdly, to what extent is the architectural use of terracotta in the western provinces different from that in Italy? Since its use in Italy had a much longer and more varied history, most of the paper will be concentrated on that area. Additionally, much more work has been done on the Italian material, and one purpose of this paper is to review some of the results of recent research. Virtually nothing has been done, however, to bring together such provincial material as there is, but it may be thought useful to point to research that could be done.

Terracotta was employed in Roman architectural decoration in two forms. First, in the shape of moulded plaques, free-standing ornament and statuary attached to a building made of other materials. Secondly, in the form of bricks and tiles incorporated within a structure which was itself of brick, but shaped to reproduce such ornamental features as columns, capitals or cornices. In some cases these served as a basis for finer detail applied in stucco, but in the early Empire unadorned brick and tile also came to be used for decorative purposes.

The first of these forms of architectural decoration was, as is well-known, related to a type of building constructed of wood and sun-dried brick, with a tile roof. In the Etruscan temple, widely-projecting eaves helped to keep the rain off the walls, and the terracotta plaques had the function of protecting exposed woodwork from weathering. As, however, such contemporary models as those from Nemi, Orvieto and Vulci also demonstrate (Staccioli, 1968), the architectural function of these terracottas was not purely utilitarian. Antefixes and cresting projecting above the eaves were decorated with masks, palmettes and other ornament which became conventional (Fig. 14.1(1)-(3)). Acroteria on the pediment and along the ridge took the form of palmettes and free-standing sculpture. Revetment plaques were decorated with figured friezes as well as with conventional floral motifs (Fig. 14.1(4)). The appeal of these brightly-painted additions was not only aesthetic: in the earlier period, particularly, the subject matter of the sculpture and the friezes was closely connected with the religious and social function of the temples and sacred buildings upon which this decoration was predominantly employed in the Archaic and early Classical periods.

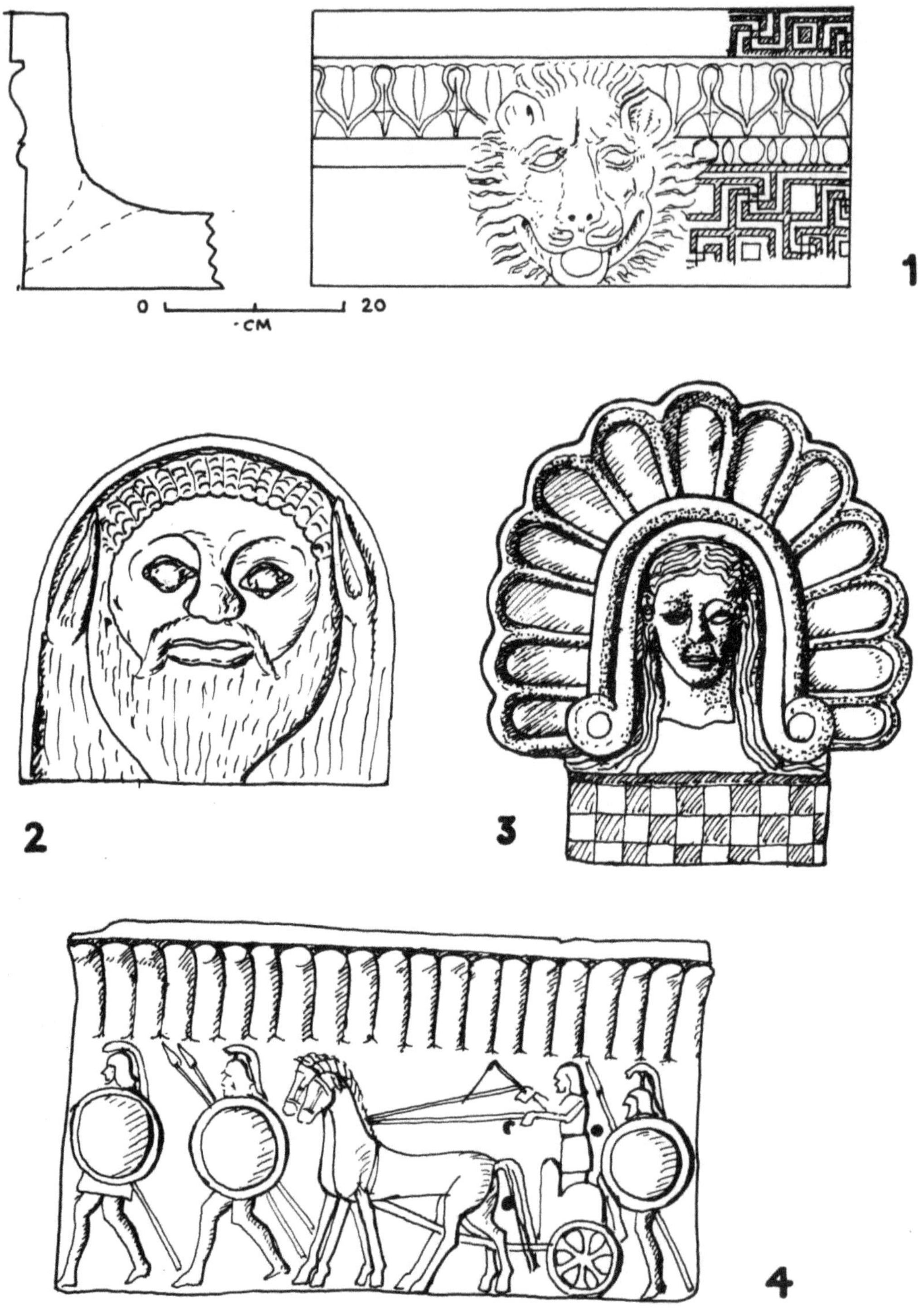

Fig. 14.1 (1) Lateral Sima, Temple of Hera, Croton
(2) Antefix, Tarentum (Trieste, Museo Civico)
(3) Antefix, Capua (East Berlin, Staatl. Mus., Antikenslg. no. 7898)
(4) Frieze, Toscanella (Munich, Antiquarium).

Numbers (2)-(4) not to scale.

It is necessary to go back to before the time during which Rome's influence in Italy was paramount architecturally, in order to take account of recent discussion of the origins of this form of decoration and the date of its introduction into Italy, and to isolate one or two particular examples of types which survived into late republican or imperial times.

The roofing system of which these decorative elements formed part was the familiar one of tegulae and imbrices, which is basically a hybrid of two systems of tile roofing developed in Greece by the seventh century BC. These are the laconian, in which semi-cylindrical tiles were employed with their upturned edges covered by similar tiles inverted; and the corinthian, in which ridged cover-tiles were laid over the adjacent sides of flat pan-tiles with raised edges. Several hybrid versions are attested on the mainland of Greece before the end of the seventh century, notably at Corinth and Isthmia (Roebuck, 1955, 156 and pl. lxii; Broneer, 1955, 112; Andrén, 1974, 2). Of these, it was the combination of the laconian cover-tile and the corinthian pan-tile which became characteristic of Italian tile roofing.

The covering tiles of the Etrusco-Italic roof have received rather little scholarly attention. A paper by Wikander (1974), discussing the important finds from a house at Acquarossa, has initiated a redress of the balance, but so far the evidence from archaeological excavations has played little part in establishing the date of the introduction of tile roofing of this hybrid form. Attention has almost exclusively been concentrated on the decorative aspects associated with it, and discussion of dating has depended to a large extent on stylistic criteria.

Arvid Andrén, in his fundamental work on Etrusco-Italic Architectural Terracottas, originally took the view that the most ancient of these pieces went back to the middle of the seventh century BC (Andrén, 1940, p. CXLIIIff.). In a recent reconsideration of the figured friezes (1974), he basically reaffirms this view against that of Akerström, who preferred to see them beginning a full century later. This opinion derived from the latter's study of architectural terracottas from Asia Minor (1966). It would seem that the decoration of early Etrusco-Italic friezes should be seen as strongly related to those of the Ionian coast, but mainland sites, notably Corinth, played an important part in the development. Akerström pointed particularly to the influence of the designs on Corinthian pottery imported into Italy in the first half of the sixth century. One should consider the possibility that ceramics, particularly those with plastic or applied decoration, and architectural terracottas were modelled by the same craftsmen.

There is now, however, some archaeological evidence from excavations to support Andrén's higher chronology. Stylistically early antefix types found at Murlo, Siena, for example apparently decorated a building which was in use during the last quarter of the seventh century but was possibly constructed about 650 BC (Nielsen and Phillips, 1974, 271 and 275 ff.). The occupation dates for the later Archaic sanctuary there suggest that its more evolved architectural terracottas are 'at least as early as about 575-570 BC' (Shoe Merritt, 1970, 15).

According to what one might call the origin myth of terracotta decoration in Italy, recounted by Pliny (Nat. Hist. xxxv, 152), Damaratus of Corinth, when he came to Etruria in the mid-seventh century, brought with him three fictores who introduced into Italy the art of modelling in clay. A development ascribed to Butades of Sicyon was the addition of masks to the outer gutter tiles on roofs, from which, according to Pliny, the ornaments on the pediments of temples originated. Butades was also said to have made the first terracotta statuary at Corinth. There does seem to be evidence for the use of architectural terracottas in Etruria in the late seventh century, and there does seem to have been come Corinthian connexion, both in the style of roofing, the style of early friezes, and the inspiration apparently derived from the decoration of pottery. To that extent the myth expresses the underlying archaeological reality. We should now turn, however, to look at some of the specifically Italian developments.

First, one may look at the antefixes and the decoration along the eaves. This takes two forms. The first is a continuous sima, that is, a vertical moulding, curved at the top, which rises from the end of the lowest of the flat roof-tiles. The form is common in Magna Graecia and Sicily, where it may be illustrated from temples at Locri, Syracuse, Caulonia, Croton and Metapontum (van Buren, 1923, pls ii-v), and in Campania, where it remained in use into imperial times (Koch, 1912, tafn. xxiii-xxv; Maiuri, 1929, 89). The main decoration was usually a frieze of lotus and palmettes, and they were pierced at intervals with spouts for the run-off of rain, often in the form of a lion's mask with open mouth (Fig. 14.1(i)). The second form is the circular or, more commonly, semi-circular, antefix attached to the end of the lowest cover-tile (Fig. 14.1(2)). In Greek Italy the decoration of this type of antefix was normally a Gorgon or Silenus head, but palmettes and other mythological figures were also used, both there and in Campania (van Buren, 1923, pls. xiv-xvi; Koch, 1912, tafn. i-xv).

The addition of the female human head to a gallery of otherwise mythological faces was probably an Italian development. In a recent paper on this type of antefix, Nancy Winter has shown that on the Greek mainland it occurs only at the Corinthian colonies in north-western Greece—Thermon, Calydon and Corfu; it is unknown at Corinth itself. Its earliest examples appear at Taranto and at Murlo, near Siena, by about 600 B.C., but its main area of popularity was in Campania, Latium and southern Etruria (Winter, 1978, especially pp. 29-31).

One early centre of production appears to have been Capua, and it was probably there, about the middle of the sixth century, that a new format was developed, in which the head was placed within a shell-like halo of squat tongues derived from the palmette antefix (Winter, 1978, 35). Later developments include elaborations to the semi-circular frame with spiral terminals which immediately surrounds the head, to form S-scrolls, lotus tendrils and subsidiary palmettes (Fig. 14.1(3)). At first the unhaloed heads were more widespread, but the shell antefix became established as a favoured version, particularly in Etruria, for about half a century from the last years of the sixth century B.C. onwards. Antefixes from Minturno and Veii exhibit technical differences in their manufacture from those of Capua, as well as a number of stylistic differences, with elongated tongues, the addition of diadems and ear-rings to the head and other details of the hair and of the moulding with volutes which surrounds it (Winter, 1978, 41).

Mention of such stylistic centres as Capua, Minturno and Veii (and also Caere) poses the question of how production of architectural terracottas was organised. Excavations of the villa rustica at Selvasecca, near Blera in southern Etruria, for example, revealed evidence for the production of ordinary roof-tiles and pieces of moulds for the manufacture of antefixes, as well as of antefixes made from them (Berggren and Andrén, 1969). Identical antefixes are known from Tarquinia, Orvieto and Talamone (Andrén, 1974, 16 and references, note 82), but it has been pointed out that their similarity does not mean that they were all made at Selvasecca. It is necessary to consider more closely the natûre of the production process, before trying to decide such a point.

There are three stages of production: the making of a prototype; the making of moulds from the prototype; and the making of antefixes from the moulds. Each stage could have been carried out at a different place from the rest, and at Selvasecca evidence for only the third of them was found. There are further implications. If the place where terracottas were produced is not necessarily the same as the artistic centre where the form of the prototype was evolved, the distribution of terracottas of a particular type tells us only about the area of influence of the first stage of production, that of the prototype. It says nothing of how within that area of influence production of the terracottas themselves was organised. Secondly, since both moulds and prototypes might remain in use for a number of years, the second and third stages of production cannot be assumed to be exactly contemporary, nor indeed can one particular antefix be assumed to have been made at precisely the same time as another of identical type.

It is necessary therefore to consider how rapidly prototypes and moulds might become unserviceable, and what is the life expectancy of a terracotta. The former is a function of the scale of production, and both questions would be worth further research, though the number of variable factors will make them difficult to answer. Evidence from such sites as Cosa suggests that the life of many architectural terracottas might be relatively short. The original decoration of the Capitolium there is dated to circa 150 B.C. Within a period of no more than a century, no less than five periods of redecoration have been identified (Brown, Richardson and Richardson, 1960, 206 ff.).

By this time, decoration had evolved into some highly elaborate forms. Along the eaves, three main elements were employed, which may be illustrated from the contemporary temple at Sassi Caduti (Plate 14,I(1)). On the pediment, it consisted of revetment plaques decorated with looped palmettes, with a small concave moulding at the top. These were pierced with holes, so that they could be nailed to the woodwork. Above them was the sima, either a projection of the cover-tiles along the edge of the roof, or separately made and seated in a groove in a tile with a moulded edge which projected over the revetment plaques below. Holes were made in the sides to contain pegs of lead or wood to join the simas together. Above the sima was the crowning moulding, often consisting of palmettes and intertwined pierced scrolling.

The reconstruction of the original decoration of the Capitolium at Cosa, the greatest of the three buildings excavated on the Arx which were decorated in this manner, shows other features of the evolved Etrusco-Italic design.

Plate 14.I (1) Revetments from pediment of temple, Sassi Caduti (Rome, Villa Guilia Museum)

(2) Le Mura di Santo Stefano, Anguillara Sabazia, from the south-west

1

2

Fig. 14.2 (1) Campana relief, Temple of Apollo Palatinus, Rome
(2) Antefix, Temple of Apollo Palatinus, Rome
Not to scale.

Along the sides of the temple, and below the pediment, there were revetment plaques with antefixes; the architrave was also revetted in terracotta.

Altogether, more than 2700 fragments of the decorative terracottas of the Capitolium were found in the excavations at Cosa, and the combination of stratigraphic evidence with the distinguishing features of style and fabric has permitted the definition of the five stages of decoration mentioned above. After the original, the next three seem to have involved no more than replacement of damaged parts, but the fifth period involved a wholesale redecoration of the main facade, though the sides appear to have been decorated with a patchwork of old and new pieces (Brown, Richardson and Richardson, 1960, 269).

This stage represents the last major flowering, perhaps the appropriate word, of the perennial formal designs of lotus, palmettes, and the like on simas and revetment plaques. During the second half of the first century B.C. there developed a wholly different style of ornament, continuing the Etruscan forms, but decorating these simas and revetment plaques with friezes of mythological subjects. These are the Campana reliefs, named for the Marchese Giampietro Campana, who published the first corpus of this particular type of architectural decoration (Campana, 1842).

Borbein's comprehensive study of these reliefs (1968) has shown that their distribution is almost exclusively restricted to central and northern Italy, ending in the north at the foothills of the Alps, and in the south round about Naples. They are not found either at Pompeii or at Herculaneum, where architectural decoration in terracotta continued to follow the older tradition (Borbein, 1968, 13). At Cosa they do not appear until after the fifth period of decoration of the Capitolium, about the middle of the first century B.C. (Brown, Richardson and Richardson, 1960, 269 and 296-300; Borbein, 1968, 21). Production continued little after the beginning of the second century A.D. (Borbein, 1968, 38-39).

The example illustrated (Fig. 14.2(1)) was found in the excavations of the Temple of Apollo Palatinus in Rome, built between 36 and 28 B.C. (Carettoni 1972, fig. 6). The pierced cresting of S-scrolls is of some incidental interest, since it appears in stone on the architectural ornament of military sites in particular in the north-west provinces: one of the pieces of evidence for saying that such sites tend to retain in a very conservative manner features of Italian ornament, even though often somewhat crudely executed, which the developing civilian traditions in those provinces tended to abandon (I hope to discuss the British evidence for this in a forthcoming paper).

Antefixes from the Palatine temple demonstrate that by the beginning of the principate far more varied subjects had joined the repertoire. On one example (Fig. 14.2(2)), a version of the Silenus, the old palmette design appears above his head in very stylised form, and scrolled tendrils provide elaborate counterparts to the handlebar moustache with which they are connected. A provincial variant of this treatment will be mentioned below.

The period during which the Campana reliefs were current was a time of rapid change in Roman architecture as a whole. A far wider use came to be made of travertine limestone and, increasingly from the principate of Augustus, of marble, in decorative superstructures. Terracotta lost its utilitarian function as a protection for wood as the traditional unfired brick was superseded, not only by tufa, which had long been in use, but also by faced concrete walls, structurally capable of carrying entablatures of hard stone which could be carved with comparable detail and finesse to that of the ornament of the terracottas which had covered timber beams and wall plates. The development of concrete vaulting in the last century of the republic, with the consequent changes in the ways in which timber was employed in the roof structure of major buildings, might logically have led to a rapid disappearance of terracotta as a decorative element in building.

In fact and, considering the conservatism of the public eye in the face of revolutionary trends in architecture, perhaps not surprisingly, terracotta was adapted, indeed, the range of its use was if anything extended, though in a manner quite divorced at first from anything seriously functional.

Archaic terracottas have generally been considered to have been made particularly for use on temples. Winter, in the context of her antefixes with female heads, has reconsidered some of the evidence, and shown that they were also used on other buildings connected with religious sanctuaries and on some tombs (Winter, 1978, 44-52). Campana reliefs and contemporary terracottas, however, are found more frequently on secular buildings than were their predecessors (Borbein, 1968, 17-19). They were also used internally; and the fact that the nail-holes on some revetments were moulded but not actually pierced through shows that some of them were probably built into a masonry wall and not fixed to timber. They were used particularly in, or rather on, tombs and in private houses, another consequence of the greater luxury in domestic building which followed from that increased exposure to Greek culture and the profits of Mediterranean expansion from the second century B.C. which so disgusted the elder Cato. Basically, as the range of buildings which were given monumental decorative treatment increased, so, initially, did the range of uses of terracotta in that decoration.

Similarly, with the increasing use of fired brick as a facing material for conrete, the practice started of using it as the basis for decorative treatment. Commonly, this was at first more structural than ornamental in itself, in that finer detail was applied over the surface in stucco, in the form of frames round and pediments over doorways, or niches, as in the Eumachia building at Pompeii (Lugli 1957, tav. clxxiv) or of pilasters and cornices, as on the Augustan Porta Palatina at Turin (Lugli, 1957, tav. clxxxi). At Pompeii and Ostia, for example, from the late second century B.C., segmental tiles were also used to build up column shafts (Lugli, 1957, 574 ff. and tav. clxxiii), and in chapter 2 of this volume Roger Wilson describes the same practice in Sicily. At the early first-century B.C. villa at Sette Finestre in the territory of Cosa, currently being excavated under the direction of Professor Andrea Carandini, we can see an example of this aggrandisement of domestic architecture in traditional materials, namely, sun-dried brick on stone footings. There, the peristyle was formed of columns made up of rather small pieces of tile, which had fluted shafts and moulded bases added in stucco. Antefixes in terracotta including palmettes in relief and lion's heads were used in the superstructure (Carandini and Settis 1979, 49ff. and Pannelli 25 and 30).

The development of various techniques for facing concrete in stone and brick incorporated a number of treatments that can be called decorative, if one's capacity for aesthetic appreciation is easily aroused, such as the combination of quoins and so-called bonding courses of tiles with opus reticulatum. Lugli (1957, tavv. cxlvii-cl) illustrates examples from Chieti, Pompeii, Ostia and other sites where the use of alternating courses of different colours of tufa with brick laid in the manner just mentioned can only be explained as having been done for visual effect, not to be concealed beneath a stucco rendering.

It is thus during the first century A.D. that we see in Italy a growing appreciation of the way in which brick might be used as a decorative material in its own right. This is distinct from earlier uses of terracotta, in that the effects were achieved by laying ordinary bricks or tegulae in particular ways, if necessary cutting them to shape, and not, initially, using purpose-made or moulded elements. The engaged pilasters and alternately triangular and semicircular pediments of the exedra of Trajan's Market, adjoining his Forum, with travertine limestone used for the capitals and bases of the pilasters, provide an early example of this application of brickwork on a metropolitan building of monumental scale (for the constructional features of this and other buildings of the period, see further, Blake and Bishop, 1973).

As such well-known examples as the Horrea Epagathiana et Epaphroditiana at Ostia show us with special clarity, this was basically a translation into brick of a manner of architectural decoration proper to stone. Even Corinthian capitals and the dentils and ovolos of cornice mouldings, as well as columns, pilasters, string courses and pediments, were built up from moulded or carved fired brick, which was also used for subsidiary ornamental features such as the aedicula in the House of the Lararium at Ostia, and for door and window surrounds.

Towards the middle of the second century A.D. we find this use of brick, and also the use of bricks of different colours for polychrome effect, in a variety of buildings in and around Rome. The site of Le Mura di Santo Stefano, two kilometres south of Anguillara Sabazia on Lake Bracciano, has recently received detailed study in a paper by Margaret Lyttelton and Frank Sear (1977). Red brick was used for the engaged pilasters and the arches of the windows on the exterior, contrasting with the yellow and reddish-orange bricks employed for the rest of the wall facing (Plate 14.I(2)). The Tuscan and, in the lowest order of pilasters, Corinthian capitals, the pilaster bases and all mouldings apart from the lower cornice were also of brick. The building, which still stands in part three storeys high, has been variously interpreted in the past, but the character of the surrounding remains suggests that it was probably a villa, and one of some distinction. That of Sette Bassi, ten kilometres south of Rome, provides another example: Piranesi's engraving, reproduced by Blake and Bishop (1973, pl. 13), shows a grand entrance, now no longer standing, with pediment, cornice and engaged pilasters with composite capitals, all in brick.

There are examples of this manner of brick decoration on second- or early third-century buildings of public or official character, notably the Theatre at Ostia, and the Excubitorium of the VII Cohort of Vigiles in

Trastevere and the Amphitheatrum Castrense in Rome (Kammerer-Grothaus, 1974, 200 and 226-227). Most of all, however, it was on a number of rather grand funeral monuments that what had become a feature of private rather than of public architecture was employed. Substantial remains of several of these survive along the Via Appia (e.g. the Tombs of Veranius and La Conocchia), the Via Latina (e.g. the so-called Barberini Tomb) and other roads leading out of the city, as well as in the Isola Sacra cemetery at Ostia (most of these are discussed and illustrated in Blake and Bishop, 1973, where further bibliographical references are given).

The most recently reconsidered of these, and perhaps the most familiar, is the now-restored Tomb attributable to Annia Regilla, the wife of Herodes Atticus, a building formerly known as the Temple of the Deus Rediculus. Kammerer-Grothaus' extended discussion of this mausoleum (1974) is followed by a review of the use of polychrome brickwork and terracotta mouldings, the latter being a later second-century development which continues into the Severan period, and a consideration of the workshop organisation which produced this distinctive group of buildings in Latium.

When we come to consider the western provinces, we find that there is evidence for only a limited number of the types of architectural decoration in terracotta that were used in republican and early imperial buildings in Italy. To a large extent, this is the result of the replacement of the timber and mud-brick architecture of the Etrusco-Italic tradition by one in which stone was the principal medium for the decoration of monumental buildings. Occasionally, terracotta plaques were employed, in the manner if not precisely the style of the Campana reliefs, such as one recently found at Cannes, decorated with garlands in a rather rudimentary manner (Gallia, xxxv, 1977, 506 and pl. 18). For Britain, one may cite a plaque found at Colchester outside the south-west corner of the <u>colonia</u> and now lost, which illustrates a gladiator not unlike one of those on the Colchester vase, in ignominious retreat from an aggressive sea-monster (Hull, 1958, 170 and pl. xxx, A). Basically, however, one cannot consider this use of architectural terracotta to have been of much significance within the context of western provincial architectural decoration as a whole.

Antefixes and acroteria are rather more common, within limits. There does not, however, seem to have been much place for them in civilian architecture. At Glanum, a terracotta antefix bearing a palmette was excavated in the nymphaeum (Rolland, 1946, 46 and fig. 35), but much more impressive is the limestone acroterion which came from the same building. It is shaped underneath to sit over the ridge-pole, and decorated with a small palmette surrounded by a larger one (Rolland, 1958, 96 and pl. 32.3). An almost identical acroterion was used on the corner of the roof of the Temple of Valetudo, which was dedicated by M. Agrippa, and the ridge was adorned with one of a female bust set within a palmette, an ultimate descendant of the terracotta shell antefixes of sixth- and fifth-century B.C. Italy. The tradition has now, however, been transformed into stone, less fragile and in keeping with other aspects of the building's construction and decoration. We can exemplify exactly the same transformation in a limestone acroterion from Gloucester, showing a male head set against a palmette (Rhodes, 1964, 19-21), and in simpler form, in an antefix and an acroterion from Bad Deutsch-Altenburg and Carnuntum respectively (Krüger, 1972, nos. 584 and 586).

Further north, the use of terracotta antefixes seems to be predominantly a feature of military sites, or at any rate of areas in which the army played a dominant cultural role so far as architecture is concerned. A rather fine collection from Neuss, for example, shows the continued regular use of the palmette, either as the sole decoration, or with a lion's or a human head superimposed. Terracotta faces also appear to have been used by themselves (Novaesium, 1904, taf. xxii). The antefix plaques were modelled on the end of an imbrex tile. Other examples, with Gorgon's heads, again illustrate the continuation of a repertoire first established in Italy seven centuries earlier.

A similar instance in which the provinces reflect earlier Italian patterns was anticipated above, in discussion of an antefix from the early Augustan Temple of Apollo Palatinus at Rome, on which a Silenus mask was placed beneath a palmette. Another feature is the organic ambiguity with which his twirled moustaches were conjoined with spiral tendrils of foliage. The same conceit may be seen on the marble mould for an antefix which was found at Pfaffenhofen, near Brumath in Alsace, of which the published photograph illustrates an impression (Espérandieu, 1918, no. 5551). The palmette occupies a similar position in relation to the head, but the ringlets of hair are replaced by a wreath of oakleaves, and sprays of leaves ending in rosettes are propagated from the beard rather than from the moustache. I have not seen the piece in question, and it is not clear to me from Espérandieu's description, which writes of an 'épreuve' made from it, whether it is a prototype, from which a mould and from that an impression would have been made, or whether the marble itself was carved as a negative matrix. A much cruder antefix, found at Brumath itself, follows the same basic type (Espérandieu, 1918, no. 5568).

Jocelyn Toynbee collected most of the evidence available in the early 1960s for terracotta antefixes in the province of Britain, and also made the point that they were much more common on military sites than civilian (Toynbee, 1964, 428-431). This can now be seen as a counterpart to the substantial differences in architectural decoration which are abundantly clear from my own study of that aspect of stonework. Among the few examples from civilian sites, an antefix in the form of a horned bearded mask from Dorchester deserves mention (JRS xlvi, 1956, pl. xvi, 2), and also one found associated with a first-century house at Silchester (Boon, 1957, 135, 148 and fig. 28), a pentagonal plaque bearing a horned but beardless deity with a rudimentary palmette above the head and tendrils below the chin which, though interpreted by Professor Toynbee as a Celticised Medusa, may rather be thought to be a relation, distant perhaps and somewhat on the wrong side of the blanket, of the heads on the Palatine and Brumath antefixes. Moulds for two antefixes were found associated with third- and fourth-century pottery from three kilns at Stibbington in the Nene Valley; one of these shows Mercury and the other an archer (JRS xlviii, 1958, p. 139 and pl. xx, nos. 3 and 4; Toynbee 1964, 426-427).

In the context of architectural decoration in Britain as a whole, it may be thought misleading to regard the series of antefixes from the colonia at York as representative of civilian building practice, in the sense that the town's architectural traditions stemmed essentially from the adjoining military presence.

As well as one or two interesting figural scenes, one showing a man taking a fish from a net, the regular types consist of a mask set within a surround like a lion's mane, and a long-necked female framed in a sort of canopy which can best be understood as a version of the moulding with spiral terminals around the heads on the Etrusco-Italic shell-antefixes (discussed above), with a border of stylised foliage (RCHM, 1962, 114 and pls. 38 and 39).

George Boon, in the first edition of his Roman Silchester, appears to have been the first to make the point that such antefixes have not been found in sufficient quantity to demonstrate that they were placed along the eaves, in the manner of the antefixes considered at the beginning of this paper, but rather that they decorated the gable ends of a building (Boon, 1957, 148). As he has since pointed out, this applies also to military sites, and the adoption of the pentagonal form, as at Neuss, Silchester and York, or, omitting the vertical sides, a triangular shape, as at Caerleon and Chester, might not be thought inappropriate for a tile covering the end of the ridge-pole at the apex of the roof (Boon, 1972, 125-126, note 75).

At Caerleon we find a number of inventive treatments, with the use of various sorts of head, surrounded by a variety of signs and a cable or reeded border (Boon, 1972, 27 and fig. 14). On two, the name of the Second Legion Augusta appears, but none has the legionary symbol which, in the form of the boar, decorates the series from Chester which was manufactured in the legionary tilery at Holt (Grimes, 1930, 136-139 and fig. 58). The Caerleon heads include horned faces, and the overall impression is one of Celtic inspiration, though that cannot really be said of the fishwife, an apparently female figure holding two dolphins by the tail.

These dolphins, in fact, first appear without the intermediate figure, balanced above a rosette. Antefixes from the same mould have also been found in excavations of the legionary baths at Exeter. At both fortresses they were discovered in early construction levels—at Exeter dated to about A.D. 60—as well as in later layers. Bidwell and Boon have pointed out, in a joint note (1976), that the implication of the finds is that the mould from which they were made was taken from one site to the other, to be used at Caerleon some fifteen years later than it was at Exeter. The discoveries document precisely the sort of circumstances outlined above in discussion of the antefix moulds from Selvasecca. The possibility should also be mentioned that the actual mould was not necessarily the same, but that separate moulds made from the same prototype might have been used.

Antefixes found associated with military buildings at the Second Legion Augusta's other fortress, Gloucester, are not paralleled at Caerleon (Hassall and Rhodes, 1974, 80 and fig. 31). They, as also those from York, were applied to the end of an imbrex tile, though that does not determine whether they were used along the eaves or, as Boon suggests, at the end of the ridge, since imbrices would have been employed in both positions.

Another possible element of terracotta roof furniture which in Britain is found mainly on civilian sites is the cylindrical or tower-like object with pierced side-openings, which has been variously interpreted as a finial, a chimney or a lamp-cover. While in certain cases a non-architectural function cannot be ruled out or may even be preferable, there can be no doubt that such examples as those from Norton, East Yorkshire, and Silchester were either chimneys or finials, since they are attached to part of a ridge-tile (Lowther, 1976, 39 and pls. ii-iv). Lowther's review of the British material (1976) also cites and gives bibliographical references for similar products of the Rhine and Danube provinces.

Finally, we may turn to the decorative use of brickwork in the western provinces. In a sense, as has already been seen, the columns composed of segmental bricks can be described as a decorative as well as a structural use of terracotta, though we may doubt whether the brick was left exposed. It is not surprising to find bricks being used for this purpose at Verulamium, a town where building stone was not immediately to hand (Wheeler and Wheeler, 1936, 132), more so that they were employed in fourth-century work at Caerleon (Boon, 1972, 68-69), a use which may be explained as reflecting the absence of competent stonemasons at that date rather than a scarcity of building stone. It would not be difficult to multiply examples, but what is really needed is a much less eclectic exercise, one which would study in detail the geographical and chronological range of this aspect of construction and the types of building in which it found favour.

The use of engaged columns, pilasters and pediments of brick in the manner of Trajan's Market and other second-century buildings in Latium does not seem to have been provincial practice, though in the amphitheatre at Bordeaux, for which Etienne considers a Severan date probable (1962, 193), brick was also used for pilaster capitals and cornice mouldings. The feature of the amphitheatre's construction more characteristic of provincial architecture, however, is the combination of coursed rubble masonry, petit appareil, with two or three courses of tile at intervals of about a metre.

The origins and distribution of this type of construction have yet to be investigated systematically, and such a study would be a valuable contribution to the history of western provincial architecture. Its use in Italy was a variation of the use of courses of brick in conjunction with opus incertum and opus reticulatum. Double courses of brick alternate with sextuple courses of rubble in the walls of the Vespasianic capitolium at Brescia (Lugli, 1957, tav. clix, 3), and with nine to twelve rubble courses in the theatre at Ventimiglia, not well-dated but probably later, possibly considerably so, than the sack of the town by Otho's troops in A.D. 69 (Lugli, 1957, tav. cxcii, 1). Such north Italian sites probably mark the route by which the technique was carried into the western provinces. In third- and early fourth-century buildings in Italy, however (and their medieval successors), the alternation of brick with only one or two courses of rubble is more characteristic (Lugli, 1957, 644).

In the provinces, the use of petit appareil with brick courses become established during the second century, though the details of the process await full study. It was employed in the Odeum at Lyons, thought to have been built in the reign of Antoninus Pius (Audin, 1965, 109 ff.). Most of all, however, it is typical of third-century fortification walls. Those of London, dated to about A.D. 200, provide a relatively early example (Merrifield, 1969, 117-120), but most of all it was the response to barbarian attacks during the course of the third century that led to its proliferation, both in town walls, such as those of Soissons, Jublains and Le Mans (Grenier, 1931, 509-513, see also Johnson, 1973), and in the coastal forts of north Gaul and south-eastern Britain (Johnson, 1976, chapters 3 and 5, passim).

In the monumental buildings of the fourth century that owed their stimulus to the imperial presence outside Italy, we find the Italian style of construction, with no more than three courses of stone between those of brick. The Baths at Arles (Constans, 1921, 273-277 and pl. vi) and the Kaiserthermen at Trier (Krencker and Krüger, 1929) thus reflect the imperial architecture of the Circus and Palace of Maxentius on the Via Appia at Rome (Lugli, 1957, tavv. cxciv, 3 and cxcv, 2), not only in that feature, but also in their use of brick-faced arches.

It is, however, with another building at Trier, the Aula Palatina, that we may conclude a survey which has covered in outline the use of terracotta in architectural decoration through a millenium. Its two rows of windows are set in arched recesses which rise almost to the full height of the building. Its exterior relies for its effect on the spacing and proportions of these structural elements (Wightman, 1970, 103ff. and 273). It is one of the first examples of an effective external treatment of a brick building which does not depend on the use of columns, pilasters or pediments, remnants of an older tradition of architecture in timber and stone divorced from their original structural function.

One should perhaps draw a distinction between the decorative use of brick as such and the decorative aspects of the design of a brick building; particularly since in this case the brickwork was faced with stucco, and the effect could also have been achieved in other materials—ashlar masonry, for example. Even so, the development depended on a long history of the exploration of the architectural possibilities of brick as a facing material for concrete. It was also to have a profound influence on the architecture of late antiquity, since it was at precisely this time that the needs of the newly-acceptable Christian religion for places of worship and assembly were met by the adaptation of the basilican hall for the purpose. The great brick-walled basilican churches of sixth-century and later Ravenna are in that sense the architectural heirs of a long history of the use of terracotta, even though by that time many of the former decorative uses of that material had expired without issue.

In the long span of Roman architectural prehistory and history covered by this paper it has not been possible to discuss many of these decorative uses in much detail. Nevertheless, it may be thought useful to have provided a perspective view of the changes which took place, changes which do not merely reflect the vagaries of fashion. Dependent on the more fundamental developments in building construction which are one of the great contributions of Rome and its empire to civilisation, the terracottas, tiles and bricks which decorated Etrusco-Italic and early Roman temples, Antonine mausolea, Ostian storehouses, military barracks, Gallic fortifications and Christian basilicas document the changing fortunes and social needs of the Roman world.

ACKNOWLEDGMENTS FOR ILLUSTRATIONS

Fig. 14.1(1) is after P. Orsi, Notizie degli Scavi 1897, p. 345 (= van Buren, 1923, pl. iv, fig. 13).

The remaining drawings are based on published photographs:

Fig. 14.1(2) = van Buren, 1923, pl. xv, fig. 62.
1(3) = Winter, 1978, taf. 18.2.
1(4) = van Buren, 1921, pl. xxix, fig. 2.
2 = Carettoni, 1972, figs. 6 and 10.

Plate 14.I(1) is the author's photograph; plate 14.I(2) was kindly supplied by Dr. M. B. Lyttelton.

BIBLIOGRAPHY

Akerström 1974: A. Akerström, Die Architektonischen Terrakotten Kleinasiens (Acta Instituti Atheniensis Regni Sueciae 4, xi), Lund, 1974.

Andrén 1940: A. Andrén, Architectural Terracottas from Etrusco-Italic Temples, Lund-Leipzig, 1940.

Andrén 1974: A. Andrén, 'Osservazioni sulle terrecotte architettoniche etrusco-italiche', Opuscula Romana, 8 (1974), 1-16.

Audin 1965: A. Audin, Lyon, Miroir de Rome dans les Gaules, Paris, 1965.

Berggren and Andrén 1969: E. Berggren and A. Andrén, 'Blera (Località Selvasecca): Villa rustica etrusco-romana con manifattura di terrecotte architettoniche', Notizie degli Scavi, 23 (1969), 51-71.

Bidwell and Boon 1976: P. Bidwell and G. C. Boon, 'An Antefix Type of the Second Augustan Legion from Exeter', Britannia, 7 (1976), 278-280.

Blake and Bishop 1973: M. E. Blake, edited and completed by D. T. Bishop, Roman Construction in Italy from Nerva through the Antonines, Philadelphia, 1973.

Boon 1957: G. C. Boon, Roman Silchester: the Archaeology of a Romano-British Town, London, 1957.

Boon 1972: G. C. Boon, Isca, Cardiff, 1972.

Borbein 1968: A. H. Borbein, Campanareliefs: typologische und stilkritische Untersuchungen (Mitteilungen D.A.I. Röm., Erganzungsheft 14), Heidelberg, 1968.

Broneer 1955: O. Broneer, 'Excavations at Isthmia, 1954', Hesperia, 24 (1955), 110-141.

Brown, Richardson and Richardson 1960: F. E. Brown, E. H. Richardson and L. Richardson, Jr., Cosa II, The Temples of the Arx (MAAR 26), Rome, 1960.

Campana 1842: G. B. Campana, Antiche opere in plastica, Rome, 1842.

Carettoni 1972: G. Carettoni, 'Terrecotte "Campana" dallo scavo del tempio di Apollo Palatino', Rendiconti, 44 (1972), 123-139.

Constans 1921: L. A. Constans, Arles Antique, Paris, 1921.

Espérandieu 1918: E. Espérandieu, Recueil-Général des Bas-Reliefs, Statues et Bustes de la Gaule Romaine, vol. 7, Paris, 1918.

Etienne 1962: R. Etienne, Bordeaux Antique, Bordeaux, 1962.

Grenier 1931: A. Grenier, Manuel d'Archéologie Gallo-Romaine, vol. V(1), Paris, 1931.

Grimes 1930: W. F. Grimes, Holt, Denbighshire: the Works Depot of the Twentieth Legion at Castle Lyons, Y Cymmrodor, 41, 1930.

Hassall and Rhodes 1974: M. W. C. Hassall and J. F. Rhodes, 'Excavations at the New Market Hall, Gloucester', T. Bristol and Gloucester A.S., 93 (1974), 15-100.

Hull 1958: M. R. Hull, Roman Colchester (Research Report of the Society of Antiquaries of London XX), Oxford, 1958.

Johnson 1973: S. Johnson, 'A Group of Late Roman City Walls in Galla Belgica', Britannia, 4 (1973), 210-223.

Johnson 1976: S. Johnson, The Roman Forts of the Saxon Shore, London, 1976.

Kammerer-Grothaus 1974: H. Kammerer-Grothaus, 'Der Deus Rediculus im Triopion des Herodes Atticus', Mitteilungen D.A.I. Röm., 81 (1974), 131-252.

Koch 1912: H. Koch, Dachterrakotten aus Campanien, Berlin, 1912.

Krencker and Krüger 1929: D. Krencker and E. Krüger, Die Trierer Kaiserthermen, Augsburg, 1929.

Krüger 1972: M. -L. Krüger, Die Reliefs des Stadtgebiets von Carnuntum, ii: Die dekorativen Reliefs (Corpus Signorum Imperii Romani, Osterreich I.4), Vienna, 1972.

Lowther 1976: A. W. G. Lowther, 'Romano-British Chimney-Pots and Finials', Antiq.J., 56 (1976), 35-48.

Lyttelton and Sear 1977: M. B. Lyttelton and F. B. Sear, 'A Roman Villa near Anguillara Sabazia', PBSR, 45 (1977), 227-251.

Maiuri 1929: A. Maiuri, Pompeii, Novara, 1929.

Merrifield 1969: R. Merrifield, Roman London, London, 1969.

Nielsen and Phillips 1974: E. Nielsen and K. M. Phillips, Jr., 'Bryn Mawr College Excavations in Tuscany, 1973', AJA, 78 (1974), 265-278.

Novaesium 1904: 'Novaesium', plates to Bonner Jhrb. 111, 112 (1904).

RCHM 1962: Royal Commission on Historical Monuments, An Inventory of the Historical Monuments in the City of York, I: Eburacum, Roman York, London, 1962.

Rhodes 1964: J. F. Rhodes, Catalogue of the Romano-British Sculpture in the Gloucester City Museum, Gloucester, 1964.

Roebuck 1955: M. C. Roebuck, 'Excavation at Corinth: 1954', Hesperia, 24 (1955), 147-157.

Rolland 1946: H. Rolland, Fouilles de Glanum I (Gallia, 1e suppl.), 1946.

Rolland 1958: H. Rolland, Fouilles de Glanum II (Gallia 11me suppl.), 1958.

Shoe Merritt, 1970: L. Shoe Merritt, 'Architectural Mouldings from Murlo', Studi Etruschi, 38 (1970), 13-25.

Staccioli 1968: R. A. Staccioli, Modelli di edifici etrusco-italici: i modelli votivi, Rome, 1968.

Toynbee 1964: J. M. C. Toynbee, Art in Britain under the Romans, Oxford, 1964.

Van Buren 1921: E. D. van Buren, Figurative Terracotta Revetments in Etruria and Latium in the VI and V Centuries B.C., London, 1921.

Van Buren 1923: E. D. van Buren, Archaic Fictile Revetments in Sicily and Magna Graecia, London, 1923.

Wheeler and Wheeler 1936: R. E. M. Wheeler and T. V. Wheeler, Verulamium: A Belgic and Two Roman Cities (Research Report of The Society of Antiquaries of London XI), Oxford, 1936.

Wightman 1970: E. M. Wightman, Roman Trier and the Treveri, London, 1970.

Wikander 1974: C. Wikander, 'Etruscan Roofing-tiles from Acquarossa', Opuscula Romana, 8 (1974), 17-28.

Winter 1978: N. A. Winter, 'Archaic Architectural Terracottas decorated with Human Heads', Mitteilungen D.A.I. Röm., 85 (1978), 27-58.

Addendum:

Carandini and Settis 1979: A. Carandini and S. Settis, Schiavi e padroni nell'Etruria romana, Bari, 1979.

TILES AS A STRUCTURAL COMPONENT IN BUILDINGS

Graham Webster

When I saw a draft programme for this Conference, I realised that many kinds of tiles were to be discussed except those most commonly found. When I rashly expressed my misgivings over this lamentable gap, Alan McWhirr immediately suggested that I should fill it. Thus I have become responsible for a rather mundane paper on the tile as a structural rather than a decorative unit. In near-eastern countries it was customary from the earliest times to build with sun-dried mud and clay, especially in ancient Babylon and Assyria from the third millennium B.C.[1] The materials were usually ready to hand, little skill was required and the hot sun did all the drying and baking. This time-honoured technique is still practiced in rural areas in many countries today and one often sees the wooden moulds being filled, usually by women, with stiff clay and slurry from river banks. It is so much easier and cheaper than hewing blocks from stone in quarries, trimming and dressing and transporting them to the building site. In many countries too there was little available building stone. This especially applied to Italy, where the Etruscans adopted the idea of using tiles from the Greek colonists. By the late Roman Republic there were many large tile works supplying the needs of the growing Roman cities and several leading families became very rich on the profits.[2] One has only to visit the large town of Ostia, the port of Rome, at the mouth of the Tiber to appreciate the full extent to which the tiles were used; almost all the buildings are so constructed throughout and the same is true of Pompeii and Herculaneum, whereas in Rome, the wealth of the early Empire made it possible to disguise the tilework with stone and marble cladding, hence that famous remark of Augustus, 'I found Rome built of sun-dried bricks; I leave her clothed in marble'.[3]

The advantages of tiles are very great, one can build easily to line and level and form precise corners and by flat striking of joints create a smooth enough surface to take an economically thin layer of plaster in comparison with the four of five inches thickness needed for undressed stonework, as so often seen in Britain.

In view of this wide-spread use of mud-bricks and tiles in Italy and many of the provinces, one may well ask why the same is not true of Britain, although London may have been an exception.[4] The answer to this is that most of the construction work of the first century was in timber throughout, and even when these buildings were replaced in the second century, almost all the houses remained in this material, the only difference being that they were then built on stone or clay foundations. Only the larger public buildings were constructed differently since their size demanded stronger material, but stone was preferred to tiles, presumably because it was plentiful. An attempt was made to introduce tiles in Britain in the first century, as the

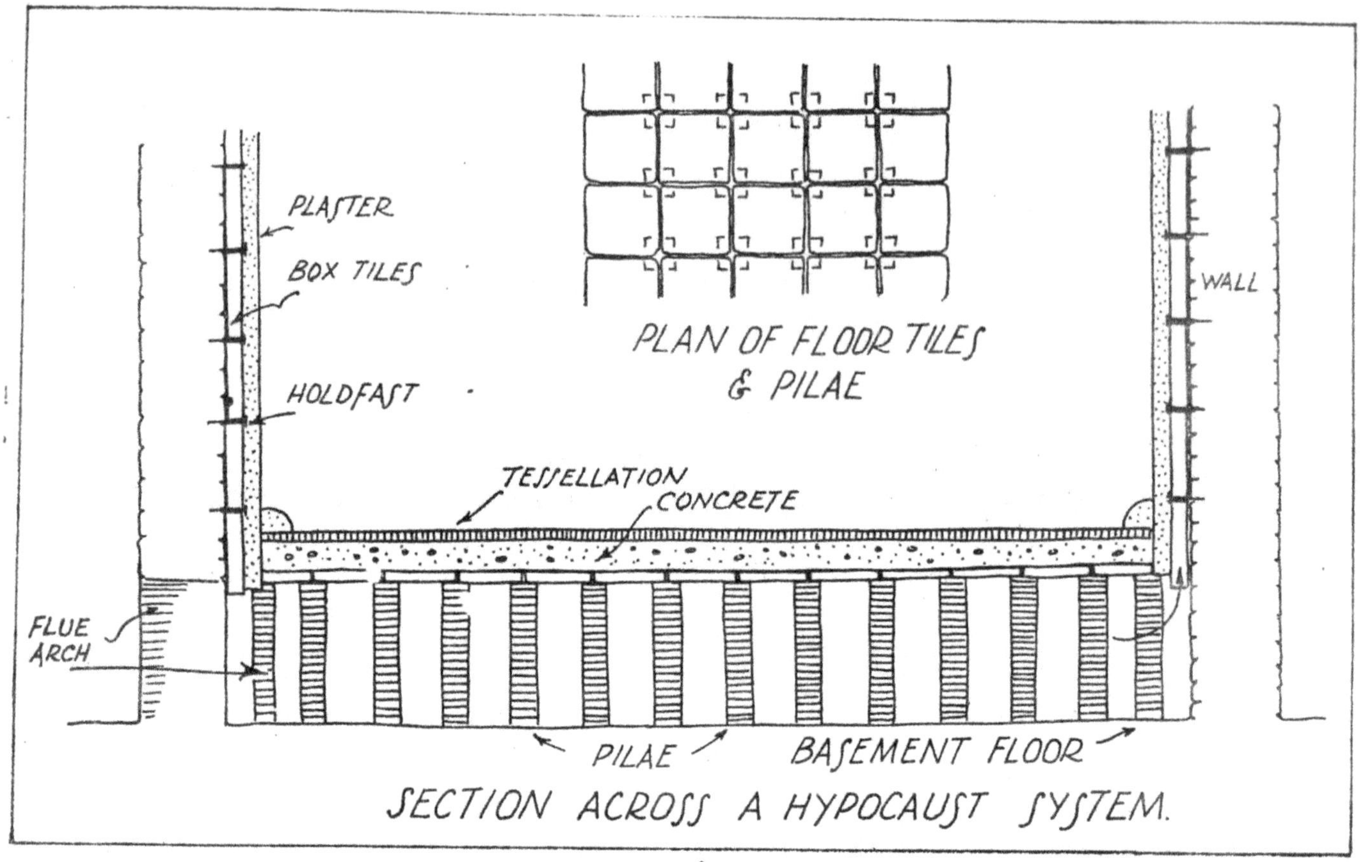

Fig. 15.1

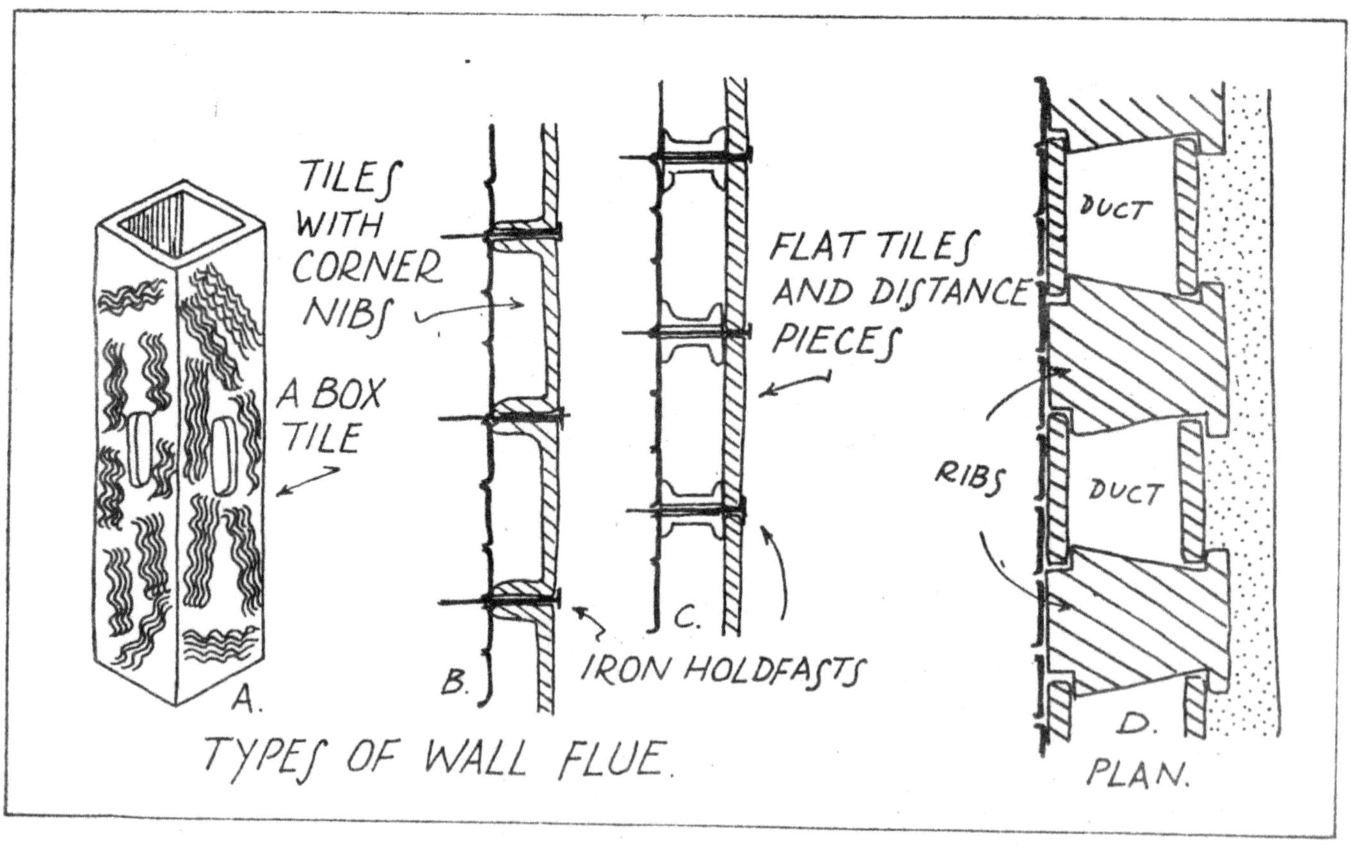

Fig. 15.2

Neronian tile-stamps from Silchester indicate.[5] The only large building in Britain of this early period of development to be extensively studied is the Fishbourne Palace.[6] The building materials were mainly nodular flints in mortar for the foundations and stone for the ashlar blocks which came from the limestone beds of the Mixon reef seven miles away; they also obtained sandstone blocks from the Weald. For the dressed stone features, the Bembridge limestone from the Isle of Wight was selected, although stone for some of the columns may have been brought from Caen.[7] Imported marble was used for fine details and wall veneers and inlays. Tiles were also used extensively, but only in subsidiary structures such as the bath-house, the floors, roofs and vaulting.[8] This clearly demonstrates that the use of tiles in Britain was on a more limited scale than in Italy, and that this was due entirely to the presence of fine building stone in this country.

My brief is to consider tiles as a basic construction component. There are two main advantages which tile had over stone, firstly, units could be made to predetermined shapes and sizes and secondly, the material can withstand frequent and considerable changes of temperature, under conditions when most stones disintegrate. It is noteworthy that with the possible exception of the Exeter bath-house[9] the few military tiles known in Britain before the late-Flavian period appear to have been used entirely for hearths. Tiles were designed for very precise functions and nowhere is this better illustrated than in the bath-house, where they also offered the second advantage. They were an essential material for hypocaust construction and this is proved by the extensive robbing of hypocausts and tile-lined drains in the late fourth and probably early fifth centuries, when, for reasons not yet understood, the industry like that of the potters', had ceased to function. Most disused buildings in town and country all over Britain were subjected to this intensive and often difficult campaign of tile-recovery, in order to provide the necessary materials for repairs to old, and the erection of new bath-houses, although it must be admitted the archaeological evidence is as yet sparse.[10] Great precision was needed in setting out the supporting columns of pilae for the hypocaust-basement floors, so that the corners of four sub-floor tiles could be placed symmetrically on the top tile of each pila to form a perfectly flat surface without any gaps between the tiles. This was the base on which the concrete was poured, creating the thick floor which was normally surfaced with stone slabs or tessellation. This kind of construction could only be done with tiles made to exact standards of size and shape. In large public bath-houses such precision was even more necessary, since the sub-floor tiles were normally at least 2 feet square and needed an extra tile at the top of each pila, larger than the other pila tiles to give a great area of support (Fig. 15.1). One imagines a considerable wastage at the factory in the manufacture of these very large tiles to ensure that each was completely flat, no distortion could be tolerated.[11]

The subject of hypocaust flues deserves a separate paper, since the wide variety of forms which have been found demonstrates the great ingenuity employed in overcoming the problems of creating vertical air channels in an efficient and economical manner (see Chapter 17, Ed.) They were needed to prevent condensation on the walls and roof and to give an even distribution of heat throughout the room. The most expensive type was undoubtedly the box-flue (tubulus) which was fixed in position with long hold-fast nails to form a complete jacket (tubulatio)[12] to each wall of a caldarium and the very hot room

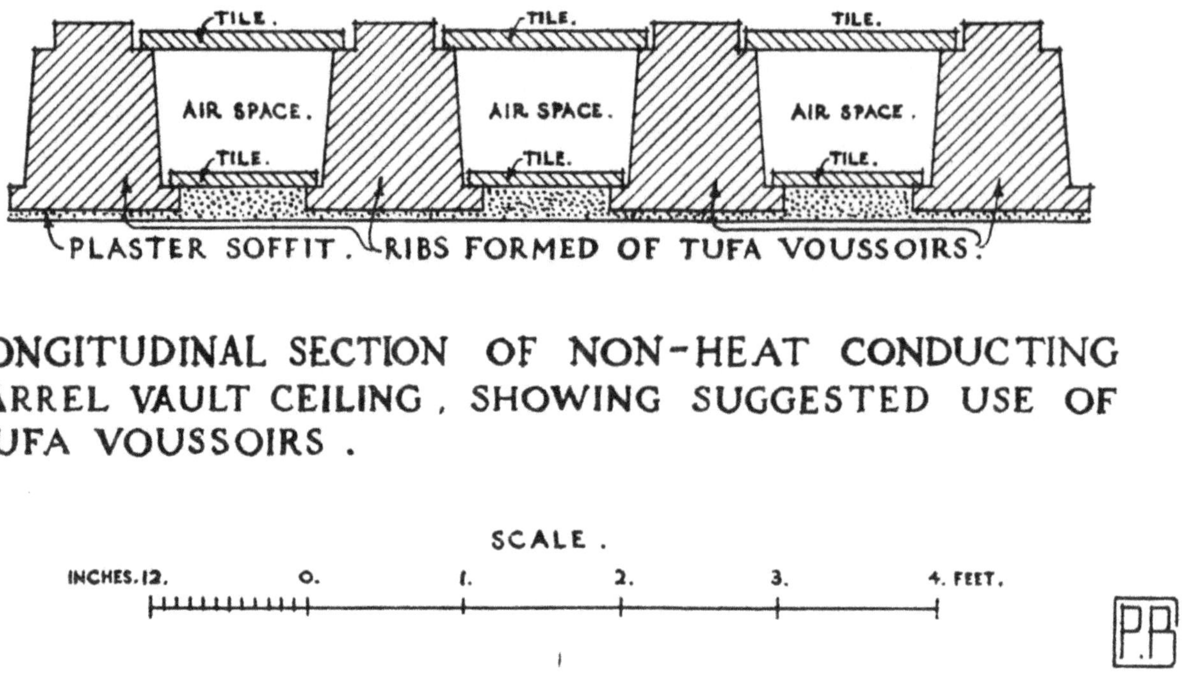

Fig. 15.3

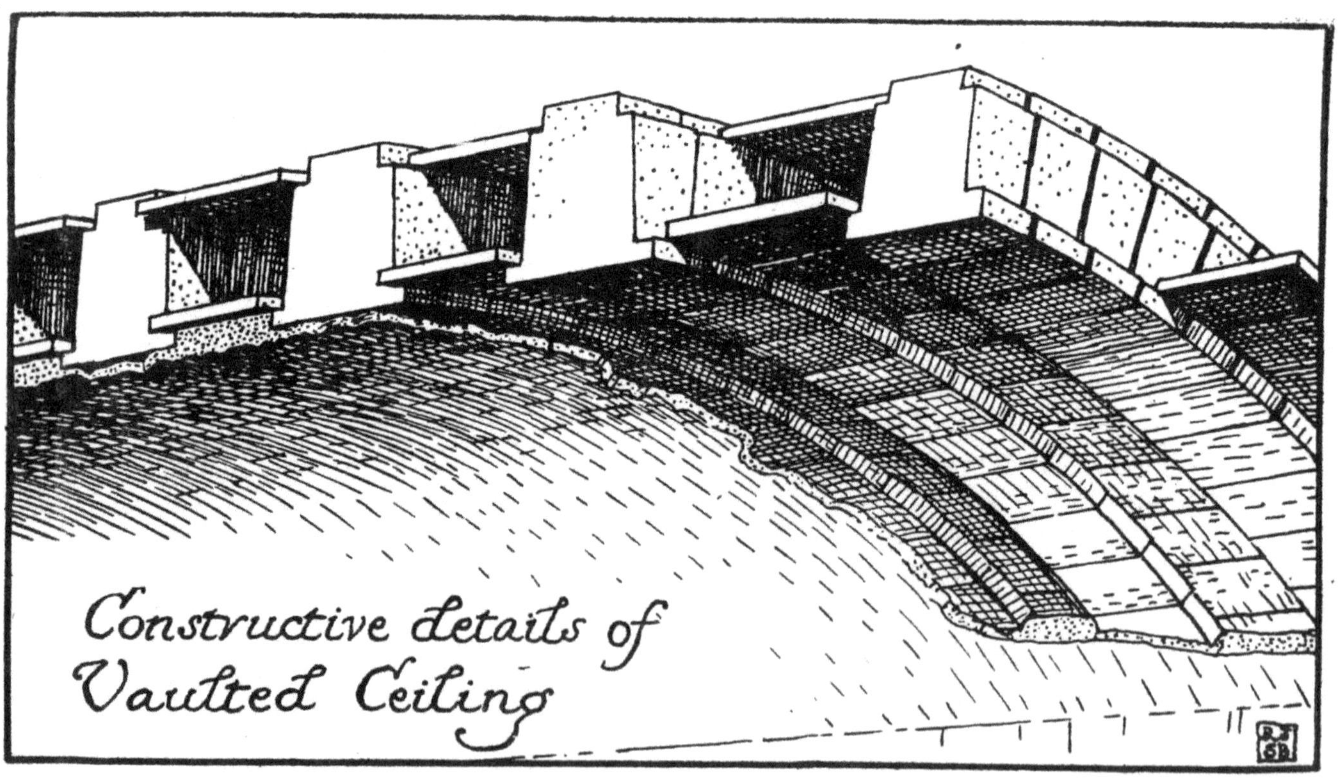

Fig. 15.4

Both drawings by Parker Brewis and reproduced by kind permission of the Society of Antiquaries of Newcastle-upon-Tyne.

in a laconicum suite (Fig. 15.2). The hot air entered the system immediately below the floor and rose in the vertical flues to the spring level of the barrel-vault or semi-dome where all the flues connected with a horizontal channel from which it found its way into roof flues spaced at intervals, as described below. The same effect was achieved far more economically with flat tiles held away from the wall by spacers (known as bobbins from their shape (Fig. 15.2)) to provide the hot-air jacket; this system may also have been more efficient since it would have allowed the hot air to circulate more freely.

An alternative to the bobbin spacer were small lugs, or lumps attached to the corners of the tile (mammatae) but they provided a much narrower air space than the tubulatio.[13] In tepidaria and heated living-quarters in a house, the hot air rose in the walls by means of channels cut into them at spaced intervals and at corners. Box-flues were inserted, but more often the plasterers merely fixed an upright piece of wood in the open side of the channel to act as form-work. Apart from the problem of condensation, the vertical flues were necessary to provide a draught to draw the gases from the praefurnium evenly throughout the sub-floor space, and this in itself was no easy achievement; as witnessed by the blocking and other alterations to flues one often finds. The main difficulty in heating a bath-house was the roof, where tiles were extensively used for vaulting and arches. One can but sympathise with the bathers in the old Republican period bath-houses of Rome suffering the cold drips of condensation falling on their hot bodies from above. At Pompeii, this problem was partly solved by the barrel-vaulting having a corrugated surface providing channels for the water to run down into a drainage gutter at the level of the springing of the arch, It was a serious drawback, as Seneca noted in one of his letters to Lucilius, when he wrote of 'that dark bath covered with a common sort of roof' (Ep. LVI) and in another (Ep. XC) enthused over the improvements to insulation with the introduction of the tubulatio, after which one only visited the old establishment for the romantic thrill of bathing where Cato bathed. But to heat the ceiling of the barrel-vault was a difficult technical problem, since the normal box-flues could only be used where the span was great enough to accommodate easily a semi-circle of the straight 2 feet lengths. With smaller spans special box-flues would have been needed and even made to fit the curve, and these were often voussoir box-flues, built in the form of projecting ribs in the vaulting. Greater ingenuity was shown with the so-called armchair tile which was wedge-shaped with the projecting lugs at the narrow end and small rectangular notches cut out of the corners of the wider end.[14] These tiles were constructed as solid ribs alternating with spaces to provide the hot-air ducts which were formed by a pair of flat tiles seated in the niches provided (Fig. 15.3); although in fact against a smooth surface, only the outer one was needed. Roof ducts also posed problems of position for the vents through which the hot gases escaped into the air and how they were controlled. This is a very important aspect, since if the draught became too fierce the caldaria could rapidly be brought to an unbearable heat. For comfort and safety, one can but echo the words of Statius,[15] 'The languid flame wanders slowly and aimlessly through the hypocaust', rather than those of Ausonius who indulged too freely in poetic licence.[16] Nonetheless stoking up the great city thermae at the beginning of the day may have produced a cloud of smoke, a point used by Martial in one of his epigrams.[17] It is very rare that a Roman bath-house has survived with

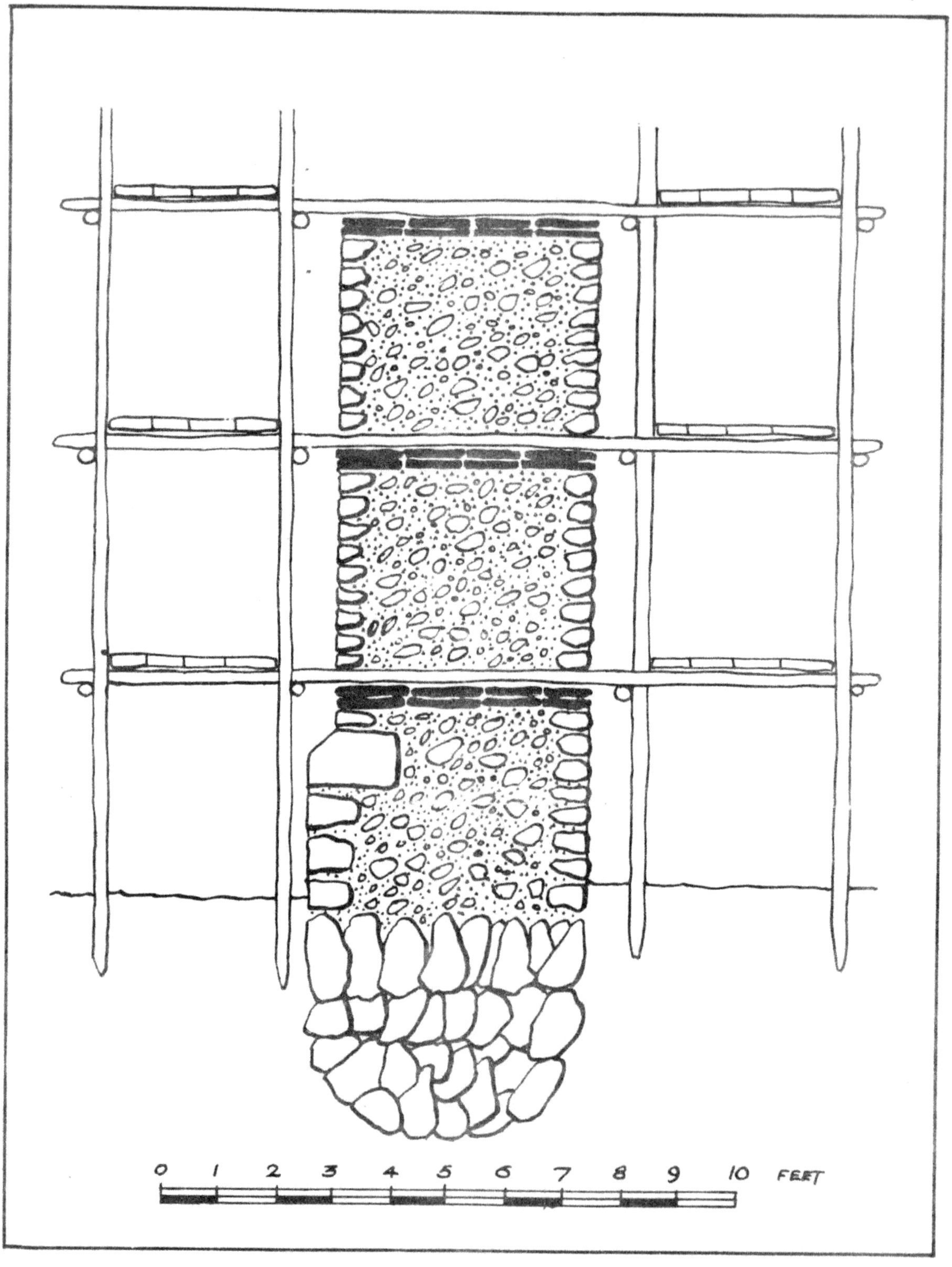

Fig. 15.5 Section across a large stone wall showing tile courses and building staging

its roofs intact, together with the ducts and vents to demonstrate how the system was organised and controlled;[18] there are, of course, late Byzantine examples still standing in the Eastern provinces, which could still perhaps be studied with profit.

The most extensive use of tiles in Britain was for roof construction with the flat (tegulae) and curved tiles (imbrices). This has been so well and so often documented, that there is no need to dwell upon it except for two points. First there is the problem of how the tiles were attached to the roof timbers. Their great weight would have given them a tendency to slide, so some firm anchorage was necessary, even in the lowest pitch. According to the contract specification surviving from Puteoli[19] iron was used for fixing the first row of tiles to the fascia, and the most sensible means would have been the use of stout brackets rather than nails. The second point to be made is that one imagines difficulties in tiling curved surfaces such as domes and vaults, but a study of the roofs of churches at Ravenna and elsewhere, shows how skilfully and ingeniously these problems were overcome, proving that the tegulae-imbrices provided a very flexible system (for a further discussion on roof tiles see Chapter 16, Ed.). Two other important uses of tiles were in drains, since it was difficult to make pipes of adequate diameter, and also one presumes tile would be more durable than stone. Tiles were used extensively to provide the floors and linings for the sides of drains, but the cover-slabs are normally of stone. Secondly, tiles were very useful as floor covering in the form of small tiles set on edge in a herring-bone fashion.

Finally, one must consider another major use of tiles in thick wall construction, of which there are many examples in Britain. Tile-laced walls in which this method was adopted were either for defences, or as main structural elements in large public buildings. Tiles were laid in courses of two or three thick at vertical intervals of three to four feet (Fig. 15.5). Their function was two-fold; first they marked the staging levels where the masons could have completed a convenient height, without undue bending and stretching, then the wall was levelled off with tiles before the staging was next raised. If enough of these walls survive, as in the case of 'Old Work' at Wroxeter,[20] the tile courses and putlog holes should be studied in conjunction, and then the whole building scheme becomes clear. At Wroxeter there is even an odd hole at the upper stage to take an individual post from which the pulley block and tackle must have been suspended. The most important function of these tile-lacing courses was to bind the wall together especially when the concrete was still 'green'. These large walls consisted of two thin skins of dressed masonry, often of rather small stones and the core of mass concrete, the tile courses being to the full thickness of the wall and holding the whole together laterally. This was even more essential in walls which were wholly of mass concrete, surfaced on the outside by sliced-flint nodules, as at Silchester, Verulamium and elsewhere.

It must be clear from this Conference that the tile factories of Roman Britain were a large and important section of the building industry, which was variously organised by private firms, civic authorities and the army. It is a subject which has long been neglected and now we hope it will begin to receive the serious detailed studies it merits, especially with regard to its social and economic aspects and the fascinating problems of transporting such heavy and awkward materials.[21]

NOTES

1. Davey, 1961, 67-69.
2. <u>CIL</u> iii and v for the stamps of their names.
3. Suetonius, <u>Augustus</u>, 28.
4. Lethaby, 1923, 15-16.
5. <u>Archaeologia</u> 59 (1905), 366.
6. Cunliffe, 1971.
7. <u>Ibid</u>. i, 80-82; ii, 1-16.
8. <u>Ibid</u>. ii, 43-49.
9. Bidwell, 1979.
10. For evidence from Verulamium, Frere, 1958, 8-9; Dorchester-on-Thames, Frere, 1964, 121-122 and Catterick, Frere 1966, 96-97.
11. Vitruvius recommended sloping the basement floor towards the <u>praefurnium</u> to assist the upward passage of the hot air (<u>de architectura</u>, Bk. V).
12. Davey, 1961, 200-202.
13. Ward, 1911, Fig. 85.
14. This system was first understood by Parker Brewis at Chesters where the tiles had been cut from tufa (Macdonald 1931, 280-284 and Fig. 8).
15. <u>Silvae</u>, i, 5, 58; <u>ubi languidas ignis inerrat dedibus tenuem volunt hypocausta vaporem</u>.
16. <u>Mosella</u> 337 '... Vulcan drawn by the glowing flue pants forth the flames and whisks them up through the channelled walls rolling in masses the imprisoned smoke before the scorching blast...' (Loeb trans.).
17. <u>Epig</u>. x, 48; 'It is 11 o'clock...for smoke is pouring from the baths of Nero'.
18. Although some of these details survived at the Lepcis Hunting Baths they were not fully described in the report (Ward-Perkins 1949).
19. <u>CIL</u> i, 577: Donaldson, 1833, 32 ff.
20. Webster and Woodford, 1966, 238-239 and Fig. 2.
21. A recent C.B.A. Conference has drawn attention to the greater use than has been recognised of small rivers and streams with the aid of the narrow punts and barges (Boe, 1978).

BIBLIOGRAPHY

Bidwell, P. T., 1979. The Legionary Bath-House and Basilica and Forum at Exeter, Exeter Archaeol. Rep. vol. i.

Boe, G. de, 1978. 'Roman boats from a small river harbour at Pommeroeul, Belgium' in J.-du Plat Taylor and H. Cleere, Roman Shipping and Trade: Britain and the Rhine provinces, CBA Research Rep. No. 24, 22-30.

Cunliffe, B., 1971. Excavations at Fishbourne, Soc. of Antiq. Research Rep. No. 26.

Davey, N., 1961. A History of Building Materials.

Donaldson, 1833. 'A Collection of the most approved examples of Doorways.

Frere, S. S., 1958. 'Verulamium Excavation Committee, Third Interim Report', Antiq. J. 38, 1-14.

Frere, S. S., 1962. 'Excavations at Dorchester-on-Thames, 1962', Archaeol. J., 99, 114-149.

Frere, S. S., 1966. 'The End of Towns in Roman Britain', in J. Wacher (ed.), The Civitas Capitals of Roman Britain, 87-100.

Lethaby, W. R., 1923. Londinium: Architecture and the Crafts.

Macdonald, Sir G., 1931. 'The Bath-House at the Fort of Chesters (Cilurnum)', Archaeol. Aeliana 4th ser., 8, 219-304.

Ward, J., 1911. Romano-British Buildings and Earthworks.

Ward-Perkins, J. B., 1949. 'The Hunting Baths at Lepcis Magna', Archaeologia, 93, 165-196.

Webster, G. and Woodfield, P., 1966. 'The Old Work at the Roman Public Baths at Wroxeter,' Antiq. J. 46, 229-239.

TILED ROOFS

Tony Rook

a) A NOTE ON PITCHED ROOFS

This brief note is prompted by a common attitude, apparently even shared by some people at the conference, that "the past is a foreign country—they do things differently there". This sentiment of L. P. Hartley is excusable in sociological affairs, and the world of the novel, but in technological areas it holds the inherent dangers of the cult of Von Danniken, where the past is only possible if we suspend all belief in physical laws and common sense. With Von Danniken we can ask pseudo-rhetorical questions which imply by their very nature that no answer is available. Of tiled roofs such questions are "why don't we find traces of the method used to fasten tiles on the roof? Why don't we find more ridge tiles? Whatever happened to all those antefixes?".

I must say that I had always, until this conference, considered the answers to be self-evident. It is for that reason, paradoxically, that they are rarely, if ever, given, and when I came to write this note, I found little published material to help me. Therefore what follows is under-researched and interim, being based upon my own notes in the field; in particular I am unsure how representative my examples are. Pending further observations, however, I feel that some note, however brief, should be included in the conference papers with the intention of suppressing the attitude that such information is unknown, unknowable, or not susceptible to logical thought.

Roman *tegula*- and *imbrex*- roofs were shallow pitched, and even if we did not have surviving examples at Herculaneum, we have a clue from Vitruvius that the pitch was about 20°. He is describing a temple of the Tuscan order, and Roman temples generally seem to have a pediment with this pitch (e.g. Maison Carré, Portico of Octavia, Pantheon). The pitch of Greek pediments seems to be lower at about 15°.

Present day *tegula*- and *imbrex*- roofs in Italy have a slope of about 20° except where they cover a vault, as in churches, where a more usual slope is about 30°. Clearly the weight of the tiles is quite sufficient to withstand wind suction except under extraordinary conditions. They will not slip down the slope under gravity either, since friction will hold them in place. An improvised experiment with a wide range of Roman tiles on site showed that the angle of repose on planed wooden boards was about 40°, and on other tiles about 35°. Clearly the angle chosen for the roofs was such that the tiles were stable without fastening, and we should not be surprised that few tiles are nailed. In fact where nails do occur, a reason should be sought for this departure.

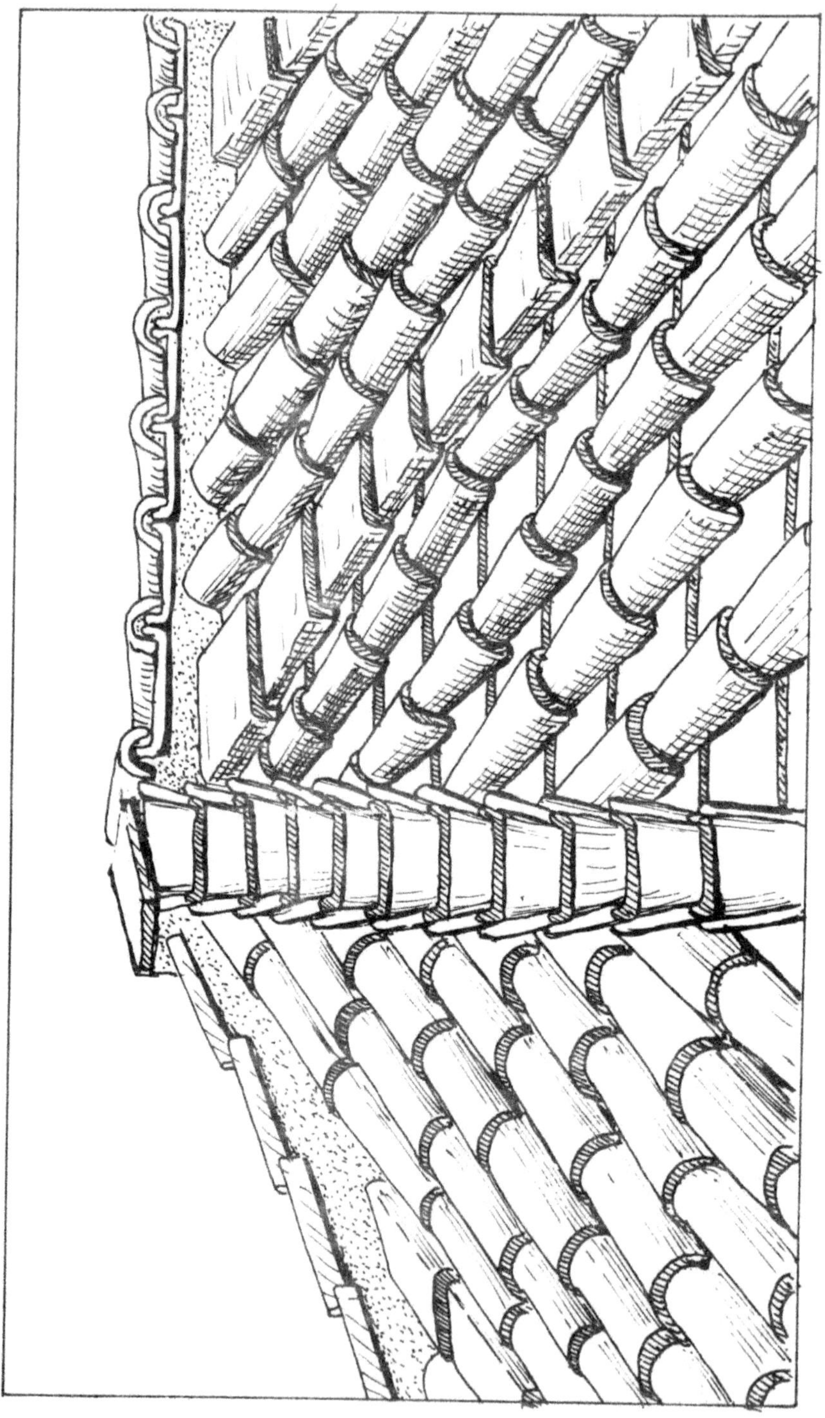

Fig. 16.1 Sketch of the angle of a hipped roof in Rome

Fig. 16.2 Sketch of a gable end at Rome

The absence of antefixes and of purpose-made ridge-tiles should not surprise anyone who has visited Italy. Antefixes are common in modern Greece, but exceptional in modern Italy. Ridge-tiles rarely occur; in some cases imbrices may be torched on with a liberal application of mortar. (This is the rule in Greece, where "double imbrex" (Morse type A5) roofs are common and slopes of 10° are usual). In Italy the details shown in Fig. 1 are usual; a small "wall" is built on the ridge (it does not continue under the tiles—see Fig. 16.2); hips are covered by a similar wall. On the ridge the final capping is of tegulae and imbrices. On the hip it is a continuous gutter of overlapping tegulae.

It is worth remarking, as illustrated by Fig. 16.1, on present-day roofs the alternation of tegulae with imbrices which is usually shown on reconstructions of ancient-Roman structure is not adhered to. Instead we see the interpolation at every fourth, fifth or sixth row on many buildings, of a line of inverted tegulae in place of a row of imbrices. This gives an interesting texture to the roof, and enables corrections of direction and length to be carried out by the tiler. We all know of Roman buildings which are out of square, or incorporate curved parts that must have been roofed; this may provide a simple technique for "fanning" the tiles round. Perhaps a count of total numbers of tiles may give a clue in specific cases.

Tiles may be laid (Fig. 16.2) on boards supported by the rafters. In some cases it seems that the tiles are directly on the rafters, which support the edges of the tegulae, and are directly under the imbrices.

Finally, I would like to explain briefly the historical reason for expecting tiles to be pegged or nailed in Britain. Why is this the case if the Romans produced low-pitch tiled roofs, whereas the traditional English tiled roof is high-pitched? At, or probably before, the end of the Roman period in Britain, the manufacture of tiles ceased. During the subsequent period, thatching became the usual roof cover. Whereas the low-pitched tegula- and imbrex-roof could dispose of water on (and under) the tiles by gravity, the capillary effect on thatch is considerable, and a much steeper angle is needed; a pitch of 50° is normal with reed thatch. Thatch was eventually replaced by shingles and tiles (often as a result of new building regulations following disastrous fires) on existing buildings which already had the high-pitched roof necessitated by thatch. Tiles were therefore made to hang on to battens nailed to the rafters, and new buildings were constructed to make use of hung tiles. There could be some lowering of roofs; in vernacular architecture one can often deduce whether a building was originally thatched from the roof angle. One reason for the success of slates when these were introduced may be that they permitted a roof angle which was more "classical" at the right time in the history of architecture. Today use is often made of large concrete tiles, of about the same weight and size as tegulae, on low-angled roofs. They are not fastened to the roofs, of course.

b) EXPERIMENTAL PRODUCTION OF TEGULAE; A PROGRESS REPORT.

A method of making tegulae by wire-cut moulding was evolved in order to produce a large number of tiles for a model roof at one-twelfth scale. The method has been scaled up to one-quarter scale and has proved easier with increased size. The process used is based upon what might pretentiously be called ethnoarchaeology, since the writer saw how roll pantiles were made in

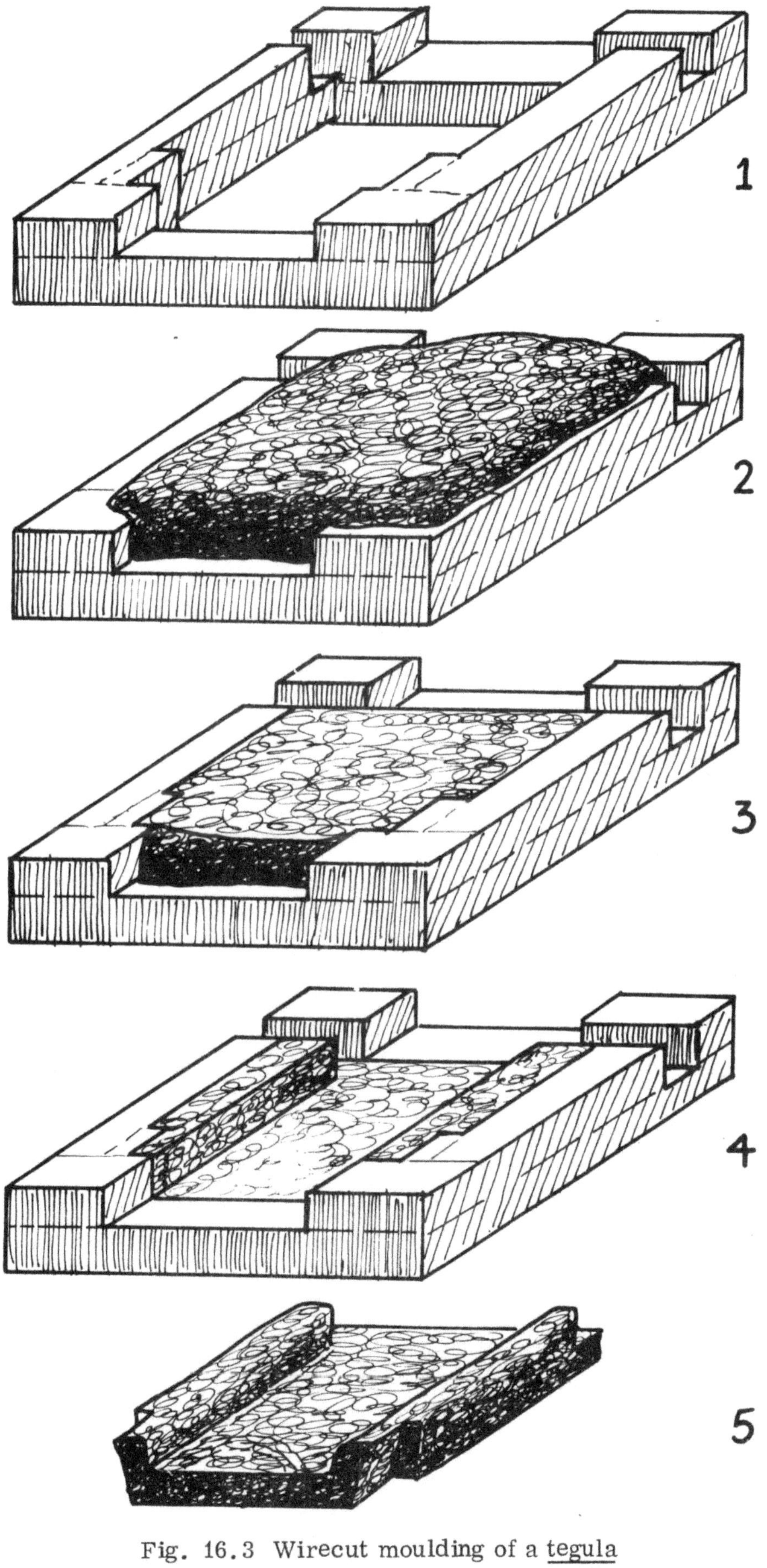

Fig. 16.3 Wirecut moulding of a tegula

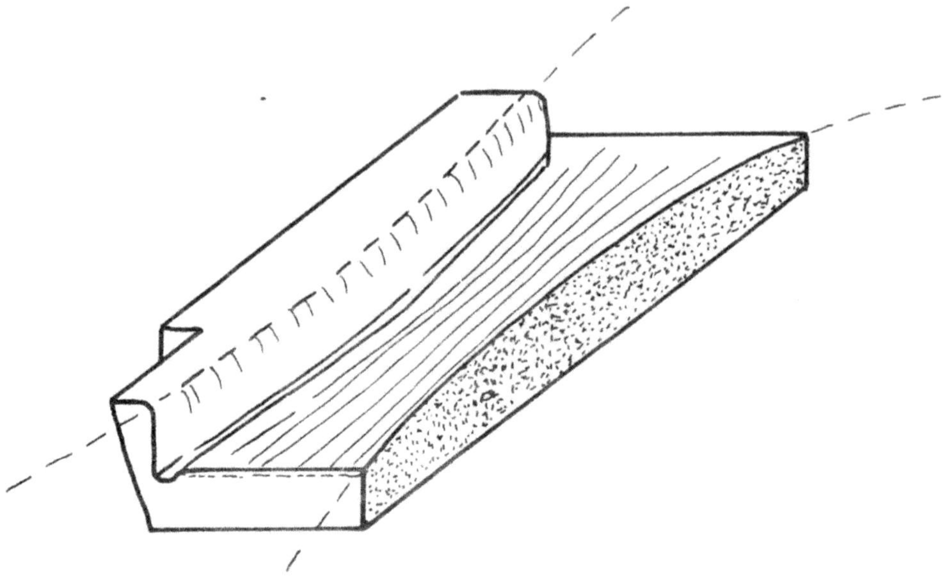

Fig. 16.4 Distortion of a tegula due to the use of an ill-tensioned cutting wire

Somerset some years ago. Only a brief summary of the method is given here, in the hope that the description, and the discussion which follows it, may provoke both further experiments by others, and also a closer examination of finds from sites.

The method is show in Fig. 16.3

1. A frame mould is constructed in scantling of suitable thickness.
2. The mould is stood on a flat base, dampened and sanded. Clay is 'batted' into it to fill the corners.
3. Excess clay is removed along the mould, using a wire tensioned in a bow.
4. Excess clay is removed across the mould in the same way.
5. After smoothing and 'signing' the clay is removed, dried and fettled.

Demoulding appears at first to be a problem, when the 'cutaways' (Rook, 1977) are overlapped by the clay. Three possible ways of overcoming this might be used: the tile can be pushed upwards from the frame, the blocks to form the cutaways could be on the baseboard, rather than the frame, or the mould could be made in two parts. Each of these might leave a characteristic signature that might be recognised.

The moulds that were in use recently in rural Britain had iron profiles to prevent the frame being worn by the wire. It would be interesting to find that these existed on a Roman kiln-site, perhaps unrecognised.

Since evolving this method of moulding, I have observed a phenomenon which supports the idea that the Roman method was similar to the one described. One other characteristic has been suggested at this conference.

Mr. Brodribb has noted that some tegulae have a line across the upper end between the ends of the upper cutaway. It is interesting that this is produced by 'the awkward squad', since it might well result from the lengthwise cut being made after the crosswise one.

My own observation is that many tiles (including one displayed at the conference) show a characteristic thickening in the middle, with the flanges similarly thickened (Fig. 16.4). This results when the wire is insufficiently tensioned.

Note that the fine lines seen on tiles are made, not by something moving along their direction, but, like saw marks, by something (the cutting wire) running across the tile.

Bibliography

Hartley, L. P. The Go Between, London, 1953. Prologue.

Morse, E. S. 'On the Older Forms of Terracotta Roofing Tiles', Bulletin of the Essex Institute, parts 1, 2, 3, 1892.

Rook, A. G. 'The Building Materials' appendix to C. Partridge, 'Excavations and Field Work at Braughing 1968-73; A Roman Bath House, Broughing 1969-73' Hertfordshire Archaeology 5 (1977), 55-6.

Vitruvius (= Marcus Vitruvius Pollio) De Architectura, 4, 7, 6.

THE EFFECT OF THE EVOLUTION OF FLUES UPON THE DEVELOPMENT OF ARCHITECTURE

Tony Rook

It is striking to students both of architecture and also of Roman society that baths play an influential part in their studies. When Seneca (a) contrasts the cramped dark baths in which Scipio had washed off the dirt of honest toil with the modern baths where people bathed and sunbathed at the same time, he was commenting, not just upon the innovations that had occurred in the design of a class of building, but also upon changes in society. Such is the interaction of buildings and their users.

The purpose of this paper is to examine the evolution of the ways in which the gases from the fires that heated baths were conducted away, and to suggest that their effect upon the structure of the baths, and hence upon the development of architecture in the Empire, was out of all proportion to their apparently modest function.

The presence of oil-flasks and strigils in burials shows that the practice of oiling and scraping the body after exercise was at least as early as the seventh century B.C., and heated sweating chambers became a normal adjunct to a palaestra of a Hellenistic gymnasium. Thus in early Greek times the sweat room was merely ancillary to the exercise yard. The development of the steam bath by the Romans changed the emphasis. With the exception that the systolic blood pressure is lowered, rather than raised, the physiological effects of a steam bath seem identical to those of physical exertion: pulse rate, respiration and perspiration are all increased. The baths grew in size, and the palaestra shrank by comparison. Fronto was carried on a litter from one room to another. A great perturbation in nature—to receive at once the benefit of exercise and the effects of resting.

Braziers were used to heat the earliest alipteria, and their use continued into the first century A.D. When the ash from Vesuvius buried the Forum baths at Pompeii in 79 A.D., the tepidarium of the men's section of the baths was still heated by a large brazier which was installed in 20 A.D. Although classical authors (Cicero, Macrobius, Nonius Marcellus, Pliny (a), Valerius Maximus) attribute the invention of the hypocaust to Sergius Orata, at least according to an interpretation by modern writers, the Greek origin of the term suggests an eariler invention elsewhere. In fact the ancients credited Orata with inventing balneae pensiles, which has been interpreted quite recently as "hanging baths" (like water-filled hammocks) "vaults" and even "shower baths". Assuming that there is some truth in the tradition, the most likely explanation is that Orata heated tanks of water over a hypocaust, and introduced the steam bath, as opposed to the dry heat of the laconica. The hypocaust existed long before Orata, who was a contemporary of Lucius Crassus, in the early first century B. C. Hypocausted baths are reported in Sicily in the fourth century B. C. (see page 11) and in the Heroön at Olympia at the end of the same century or the beginning of the third (Kunze & Schlief, 1943), at Gortys in Arcadia in the second century B. C. (Ginouvès, 1943) and in the IVth Olympic Thermae (Kunze & Schlief, loc. cit.).

It is not clear how the earliest hypocausts were vented to the outside, and one definite possibility is that the gases from the fire actually passed through the chamber occupied by the bathers. The laconicum was a circular building with a hemispherical or conical dome, and, with a hypocaust, would be analogous to a pottery kiln. Objections to this structure, on the grounds that the air in the room would become unbreatheable are quite unfounded, since to obtain the temperature needed for a sweatroom would need so much excess air over the fire that no carbon monoxide would be formed, and the dilution of the carbon dioxide would be quite acceptable. We know that braziers could be used, as stated above. In any case, there was always the option of raking out the fire before the bathers entered, and using the hypocaust as a heat exchanger. The operation would be suggested by the traditional bread-oven, and similar to that of the old Swedish sauna.

Support for the laconicum operated like a kiln is given by the description by Vitruvius (a). It is circular "so that the heat may spread from the centre around the circumference" and there is an opening in the centre of the dome with a bronze disc (clipeus) which hung on chains, and could be raised and lowered to adjust the temperature. It is difficult to see how this would function unless it controlled the draught through the heating system. The description by Vitruvius (b) of a kiln for making lampblack as being like a laconicum is also suggestive, although the specific mention of vents (nares) opening into the chamber may indicate that they are confined to this type of structure only when used to make lampblack. Unfortunately, there is no mention in Vitruvius of any kind of vent, flue, or chimney for hypocausts! Another, but chronologically late, indication that furnace gases entered laconica is given by Plutarch, who says that olive wood has a beneficial effect on the bathers' bodies when burnt, but it is injurious to the interior decoration of the laconicum. He also forbids the burning of darnel (lollium tremulentum, the "tares" of the Bible which were gathered into bundles and burnt [Matthew]) as this plant gives the bathers headaches and induces vertigo.

The earliest true chimneys with hypocausts seem to be represented by the clayware pipes set into the walls of the Central baths at Herculaneum. These pipes are identical to rainwater downpipes, but are set with their necks upwards. They are placed in narrow slots in the inside faces of the walls, and not plastered over. They open at the top into large horizontal tunnels in the thickness of the masonry between the springing of the vaults and the flat roof, and hence to the outside air like windows over the stoking passage. It may not be entirely coincidental, bearing in mind the analogy of a kiln, that these chimneys seem to be introduced at about the same time that "muffle" kilns came into use for producing consistently oxidised pottery. Pipe flues were still being installed when the Redhouse baths were built at Corbridge (Daniels, 1959).

Vitruvius (c) describes a method of dry-lining walls to prevent dampness (he does not distinguish capillary ingress from other causes). Tiles are fastened to the inside of the walls with an air space between, and vents through the wall at the top. A trench or channel is cut around the base of the walls, with pilae to support the floor and the lining tiles. Unless Vitruvius was mistaken in what he was describing (and there seems to be no surviving example of this construction) it must have been some later technician who, seeking to cure dampness in the baths, made the obvious connection—between the space under the floor, and the cavity in the wall—and revolutionised the heating system.

It is unfortunate that Vitruvius does not describe the wall tiles he is recommending in detail, and that different texts read hamatae (hooked) and mammatae (nippled). In surviving hypocausts a number of clayware units were used to produce continuous jacketing. These include flat tiles with flat 'knibs' at their corners (to which the description "hamatae" could well be applied), ordinary flanged roof tiles (which we nowadays call tegulae without epithet), and flat tiles with special spacing pieces like cotton reels or small bottles, through which the fixing nails passed. Especially notable are the tiles with nipples attached to them, through which the nails passed; more economically, the nipples were solid, and the nails passed through notches cut in the edges of the tiles. These tiles are called tegulae mammatae by modern archaeologists, who presumably take their support from Pliny (b) for a reading of Vitruvius that depends more upon archaeological revelation than textual criticism.

The various clayware units mentioned above to create continuous cavities to conduct flue-gases seem to have been introduced in the Augustan period. They are found in the Stabian baths and private baths at Pompeii, and persisted long enough to appear in sites in Britain. Their disadvantages were evident: they were physically unstable, and could not stand without wasteful use of long nails; nails which could not easily be used with some types of masonry. At Griggs Bridge (Partridge, 1977) a flint masonry bath-house was lined inside with calcareous tufa; this was most likely for providing fixing for the nails, rather than for any other advantage.

To overcome the problems of flat tiles with spacers, a new clayware unit appeared which combined the stability of pipes with the flat surface of the tile: the box flue tile or tubulus. Rectangular in cross section, tubuli could be stacked against a plumb wall with a minimum of T-shaped iron fixing-spikes, particularly when they were keyed with a comb or stamp to make them adhere to mortar and plaster. They were made in a wide range of sizes.

Cavity vaults were known to Vitruvius (d) although he does not describe their construction. A range of different units of tile and stone (especially the lightweight tufa) were used to produce hollow vaults. These include wedge-shaped voussoir box flue-tiles. It is not certain that the vault itself was heated by the furnace-gases. The apex of a vault rarely survives, of course. At Bath the displayed section of vaulting shows a cavity which is blocked by transverse flat tiles, and elsewhere the surviving barrel-vaulted structures examined by the writer have exit vents close to the springing.

The tubulus seems to be the ultimate development in the construction of flues. It found application both in continuous tubulation and as single chimneys for cooler rooms. It appears in the luxury villas that were destroyed by the eruption of Vesuvius in A.D. 79, in the Suburban baths at Herculaneum, which were built after the earthquake of A.D. 62, and the Central baths at Pompeii, which were being built at the time of the eruption on a site made available by the earthquake. Even allowing for conservation by builder and customer, and conservation of materials, the use of flues on these sites at Pompeii and Herculaneum, but not elsewhere here, suggests that the tubulus was not available until the second half of the first century A.D. in this part of Italy, Seneca (b) says that the tubulus had "come to light within our memory"; he died in A.D. 65.

It is not possible to state with any precision the date of the introduction of each improvement in flue construction, and the study is bedevilled by the insertion of improvements into existing structures. A possible sequence may be that piped chimneys were introduced in the first century B.C. (perhaps at the time of Orata?), tegulae mammatae and the like at the beginning of the Christian era, and tubuli around the middle of the first century A.D. It is against this sequence that can be set an account of the influence that flue design had upon the development of architecture.

One effect of the introduction of jacketed walls has been suggested by Dr Webster in an earlier paper (see page 289). The floor tiles would need to bridge the pilae of the hypocaust with fair accuracy, and an accurate narrow gap around the walls would have to be produced to connect with the tubulation. This might well result in a standardisation of tile sizes which had not hitherto been necessary. The point needs verification by statistical measurement at different times and in different places. The writer has observed (Rook, 1975) that there does seem to be a modular convention in the construction of domestic bath-buildings.

A more important effect was upon architecture itself. Domes and vaults were ideally suited to the function of the baths, and changes in the baths had a profound effect upon the development of structures using these elements. The laconicum was traditionally a domed building, and the earliest examples of such structures must be the laconica later converted to circular frigidaria at Pompeii and Herculaneum; these are probably of the second century B.C. When steam baths came into fashion, in the first century B.C., it was found that the vault was a sine qua non, since, especially when ribbed or corrugated, it tended to guide the condensation to the sides, rather than permit it to drip on to the backs of the unsuspecting bathers!

With brazier heating, or even with piped chimneys and a floor thick enough to support baths, transfer of heat into the room was inefficient. Doors and windows were kept small to reduce heat losses, and the size of the room was also limited. The bathroom resembled a tunnel—the balneolum angustam, tenebricosum ex consuetudine antiqua described by Seneca (a).

When heated cavity-walls replaced piped chimneys, the situation changed. Heat transfer through the thick floors was slow, but wall tiles, particularly tubuli, were thin, and the covering of plaster over them could be thin. Heat transfer from the chimney gases was much more efficient, and the walls have a much larger area than that of the floor. The improved heat-transfer meant that much larger volumes could be heated. Quite as important, but less understood, is that the walls became heating elements, supplying radiant heat to the bathers, where hitherto it had been absorbed. The increased "radiation temperature" meant that lower air-temperatures could be experienced with comfort. It was found that large unglazed windows could be provided in some rooms, and a new class of room was introduced: the sunbathing lounge, heliocaminus or cella solaris, which was the principal room of many baths, and is usually described, mistakenly, as the caldarium. The main axis of the imperial baths was aligned so that this room had its great windows to the south-west, since the Romans bathed in the afternoon.

That these windows were unglazed can be seen from the surviving remains, as for example in the Forum baths at Ostia, and can also be deduced both from the name and shape of this element of the baths, and from the descriptions of people bathing and sunbathing whilst looking out at the view (Seneca (a), Pliny, (b)) since Roman glass was translucent rather than transparent.

The possibility of heating very large bathrooms with wide windows came at the very time that the potential of concrete was realised. It was in the baths, that the flexibility of space afforded by the use of concrete was mainly exploited. The importance of baths in the history of architecture lay in the fact that they were the only large public buildings for which no long-standing conventional structure existed at the time of the concrete revolution. Large baths were made possible by the invention of the modest flue-tile "devised by the lowest slave" (Seneca, (c)). Thus the effect of flues on the design of buildings, and ultimately upon the users of those buildings, was far greater than might be suspected from their apparent function.

BIBLIOGRAPHY

Cicero: Marcus Tullius Cicero, Fragmenta Librorum de Philosphia Depertorum (Müller Ed.), 5.76.

Daniels 1959: C. M. Daniels, 'The Roman bath house at Red House, Corbridge', Archaeologia Aeliana, (Fourth Series), 37 (1959), 85-117.

Fronto: M. Cornelius Fronto, Epistulae 5.44 (ad M. Caes.).

Ginouves 1953: R. Ginouves, 'Gortys d'Arcadio' Bulletin de Correspondence Hellenique, 77 (1953), 263-271.

Kunze & Schlief (1943): E. Kunze and H. Schlief, Berichtüber die Ausgraben in Olympia 1940 und 1941, Berlin, 1943, iv.

Macrobius: Ambrosius Theodosius Macrobius, Saturnalia, 3, 15, 3.

Matthew: The Gospel According to St Matthew, 13, 24-30.

Nonius Marcellus: Nonius Marcellus, De Compendiosa Doctrina, 193-194.

Partridge, 1977: 'Excavations and Fieldwork at Braughing 1968-73; A. Roman Bath House, Braughing 1969-73', Hertfordshire Archaeology, 5 (1977) 24-64.

Pliny (a): Gaius Plinius Secundus, Historiae Naturalis, 9, 168.

Pliny (b): Gaius Plinius Secundus, Historiae Naturalis, 35, 46.

Pliny (c): Gaius Plinius Caecilius Secundus, Epistulae, 2, 17, 12.

Plutarch: L(?) Mestrius Plutarchus, Quaestiones Conviviales, 658E.

Rook, 1975: A. G. (Tony) Rook, Aspects of the Design and Operation of Hypocausted Baths in Rural Britain, Unpublished M.Phil Thesis, University of London Institute of Archaeology, 1975, 74-76.

Seneca (a): L. Annaeus Seneca, Epistulae Morales, 86.4.

Seneca (b): L. Annaeus Seneca, Epistulae Morales, 90.25.

Seneca (c): L. Annaeus Seneca, Epistulae Morales, 90.26.

Valerius Maximus: Valerius Maximus, Factorum et dictorum memorabilium, 9, 1, 1.

Vitruvius (a): Marcus Vitruvius Pollio, De Architectura, 5, 9, 5.

Vitruvius (b): Marcus Vitruvius Pollio, De Architectura, 7, 10, 2.

Vitruvius (c): Marcus Vitruvius Pollio, De Architectura, 7, 4, 2.

Vitruvius (d): Marcus Vitruvius Pollio, De Architectura, 8, 2, 4.

A PETROLOGICAL STUDY OF LHS AND TPF STAMPED TILES FROM THE COTSWOLD REGION

Tim Darvill

Introduction

Over the past few years the direction of archaeological research has been realigned, away from the well known or unique aspects of material culture to those which although not seeming to hold immediate appeal do offer much new archaeological information (Renfrew, 1978). The Roman ceramic tile, found on almost every Roman site, seems to have been swept up in this change. The increasing number of papers about tiles appearing in journals such as Britannia in recent years, and the organisation of the conference at which this paper was given emphasise the rise of the tile in the league table of "in" subjects to study.

Fig. 18.1 shows a possible scheme of studying tiles. Many of the avenues shown are explored elsewhere in this volume, while this paper attempts to outline something of the economic organisation of the tile industry.

The organisation of the military sector of the industry has already been studied by Peacock for the CLBR tiles of the Channel Fleet (1977) and by Grimes for the legionary production centre at Holt (1930). One feature of the military brick-making centres was that the tiles were often stamped. It has also been realized for a long time that within what may be loosely termed the Cotswolds there is a series of discrete groups of stamped tiles, which seem to come from civilian production centres (Clifford, 1955). A detailed examination of these groups of tiles shows that several levels of production existed (for a provisional discussion see McWhirr and Viner, 1978). The lowest identifiable level was a form of estate production, with products confined to a limited geographical area, and possibly all from buildings within one estate. The lower level units of production will be discussed at a later date, while this paper will deal only with the largest identifiable production network. The products of this system were stamped either LHS or with one of the TPF series stamps (Figs. 18.3; 18.7; 18.8; 18.11). Tiles stamped with these groups of letters constitute some 58% of the known 240 stamps from the Cotswold region. Their distribution is centred on Cirencester and the area within a 30 kilometre radius of the town, although outliers are known from Old Sarum and Silchester (McWhirr and Viner, 1978, figs. 6 and 7) and most recently 3 examples of the LHS stamp have come to light at Kenchester. The wider spatial distribution of these groups of tiles, compared with other groups not discussed here such as ARVERI, VLA, LLH or TCM stamped tiles, is suggestive of a more complex, integrated economic organisation. This contrast between the size and distribution of the groups of stamps was not really brought out by Hodder when considering the tiles as examples of a "local" marketing

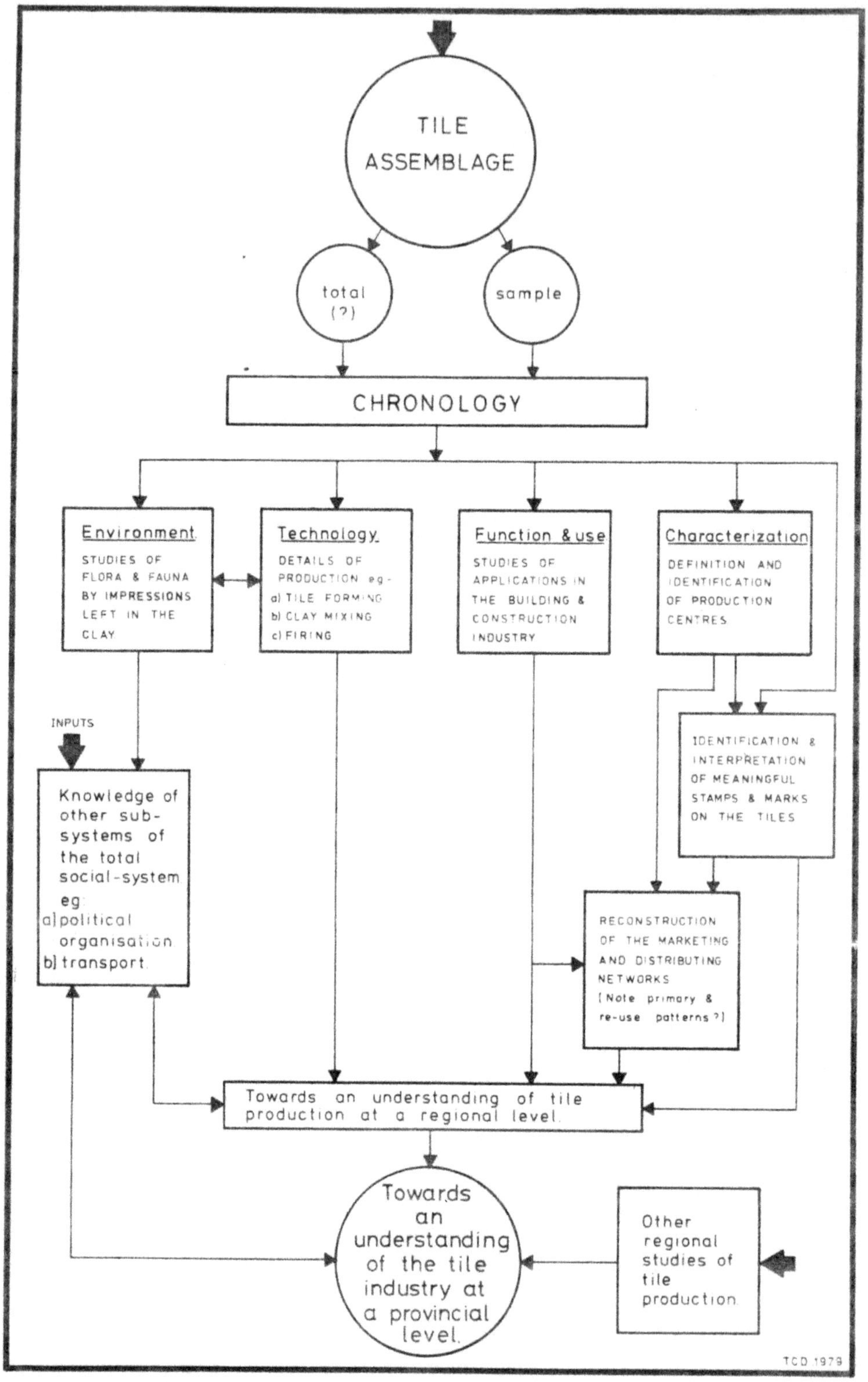

Fig. 18.1 Suggested scheme for the study of Roman tiles

pattern; a pattern which is not wholly supported by this analysis (Hodder, 1972, 901 and 1974, 42).

QUESTIONS AND OBJECTIVES

The analytical procedure used in this study was structured to answer three specific questions:

1) Where were the main tile-producing centres in the Cirencester region?

2) What was the distribution achieved by their products, and was there more than one system of production present?

3) What further light can be shed on the technology of production?

A THEORETICAL BASIS

At the outset of a study such as this it is important to consider conceptual questions which will affect the conclusiveness of the results (Clarke, 1978, 15). It is important to work from the known to the unknown, and in this case it is only the stamps about which we have any knowledge. Stamps are official marks of some kind which can be grouped by similar letter configurations, and also related to other stamps by seriation and die similarities. This allows one series of groups to be established. If the fabric type of the tile, and its source of origin, is known then a second series of groups can be built up and compared with the first set of groups to produce patterns which can then be interpreted in meaningful socio-economic terms. The use of petrological techniques alone might point to common origins, i.e. a given clay source, but would not be able to demonstrate the relationships between production centres. Likewise the use of a "signature", such as a comb impression alone will result in ambiguous evidence which is impossible to interpret since either brickcaster and comb or the tile itself may have moved in geographical terms. The use of petrological fabric analysis and stamp typology together allows valid statements to be made about the industry, applicable to the time and place in question, but hopefully giving more general insights.

The use of stamped tiles is important also because the limited life of a stamp die ensures that data produced contemporaneously are being examined. Until detailed studies of tile assemblages from the Cotswold region are available the assumption must be made that only tiles from producers stamping a proportion of their products were used on a given building project. Unfortunately, in most cases, it is only the stamped tiles which are available from past excavations. Given the above assumption (which may well prove wrong) it should be emphasised that a <u>sample</u> is being taken from the original tile population, the proportion of tiles surviving from that original total is however not clear. It is dependant not only on the proportion of tile output stamped, but also on the sampling strategy of the excavator, which is almost never stated in quantifiable terms.

One further problem in interpretation are the "cultural" and "non-cultural" <u>transforms</u> which have determined the archaeological record (Schiffer, 1976). These factors should be considered in the excavator's interpretations of what he finds, although this is often not so. For the present study it is only the tiles and findspots which matter, in order to speak of economic organisation, as it is enough to isolate demand points and to assess their relative proportions. That there was re-use of tile is well known, and many of the tiles seem to come from secondary deposits. These are Schiffer's S → S processes (Schiffer, 1976, 37-39); what we have to assume is that there was little geographical movement during this process, and that the tiles remained within the demand locus. There is certain evidence that this was the case, which will be reviewed later, although the possibility of a builder taking some raw materials in the way of rubble and used brick with him when he travelled between projects should not be overlooked.

It is possible to isolate three types of demand locus. The high demand centre will be the large towns. These have a fairly continuous demand for tiles and bricks, although fluctuations in demand will occur depending on socio-economic development or regression; the town of Cirencester can be cited as an example. The medium demand centre will still have quite a high requirement but there will be more fluctuation in that demand with changing fortunes, and also with a higher degree of conservatism in re-building, which is also on a more modest scale; sites such as Wanborough and Kingscote are good examples of this. The low demand centre will be villas, where requirements will initially be high, but will be followed by a steep fall-off, with only occasional intensification for re-building or extension, and also perhaps for other work on the estate. The Cotswolds abound with examples of this type of demand centre; Hucclecote, Frocester and Barnsley Park being well-studied examples.

DATING

The dating in absolute terms of the stamped tile series under discussion here is not really practical until a few production sites have been fully excavated. The simultaneity of production of the various stamped tiles can be seen from two important closed associations. The first comes from phase I of the Hucclecote Villa which had a ceramic tile roof. It was demolished in about A. D. 150, being sealed below the building of the second phase, which had a stone slate roof (Clifford, 1933, 353 and Williams, 1971, 99). From Hucclecote there are tiles stamped TPF, TPFA, TPFP, TCM and also examples of the RPG stamps from Gloucester tilery (Clifford, 1955, 68). The second closed group is from Victoria Road in Cirencester where on the 9th February 1922 workmen for St. Clair Baddeley found "a heap of fallen wall-tiles in one of the rooms of a R-Br. house... the deposit could be detected just shewing itself under the edge of some two feet of dark soil" (Baddeley. 1922, 117). Elsewhere he notes that the tiles were on the floor surface of the room (Baddeley, 1921, 209), and it seems likely that here was a destruction layer similar to the Hucclecote one. The Victoria Road assemblage contained tiles stamped: TPF, TPFA, TPFC, TPLF, ARVERI, and LHS.

It seems quite likely that the use of an organised system which required tiles to be stamped during production was derived from the Mediterranean through the Roman army. Greene notes how "legionary" pottery styles were introduced by military potters at Holt (Greene, 1977, 126) and a similar process undoubtedly brought brick-making too. McWhirr reviews the evidence for dating early production and sees no reason to date military brick stamping before about A. D. 100 (p. 258), although the Nero stamps from Silchester indicate first century stamping in official tileries. It is also the case that during the third century stamping with literate stamps dies out in other parts of the ceramic industry. In the Oxfordshire potteries, for example, Young notes that the stamping of mortaria is only rarely practiced after A. D. 240 (Young, 1977, 236). At present, on the flimsy evidence available the main phase of stamped tile production in this area seems to be from about A. D. 100 until about A. D. 250. This mid-third century change in the industry can also be seen in the swing to using stone slate roofing in many Cotswold villas, and the presence of large assemblages of bricks with no stamps. Building VI 5 at Cirencester for example was sealed below a fourth century courtyard, and utilized a large number of tiles in a tile-coursed partition wall, none of which were stamped (McWhirr, 1978, 77).

NOMENCLATURE AND CLASSIFICATION

Throughout this paper the words brick or tile are used to mean those ceramic products often called tegulae, imbrices or pillae. In many cases the fragments which were studied could not be strictly grouped by function and the application of Harley's formula for differentiating brick from tile was impractical because of the fragmentary nature of the evidence (Harley, 1974). The important point is that the bricks and tiles were used in routine building work for many different purposes, and some were undoubtedly made especially for specific functions (cf. Swain, 1978. 8-9).

In 1955 Mrs. Clifford published the first synthetic analysis of the stamped tiles of Gloucestershire (Clifford, 1955). She gave each identifiable letter sequence a number, followed by a sub-division on the basis of die type. This descriptive typology has been followed to a certain extent since then, but it is clearly time that a new system was evolved which incorporated all the stamped tiles in the country.

Since 1955 many new stamped tiles have come to light and have been examined by Alan McWhirr, David Viner and the author. Rubbings were made of each stamp and then compared with each other over a light table, enabling die types to be established. Figs. 18.3, 18.7, 18.8 and 18.11 show the die variations for each group of stamps studied here. They are numbered in a new sequence which does not relate to Mrs. Clifford's system. The new nomenclature utilizes the letters in the stamp to designate the group, followed by a letter in lower case which distinguishes the die type. This system, or a variant of it, might well be considered as a definitive classificatory system, and has the advantage of being simple and specific. Appendix III shows Mrs. Clifford's system correlated, where possible, with the one used here.

TILE FABRIC ANALYSIS

Samples were collected for microscopic examination from seven sites of different types in Gloucestershire. These are shown by open squares on Fig. 18.2. Clay samples were also taken, made into bricks and then microscopically examined, from most of the clay beds in the county. Samples of clay and tiles were also collected from the known kiln sites at Minety, Wiltshire, during fieldwork with Mike Stone.

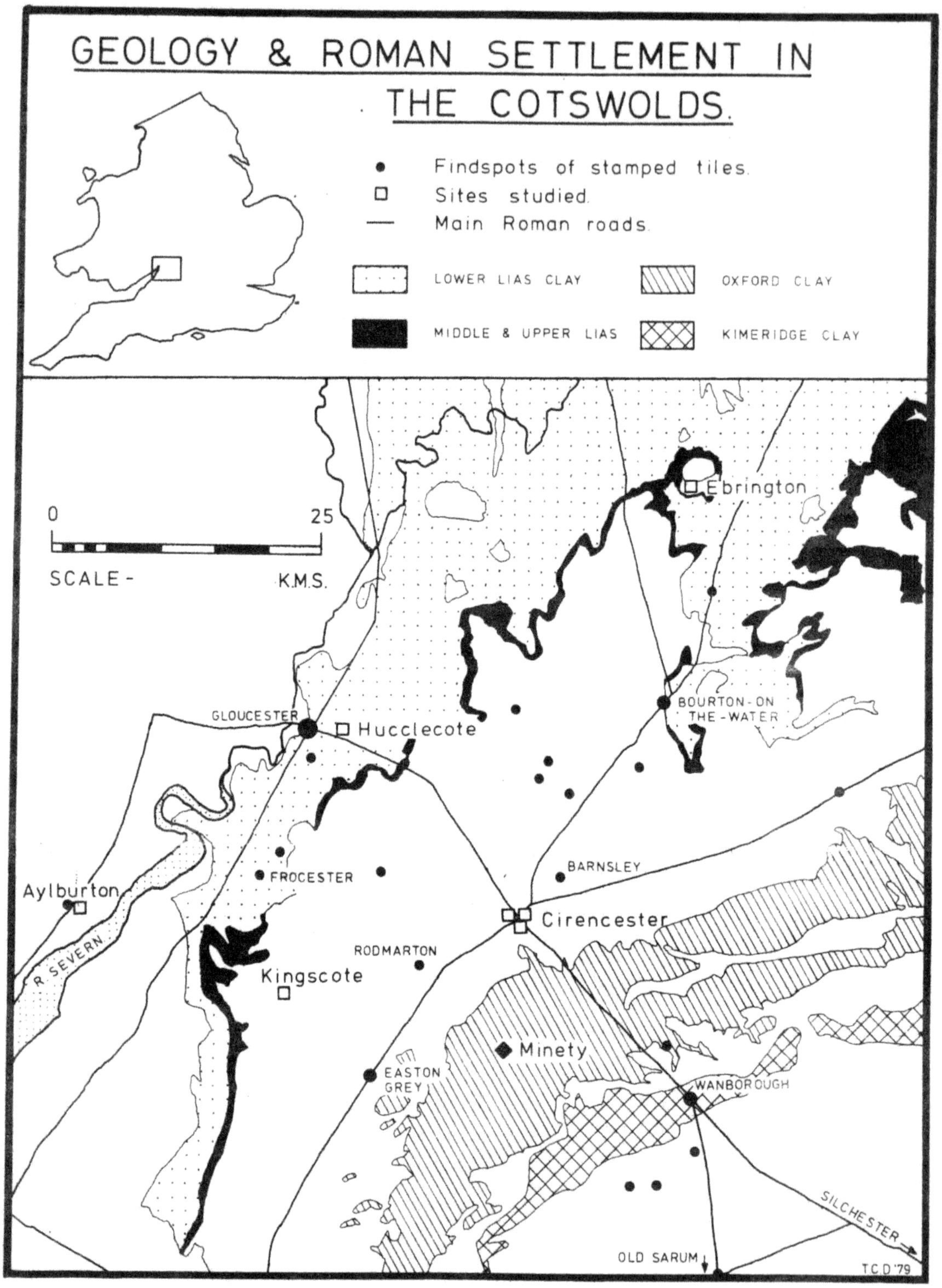
GEOLOGY & ROMAN SETTLEMENT IN THE COTSWOLDS.
Findspots of stamped tiles.
Sites studied.
Main Roman roads.
LOWER LIAS CLAY
OXFORD CLAY
MIDDLE & UPPER LIAS
KIMERIDGE CLAY
0
25
SCALE -
K.M.S.
Ebrington
GLOUCESTER
Hucclecote
BOURTON-ON THE-WATER
BARNSLEY
FROCESTER
Aylburton
R. SEVERN
Cirencester
RODMARTON
Kingscote
Minety
EASTON GREY
WANBOROUGH
SILCHESTER
OLD SARUM
T.C.D.'79

Fig. 18.2

Thin sections were made from the samples, in a technique now familiar to many archaeologists (Peacock, 1970, 397; Cornwall, 1958, 141 and Tite, 1972, 215). One face of the sample is polished and stuck on to a glass slide. The sample is then ground down on a diamond lap wheel and carborundum powder until it is 0. 03 mm thick. After affixing a cover slip and cleaning, the slide can be studied under a petrological microscope.

In the absence of a distinctive minerals or rock inclusions in the fabric, much as is to be expected in a region of sedimentary geology, other techniques of characterization were used. Since, geologically, a pot or brick is essentially a modified sedimentary rock, it seemed reasonable to apply the methods of textual analysis used by sedimentologists. In this case the sizes of quartz grains, which were abundant in the sections, were measured using a calibrated graticule in the microscope. A sampling device known as a point counter, located on the turntable of the microscope, moved the section along in parallel rows, ensuring that a representative selection of quartz grains were measured.

The longest axis of about 100 grains was measured in each sample, but in fact the number of grains counted does not seem to make much difference. This is shown in Table 18. 1 which was prepared using results of experiments by Mr. Simyu Wandibba at Southampton University. Using Iron-Age pottery from Hampshire he characterized each sample with counts of 50, 100, 150 and 200 grain measurements. It is clear that given the variance within a natural formation of clay a count of 50-100 grains is quite adequate, bearing out Pye's findings from a study of arenaceous sediments (Pye, 1943).

The grain-size measurements were grouped into classes based on Krumbein's phi scale (Krumbein 1938), (Ø units) and their cumulative percentages were plotted on probability paper. The 5th, 16th, 25th, 50th, 75th, 84th, and 95th percentiles were read off the graphs and a series of basic statistics were calculated, using the formulae of Folk and Ward (1957). These simple statistics, such as mean grain size (M_z) and standard deviation (σ_1) are used on the scattergrams which appear below. Appendix II shows the raw figures for numerical comparison, while the scattergrams used in the text give a more visual idea of the groupings. The technique has been used with success in the past by Peacock to differentiate sand-tempered pottery fabrics from Fishbourne (Peacock, 1971).

ANALYSIS OF TILE GROUPS

TPF Tiles

This is the most numerous group examined. Fig. 18. 3 shows the variety of die types known at present. There are two types with no surrounding frame, or stops between the letters, while the remaining five groups all have frames round the letters, and four of them have triangular stops between the letters and after the F. It is possible to distinguish two fabric groups among the TPF tiles. These can be seen contrasted on Plate I, where c shows a tile of fabric 1 from Beeches Road Cirencester and d shows a tile of fabric 2 from the Hucclecote Villa near Gloucester. The difference can also be seen on Fig. 4A where the scattergram shows the results of quartz grain-size analysis discussed above and shows the two groups quite clearly.

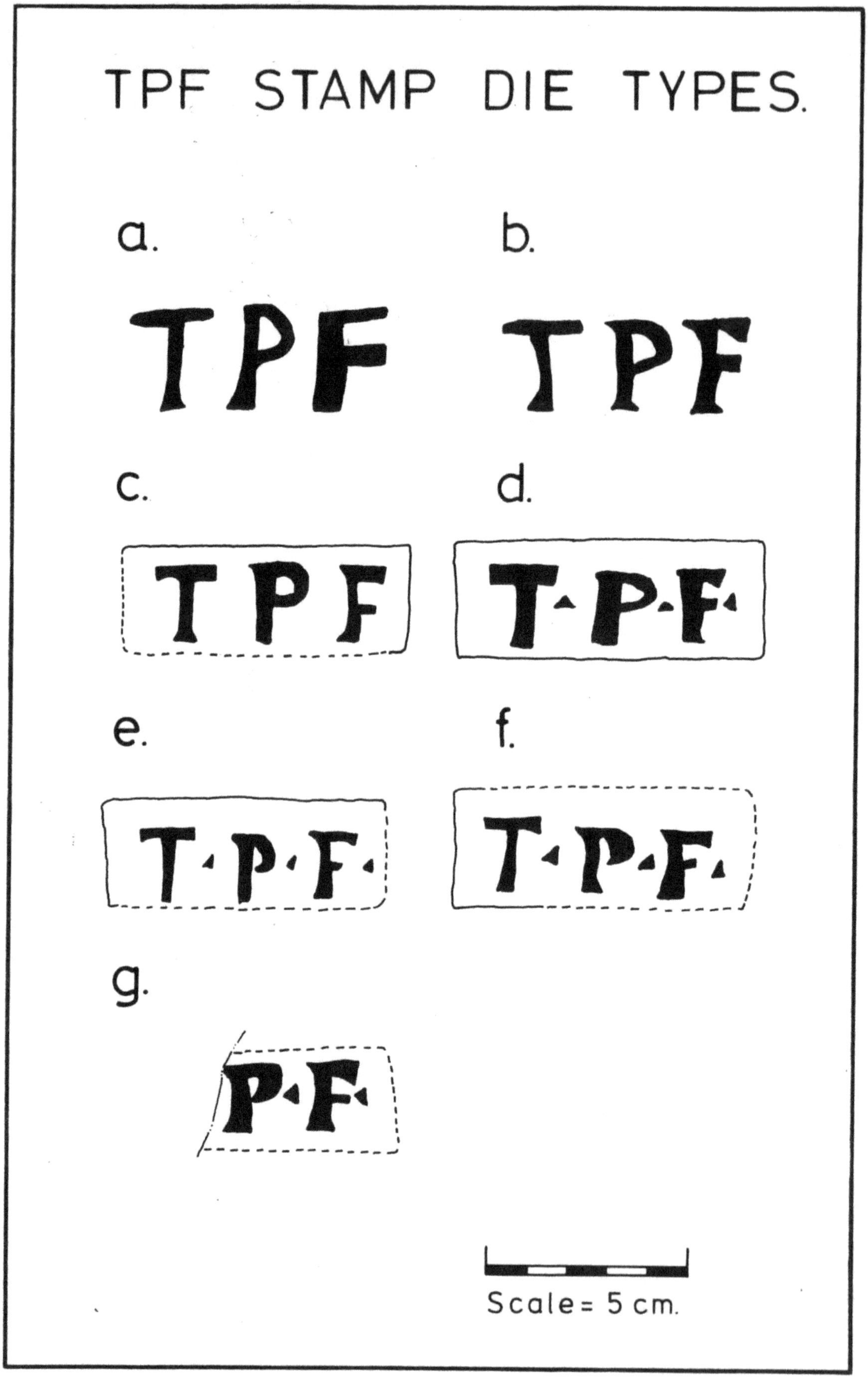
TPF STAMP DIE TYPES.
a.
TPF
b.
TPF
c.
TPF
d.
T·P·F·
e.
T·P·F·
f.
T·P·F·
g.
P·F·
Scale= 5 cm.

Fig. 18.3

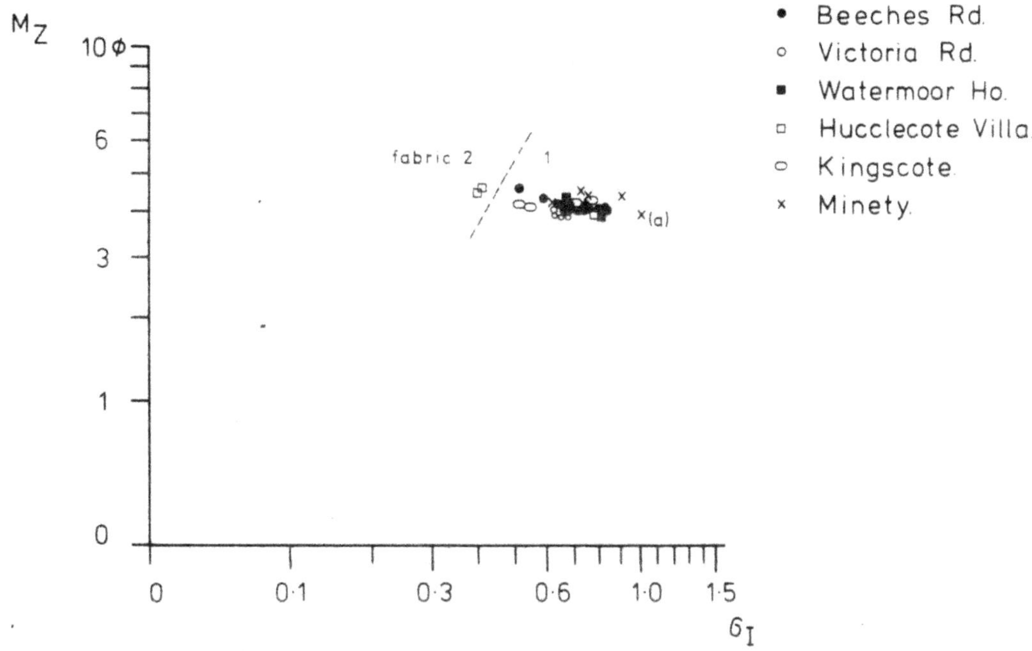

A. Scattergram of TPF material.

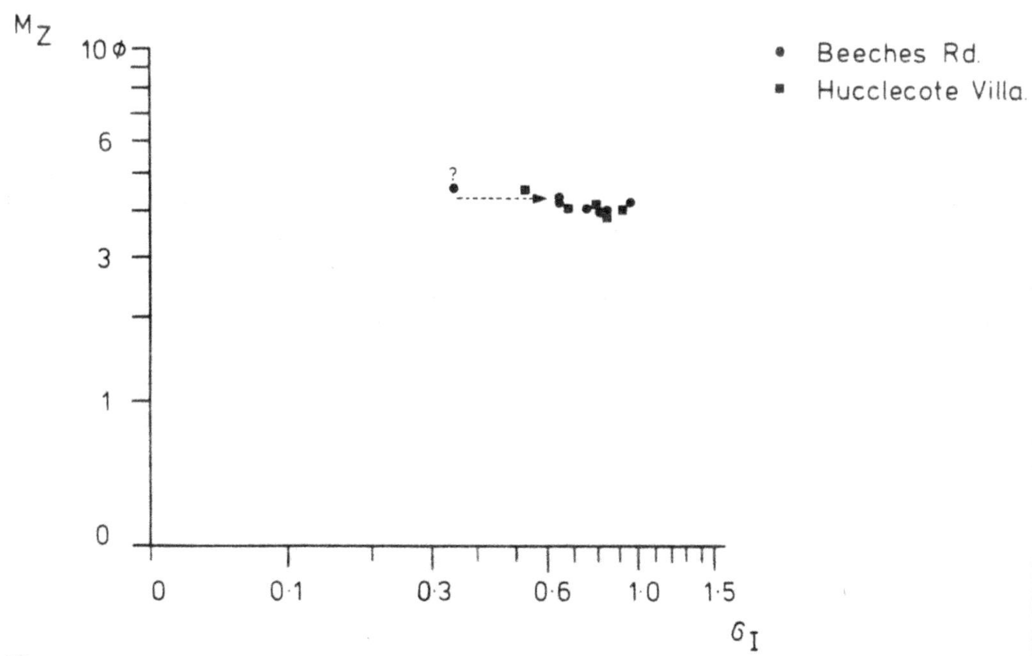

B. Scattergram of TPFA material.

Fig. 18.4

Table 18.1 Comparison of quartz grain-size sample counts

	50		100		150		200	
Sample no.	M_z	6_I	M_z	6_I	M_z	6_I	M_z	6_I
Worthy Down								
321.2.1	2.93	0.69	3.03	0.70	3.02	0.77	3.03	0.77
321.2.2	3.62	0.67	3.55	0.84	3.53	0.72	3.61	0.67
321.3.1	3.14	0.91	3.12	0.96	3.04	0.91	2.86	0.88
321.5.5	4.29	0.38	4.22	0.39	4.20	0.41	4.22	0.48
Twyford Down								
311.00.1/5	3.80	0.57	3.89	0.56	3.34	0.61	2.81	0.65
311.00.2/5	3.84	0.54	3.76	0.62	3.76	0.62	3.69	0.67
311.00.3/5	2.55	1.21	2.38	0.99	2.43	1.10	2.41	1.01

Data from experiments by Mr. S. Wandibba

Fabric 1 is characterized by quartz grains up to 0.4 mm in diameter, and very little mica. There are occasional grains of plagioclase felspar, but most distinctive are the small fragments of ferrous oxides and iron ores. The clay usually shows some streaks of yellow within the fabric, and there are often small voids, resulting from calcareous material burning out during firing, although there is some variation in the number of voids in the tiles. They are fired hard, usually not able to be scratched with a fingernail, and sometimes difficult to mark with a knife blade. Under the hand lens a fresh break appears "grainy" and grains of quartz can usually be seen. Odd pieces of angular flint sometimes occur on the surface.

Tiles of fabric 1 have been found at all three of the sampled sites within Cirencester (Beeches Road, Victoria Road and Watermoor House). Further afield the Roman settlement at Kingscote has to date produced four examples (Swain, 1978, 16) and one of the three TPF stamps at the Hucclecote villa is of this fabric. Having examined these under the microscope it was possible to visually distinguish other examples of this fabric from the area, as seen in Appendix I.

One stamp of TPF (very similar to die a) has been found at the kiln site of Minety on the Wiltshire/Gloucestershire border, see Fig. 18.2. This example is fired much harder than the other known TPF tiles and its general condition suggests that it is a kiln waster. Because of this and the fact that only a small sample could be removed without damaging the stamp, this tile falls on the right hand end of the distribution on Fig. 18.4A, where it is denoted by (a). Other more typical, although unstamped, Minety tiles do however fall squarely within the main group of fabric 1 points. Plate 18.Ia shows a photomicrograph of a tile sample from Minety, and Plate 18.Ib shows raw clay from the site, which has been fired into a brick to facilitate comparison. The similarity between these two samples and c on plate 18.I illustrates the close match of the TPF tiles with those from the Minety kiln sites.

One objection which might be raised is that during the brick-making process sand-tempering is sometimes added (cf. Richardson and Webb, 1910); if this were so surely it would affect the results of quantitatively comparing quartz-grain-size distributions. Technological studies from thin section analysis allow this procedure to be looked for (Peacock, 1970). Fig. 18.5 shows a series of curves compiled from the cumulative percentages of each grain size class. Samples of tile from the Minety kiln field, and also clay, taken from the top of the clay pits in the Minety complex are compared and show a very close correlation. If sand-tempering had been added a bi-modal curve would result. This experiment demonstrates that at least this brick-yard did not add sand for tempering, although sand was used to prevent the clay from sticking to the moulding surfaces, and can be seen adhering to underside surfaces of some tiles.

From the results discussed above it would seem that tiles stamped TPF, and made in fabric 1, were produced at or near Minety. The distribution shows that they travelled about 10 kilometres into Cirencester and to sites to the west of Minety, even as far as Hucclecote Villa some 40 km away,

TPF tiles of fabric 1 only constitute one third of the TPF tiles at Hucclecote, the other two tiles are of fabric 2. This fabric can be seen on Plate 18.Id. It is much finer grained than fabric 1, having only rare quartz grains over 0.1 mm across. It contains fine flecks of mica, and no plagioclase felspar.

Tiles in fabric 2 have only been identified at Hucclecote villa. Samples of Lower Lias clay, on which the site stands, show a very close similarity to the fabric 2 tiles, which are also softer than those in fabric 1. Plate 18.Ie shows a sample of Lower Lias clay from eastern Gloucester, not far from the villa itself, and the similarity can be seen. It is also the case that the fabric 2 tiles bear stamps (TPF f) which are found in large numbers in Cirencester, where they are of fabric 1. This demonstrates that the brickcaster or the stamp alone must have moved, a situation predicted by Mrs. Clifford (1955, 69), but until now unproven. Richardson and Webb point out that the Middle and Lower Lias clays are not very good for brick-making and suggest that "unless a considerable demand for such a class of brick be assured beforehand, it is unadvisable to attempt to make bricks out of the blue Lias clay" (Richardson and Webb 1910, 246). This seems to illustrate exactly the evidence from Hucclecote. Good quality bricks, for various functions, were brought in from a number of sources, including Minety and also RPG tiles from Gloucester. Some lower quality tiles were also required (unfortunately both surviving fragments of fabric 2 tiles cannot be accurately classed by function, although one may be part of a _tegula_) and these were more economically made on the site by brickcasters from Minety. It is likely that a clamp kiln was used for this; examples of which are claimed from this country (cf. Grimes, 1930). Ethnoarchaeological analogy of a nineteenth-century kiln at Wijk bij Duurstede in the Netherlands shows that the sort of evidence we might expect to find would be "four black tracks, each about 60 cm wide and 4.5-5.0 m long...parallel to each other about 90 cm apart" (Hollestelle 1974, 186), so that clearly very careful excavation would be required to locate and recognise Romano-British examples, even if they have survived at all.

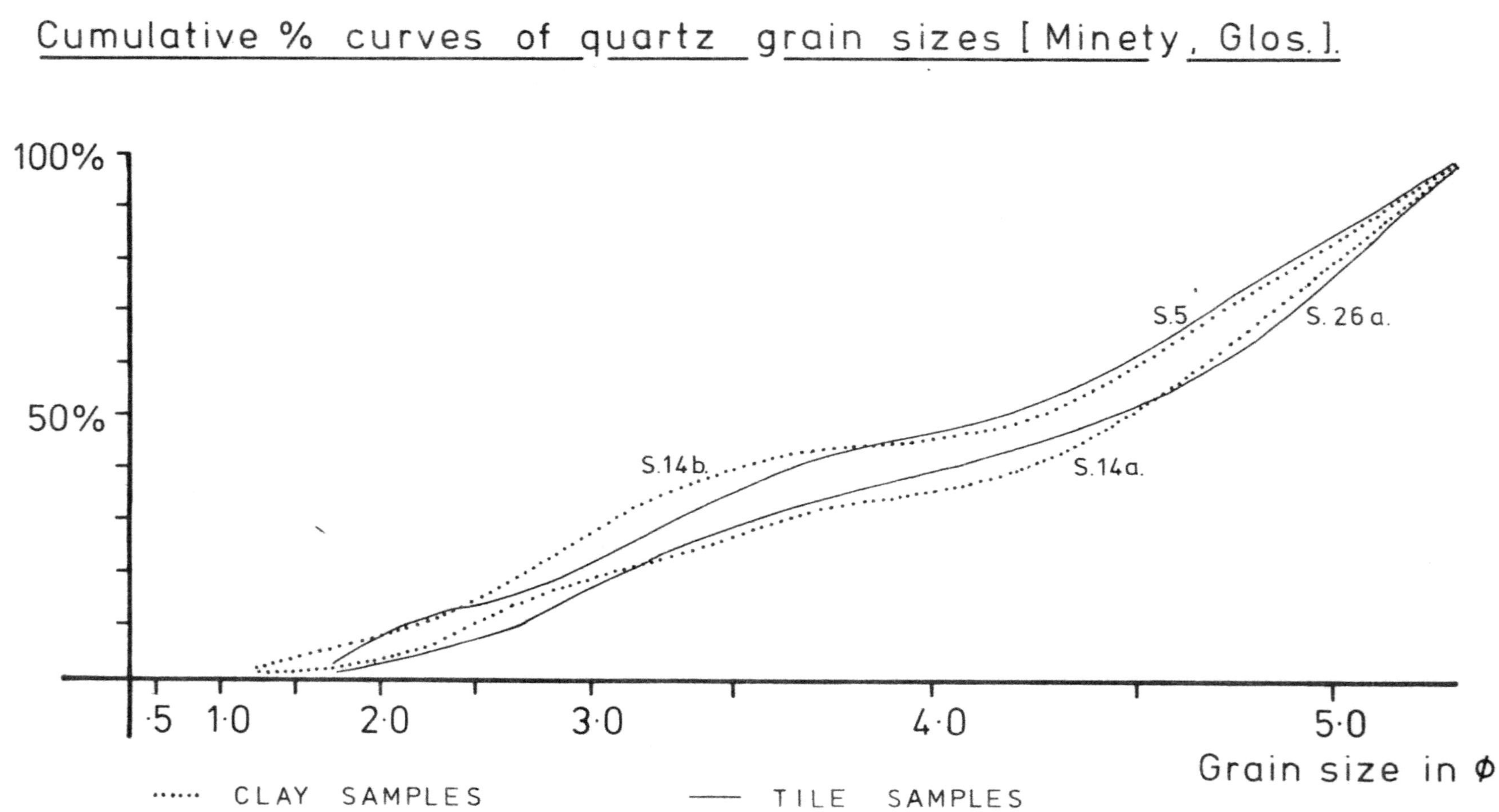
Cumulative % curves of quartz grain sizes [Minety, Glos.].
100%
50%
·5
1·0
2·0
3·0
4·0
5·0
Grain size in ϕ
S.14b.
S.5
S. 26 a.
S.14a.
...... CLAY SAMPLES
— TILE SAMPLES

Fig. 18.5

TPFA Tiles

These tiles show only one fabric group. Fig. 18.4b shows a scattergram of grain-size data. The points are rather spread out, but this is possibly because a slightly different process is being used during manufacture. The TPF brickcasters seem to have used fairly well-mixed clay, while the TPFA tiles show the use of a less well-mixed clay. This means that with small samples the standard deviation, which is represented by lateral distance on the scattergram, is not consistent. Despite this, TPFA tiles also seem to come from Minety. Plate 18.IIa shows a sample of TPFA tile from Hucclecote Villa; and it is clear that it is very similar to the Minety fabrics shown on Plate 18.I (a, b and c).

The suggestion that these tiles come from Minety is further supported by an an analysis of some tiles from Beeches Road, Cirencester by the Oxford Laboratory for Archaeology and Art History. The samples were subjected to optical emission spectroscopy (cf. Tite 1972, 260-264 and Britton and Richards, 1969), with 9 elements being used in the analysis. Fig. 18.6 shows the curves resulting from the first analysis, which was done in 1974. Samples of tiles from Minety were also examined and the curves show the similarity between all the TPF, TPFA, TPFP and Minety samples. Since 1974 techniques of analysis have improved and the apparatus has been recalibrated (H. Hatcher pers. comm.) but since all the samples were tested at the same time comparison within the group is valid. It is noticeable that the largest differences between samples is in the calcium (Ca) and the iron (Fe) values. These elements are however subject to transference either into the tile or out of it due to water percolation through the archaeological deposits, and the calcium count will also be affected by the firing temperature of the brick. Since the optical emission spectroscopy analysis is concerned with the actual chemical composition of the clay minerals (too small to be seen under conventional petrological microscopes) rather than the larger inclusions, it provides an independent check on the textural analysis. The value of integrating fabric analysis and chemical analysis is well illustrated and they should be used more widely in ceramic studies to complement each other.

The number of tiles made at Minety which have been found at Hucclecote seems to preclude interpretation as just a few odd tiles reaching there as second-hand material. It is interesting to note that two classes of the TPFA stamp, the only two which have stops between the letters (Fig. 18.7 TPFA d and e), are found almost exclusively at Hucclecote, only one other example being known, from the Lillyhorn Villa near Bisley (now in the wall of a summer-house at Watercombe) (Baker, 1846; Baddeley, 1906, 178-9 and Clifford 1955, 70, 9b). Two other classes (Fig. 18.7 TPFA a and c) without stops, are found exclusively in Cirencester, although two examples of TPFA a in the Lysons collection in the British Museum probably come from Rodmarton. The remaining group (Fig. 18.7 TPFA b) again without stops is found both in Cirencester and Hucclecote; is this perhaps indicative of special orders being made up, or is it a chronological difference?

TPFC and TPFB Tiles

No examples of the TPFB stamped tiles, of which only two are known, have been examined microscopically; in visual terms however they both appear to

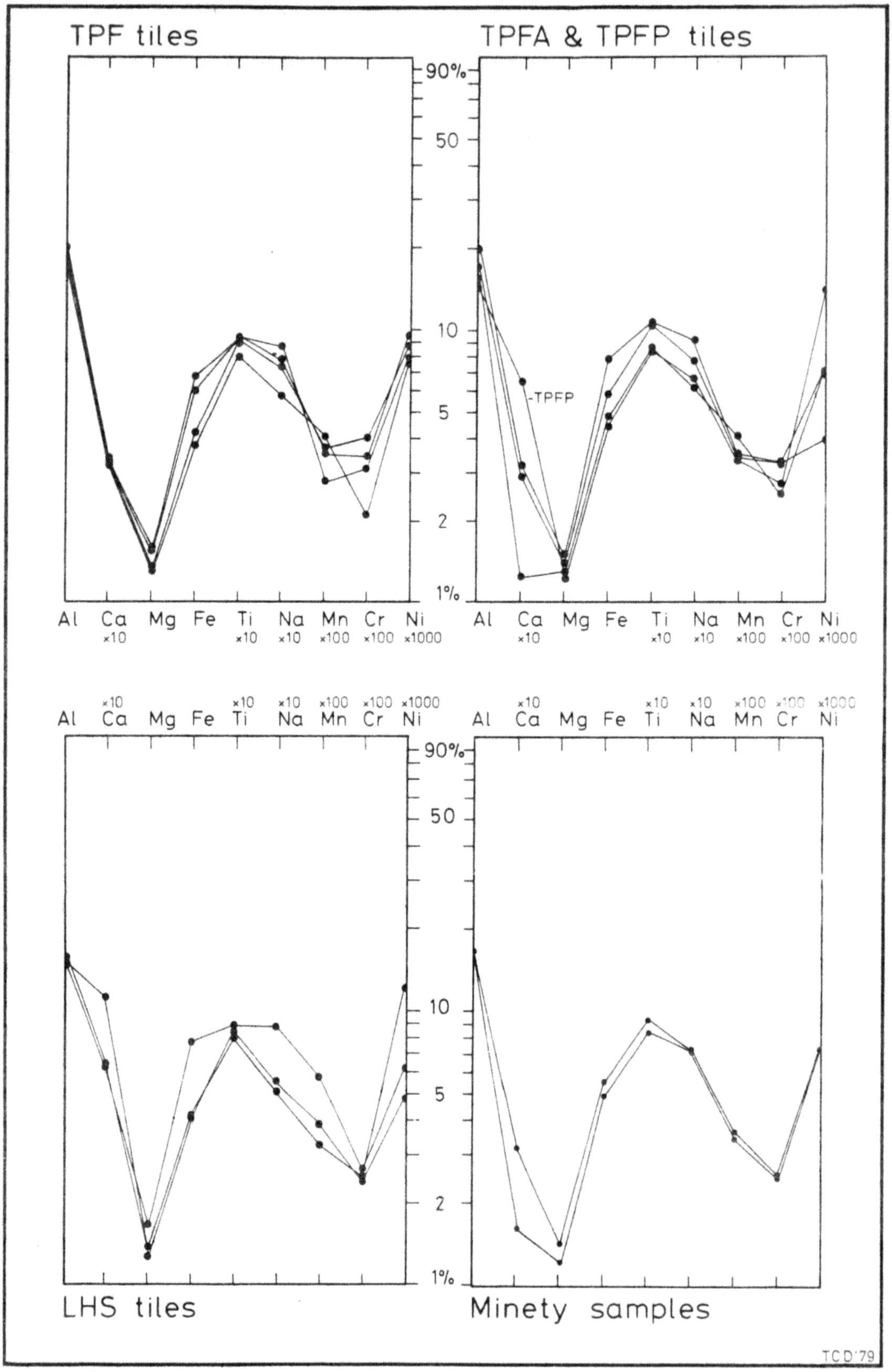

Fig. 18.6 Optical Emission Spectroscopy curves from tiles found at Beeches Road, Cirencester, and Minety

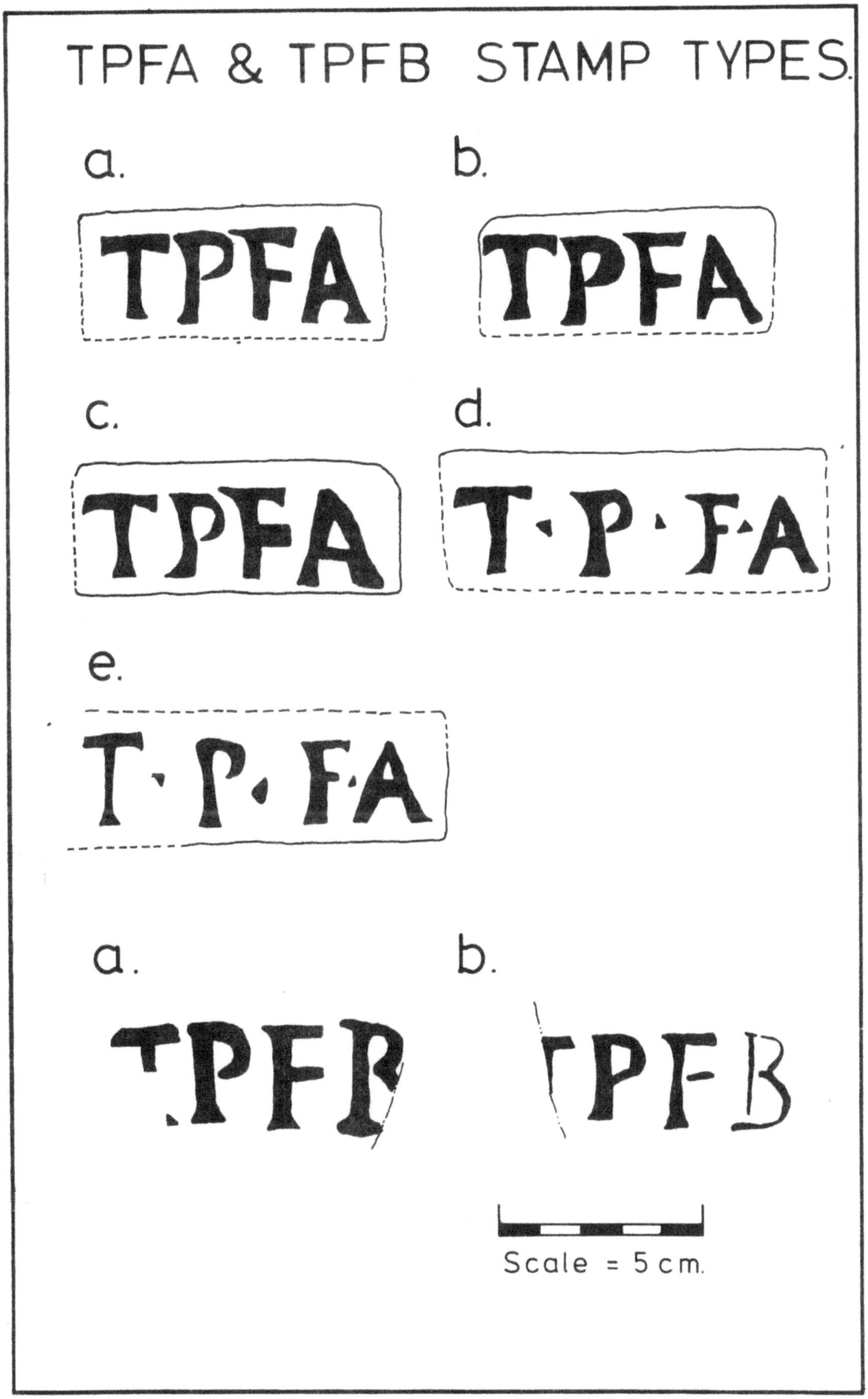
TPFA & TPFB STAMP TYPES.
a.
TPFA
b.
TPFA
c.
TPFA
d.
T·P·F·A
e.
T·P·F·A
a.
TPFB
b.
TPFB
Scale = 5 cm.

Fig. 18.7

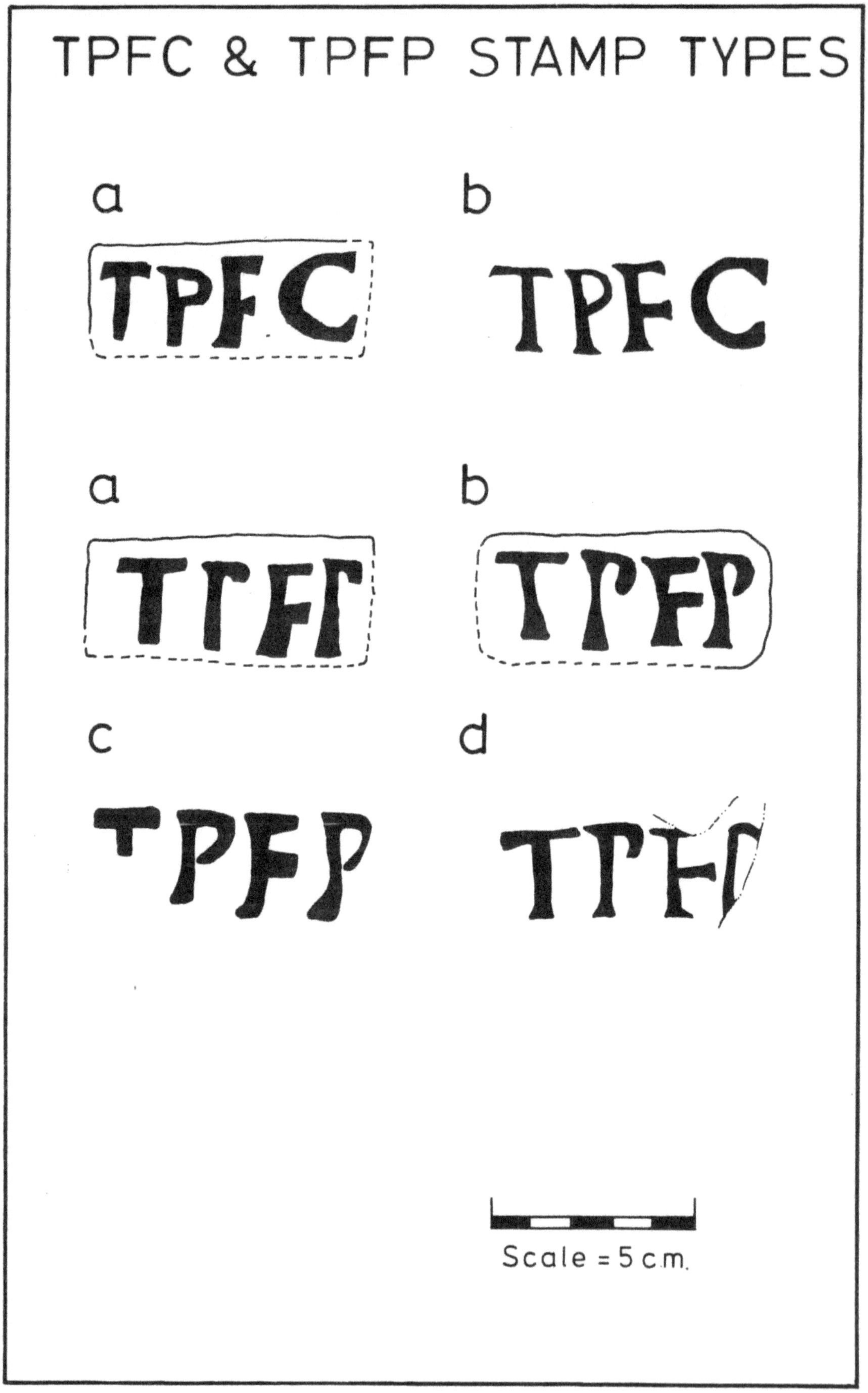

Fig. 18.8

be similar to tiles from Minety. No examples of tiles stamped TPFC have been microscopically examined either, but again the fabrics are visually similar to those from Minety. Fig. 18.7 shows the variants of the TPFB stamps and Fig. 18.8 shows the two types of TPFC stamps.

TPFP Tiles

The die types can be seen on Fig. 18.8, and Fig. 18.9 A shows the scattergram of quartz grain-size statistics. Like the TPFA tiles the graph suffers from sampling and counting problems but only one fabric can be identified. There is one example from Hucclecote which has a similar stamp (TPFP c) to one at Cirencester again showing the movement of tiles. Plate 18.II g shows a photomicrograph of a tile from Beeches Road Cirencester, emphasising the similarity to tiles from Minety. The optical emission spectroscopy analysis discussed above also shows the comparability of the TPFP fabric with those of Minety (see Fig. 18.6).

DISCUSSION OF THE TPF SERIES TILES

It is apparent that all the clusters of points on the scattergrams are not quite in the same place, yet a common origin is postulated. Indeed if all the TPF series data are put on one graph, as in Fig. 18.10, an elongated spread is seen, and even the two Hucclecote tiles of Lower Lias fabric are almost swamped (at 2 on Fig. 18.10). This is unfortunately a product of the method of analysis. The natural variations in the clay and the inherent sampling errors in removing a piece of tile and in counting the quartz grains will result in confusion round the periphery of the clusters. The patterns are however clear.

What is also evident is a change from right to left in the make-up of the cluster on Fig. 18.10. On the right are most of the TPFA samples, in the next bunch are the TPF samples and on the left are the TPFP samples. This is itself interesting as Minety is an extensive site, of about half a square mile. Recent fieldwork by Mike Stone of Swindon Museum has shown that there are at least 10 kilns at the nucleus of the distribution, first spotted by Crawford in 1921. Two stone buildings are also known from the recent work and several clay pits show as parched areas on the fields, which are at present under the plough. The kiln fields are in an area of Oxford Clay which covers a wide geographical area (see Fig. 18.2) and it is quite possible that some of the tile-production sites may not be on the site we call the 'Minety kiln-field' at present. No work has been done on the changes in grain-size distributions over a given area, or down through a clay bed so that it is difficult to assess how dispersed the production sites might be. The combined evidence of the petrological analysis showing a slight variation between each stamp type and the close similarity of the chemical evidence suggests that around Minety there was a concentration of production. The Oxford Clay beds dip towards the north and a borehole in Lewis Lane, Cirencester, drilled in the 1920s, showed that clay was 28 feet below the surface, being covered by alluvial gravel (Geological memoirs for the Cirencester region p. 92).

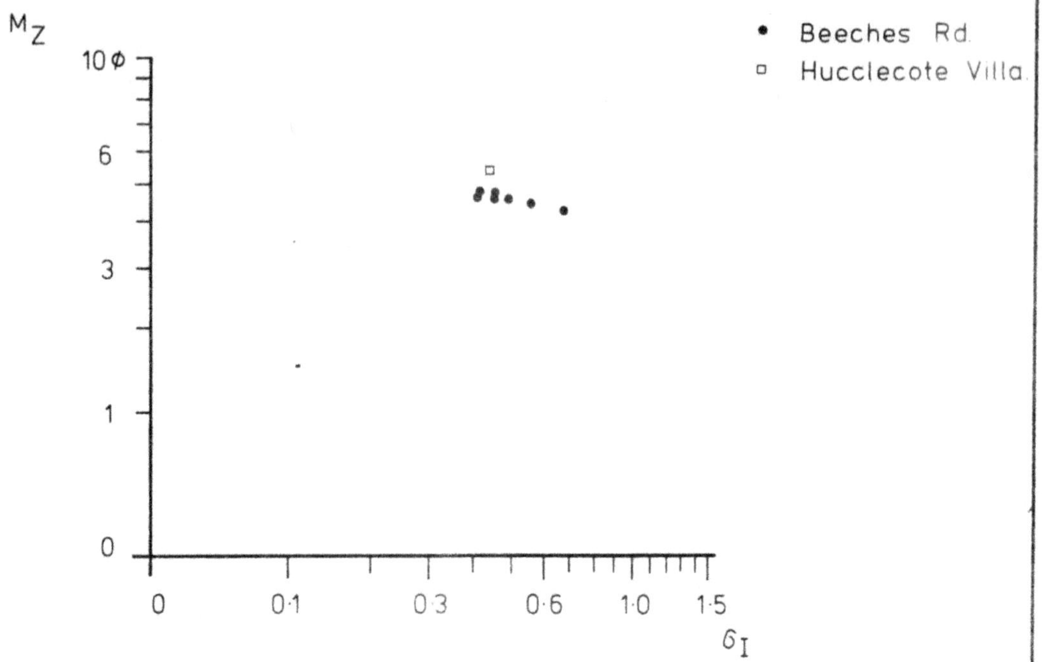

A. Scattergram of TPFP material.

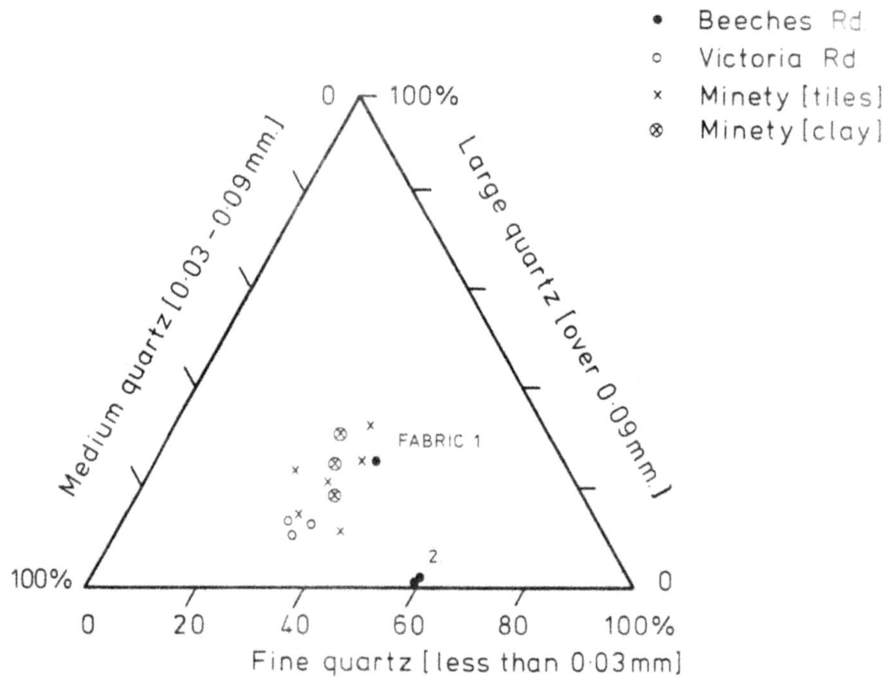

B. Ternary diagram of quartz grain sizes.

Fig. 18.9

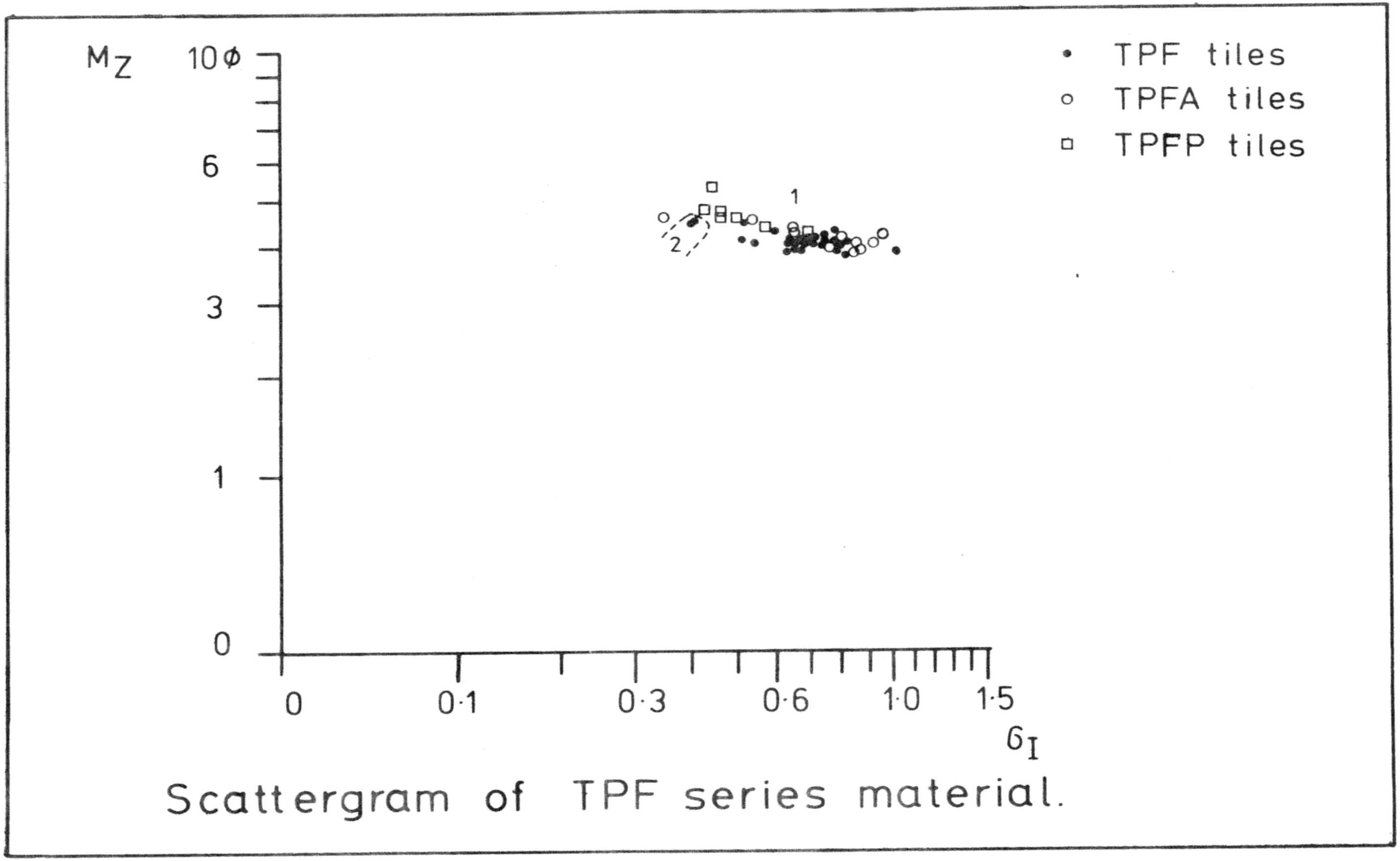

M_Z
10Φ
6
3
1
0
0
0·1
0·3
0·6
1·0
1·5
σ_I
1
2
• TPF tiles
○ TPFA tiles
□ TPFP tiles
Scattergram of TPF series material.

Fig. 18.10

LHS tiles

This is the most geographically widespread group of tiles discussed here. They are concentrated in Cirencester but there are examples to the south along Ermin Street at Wanborough and further afield at Old Sarum and Silchester. A recent find of three fragments at Kenchester extends the distribution over 75 kilometres to the north-west of Cirencester too. The stamp dies are variable and can be seen on Fig. 18.11.

It is possible to distinguish 2 fabric groups. These are contrasted on Plate 18.II where h shows a sample of fabric 1 from Victoria Road, Cirencester, and i shows a sample of fabric 2 from Beeches Road, Cirencester. There are only 3 examples of fabric 2 at present.

Fabric 1 is identical with the Minety fabric described under the TPF fabric 1 above. Four examples of the LHS stamp have been found on the kiln site at Minety, and there is no doubt that these tiles were made there. The optical emission spectroscopy curves also highlight the similar chemical composition to other groups originating at Minety (see Fig. 18.6). On Fig. 18.9 the fabric groups can be seen on a slightly different method of data presentation. A visual examination of the Wanborough tiles shows them to be of Fabric 1. The Silchester example is also visually of fabric 1 and stamped with a rough stamp which is very similar to an unprovenanced example in the Corinium Museum, although almost certainly from within Cirencester. The two examples from Old Sarum are interesting as the stamps show a total contrast to the other LHS stamps (see Fig. 18.11 f and g). These two tiles appear to be of fabric 1, although they have fragments of chalk in them. Chalk does occur in the other LHS fabric 1 tiles, and sometimes in other Minety tiles. More often there are voids however where some calcareous material, maybe chalk, has burnt out during firing. The Old Sarum chalk fragments are also rather larger than the usual pieces. The difference in stamp type may suggest a chronological break with the other LHS tiles, thus accounting for the differences in chalk inclusions, which were unlikely to have been added deliberately as this would cause cracks in the bricks, as is shown on the Silchester example. It is more likely that fragments of this sort got into the surface of the brick, which is where the large pieces were, during building or construction work at the drying site. The similarity in clay fabric and in the form of stamp in comparison to other Minety type stamps (e.g. TPF series stamps) does however suggest a common origin.

Fabric 2 is much finer than fabric 1, and contains very much less quartz and lacks the fine mica found in the Oxford clays (see Plate 18.II i). At present the source of this second fabric is unknown. What is interesting is that just one example of the ARVERI stamped tiles which, although they are mostly confined to Cirencester, in this case comes from Kingscote, is of a very similar fabric, and can be seen for comparison on Plate 18.II j. It is tempting to think of two brickcasters setting up on clays not usually used by them, and making their usual bricks, the ARVERI stamped tiles going to Kingscote and the LHS tiles to Cirencester. Further examples of LHS fabric 2 and ARVERI tiles like the one from Kingscote need to be examined to clarify this aspect of the industry.

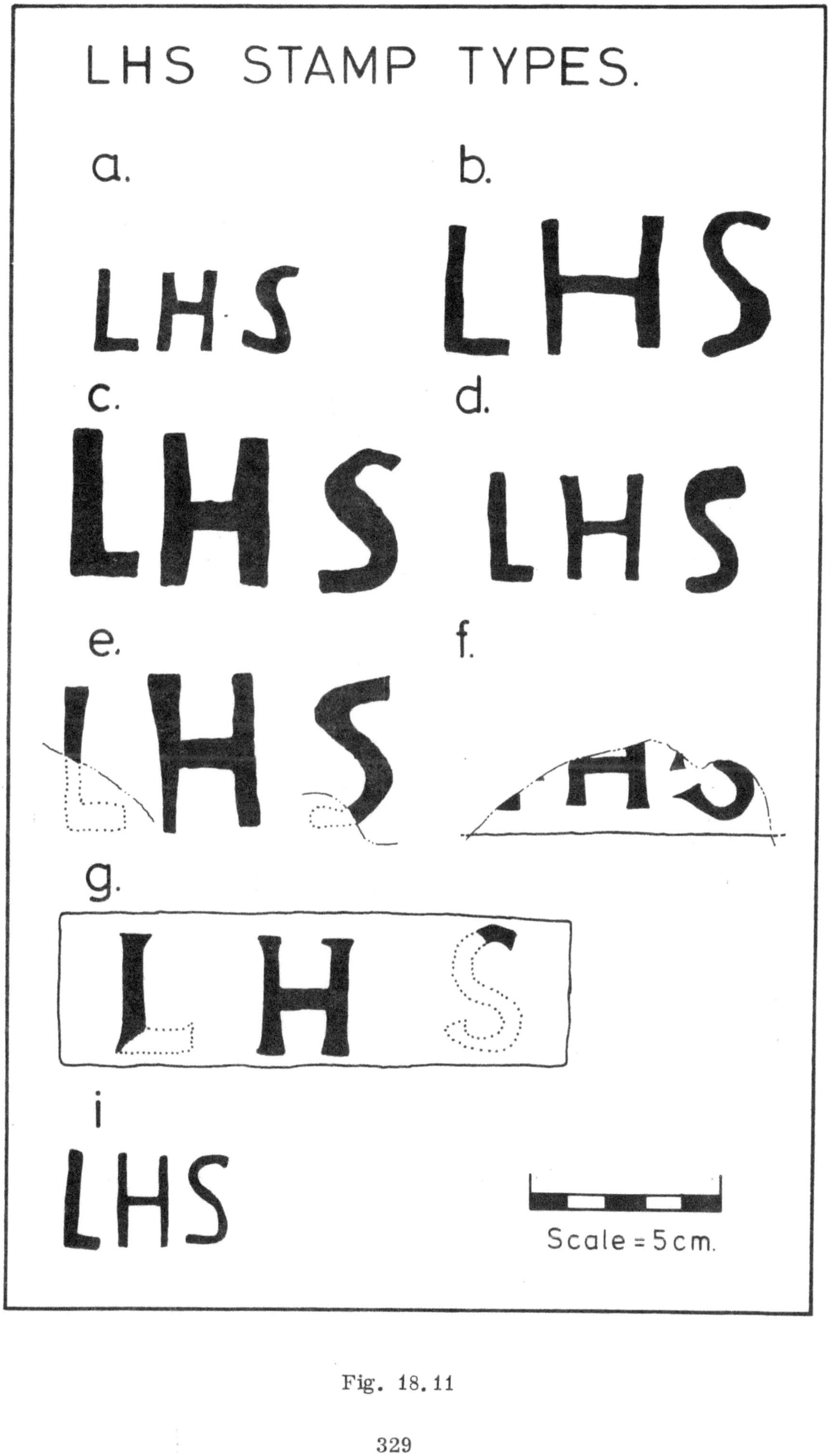
LHS STAMP TYPES.
a.
LHS
b.
LHS
c.
LHS
d.
LHS
e.
LHS
f.
g.
LHS
i
LHS
Scale = 5 cm.

Fig. 18.11

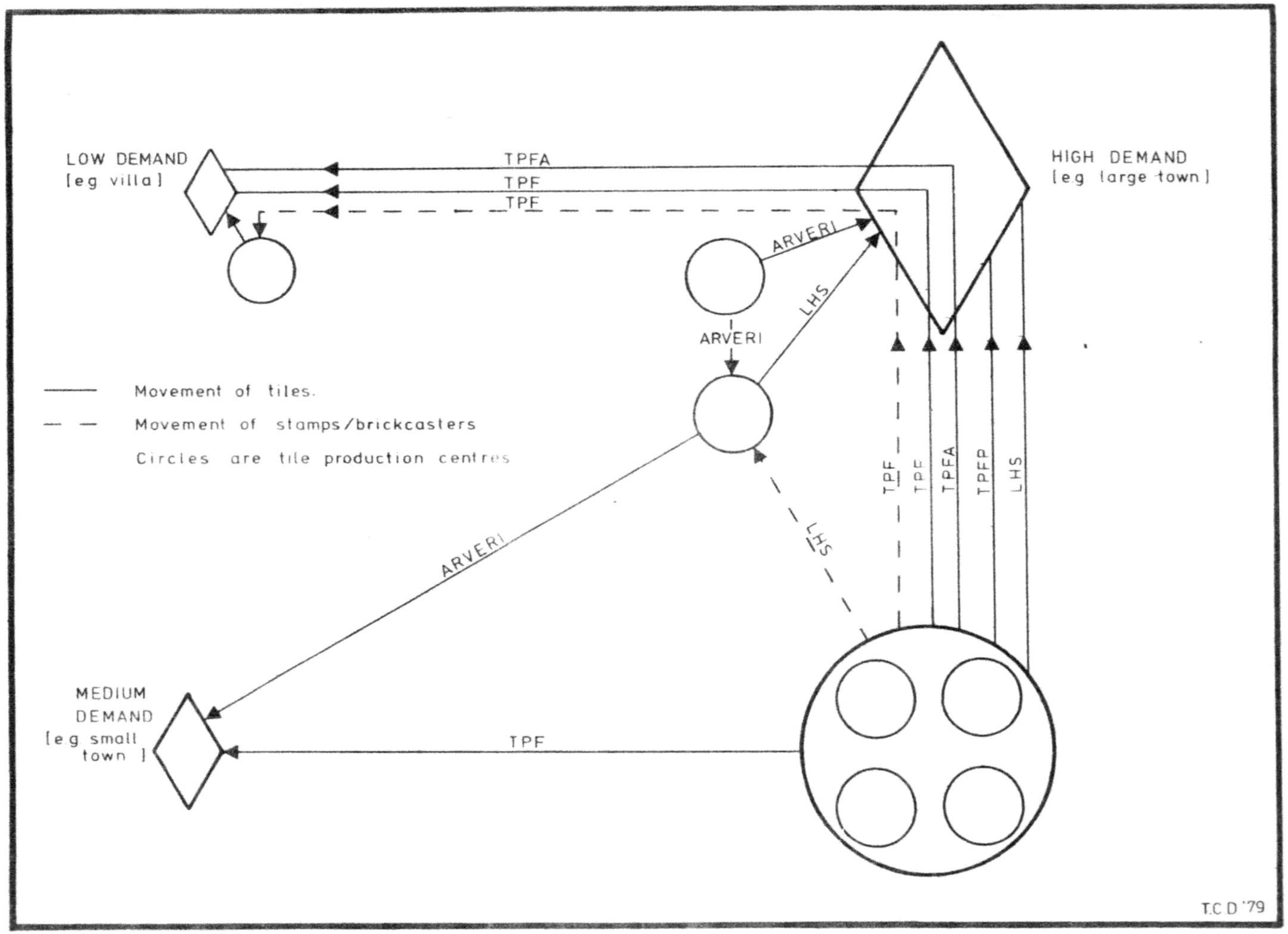

Fig. 18.12 Summary diagram of tile production and demand centre interaction

CONCLUSION

The first important conclusion is that the tile industry seems to be considerably more complex than was previously thought. The study of stamped tiles from the Cotswold region shows the presence of several levels of production. The highest level of production so far examined with large numbers of related stamps over a wide geographical area, allows several facets of this sector of the industry to be seen. The combination of petrological and chemical analysis of the tile fabrics and the study of the stamps themselves ensures validity of the results. The study of this large group of tiles also illustrates the potential, and limitations, of textural analysis.

Within the sector of the industry examined here it is possible to see three points emerging. These are summarised in diagrammatic form, as identifiable activities of this group of brickcasters, on Fig. 18.12

1). First the evidence suggests a large production centre of some kind, making large numbers of tiles. The production was concentrated in a limited, although as yet not totally defined area. This could be called a nucleated industry. The scale of the Minety site as revealed by field work shows a well-organised production-unit; the kilns are arranged in a linear group, with clay pits some distance away, but on a similar line. The presence of stone buildings suggests investment to modify the local environment to allow drying over a longer production season, as well as protection for workers, further allowing work over more of the year than could be possible in the open. The size of the kilns, as revealed by recent fieldwork and exploration, suggests large and complex structures representing considerable capital investment.

The economic benefits of this scale of production can be seen in the wide spatial distribution of the products, especially shown in the LHS stamped tiles. While the actual site of the Minety kiln fields might be thought by some to be small for a true nucleated industry, it is very likely that more kilns remain to be found in the neighbourhood. The evidence seems to suggest that there was a series of individual workshops within the complex. The connections shown by the series of TPF stamps might even be used to suggest kinship links, and if this was so the LHS stamps might show that the industry was not a totally family-based concern. At first glance there are differences between the output of each unit. The tiles in the Lysons collection in the British Museum, probably from Rodmarton, show a neat correlation of firing conditions with stamp, although the fabrics seem to be essentially the same. The tiles stamped TPFA have a central core of reduced clay, while those stamped TPFC are a lighter colour with no reduced core, and those stamped TPFP also have no core but are a darker red in colour.

It is not clear whether middlemen operated or not, but the make-up of the tile assemblages available suggests that demand was met collectively; a feature of some but not all nucleated industries. The demand from a given building project may well have been too high for any one workshop, so giving rise to products from a number of workshops being present. The case of Sandgate Castle in Kent built in 1539-40 provides a good ethnoarchaeological parallel. Some 147,000 bricks were used, and they came from at least 13 different brickworks (Rutton, 1893, 236).

The distribution of tiles emphasises the ease of transport along Ermin Street, giving rise to the long-distance movement. The distance/cost effects of arterial routeways is emphasised by Hodder and Orton when considering explanations for archaeological distributions, particularly those of heavy goods (1976, 184). These patterns are well shown in the evidence here. The cause of the pattern might well be the superior quality of the bricks produced. That they are hard has been noted above and the need for such a brick in the construction of larger buildings should not be underestimated.

Although Cirencester might seem to be the main market for this industry, relative quantities may be over emphasised by the concentration of archaeological activity in the town. It is clear that the Minety complex is well sited to provide for Wanborough and Kingscote as well as the numerous villas in the central Cotswolds and central Severn valley. That each stamped tile represents a larger quantity of tiles going to a site seems likely, and it would be interesting to examine larger assemblages at these sites.

The location of the Minety complex might well be determined by economic factors of supplying a range of markets, both in scale and space. Legal and social pressures on the location are heavy and being an anti-social industry may have prevented it from being nearer, and so more dependent on, Cirencester. The availability of raw materials such as clay, wood and water are important too. The pressure on land use for agriculture could also preclude the use of such a large area of land nearer Cirencester. Once established, the presence of a pool of skilled labour also tends to sustain nucleation (cf. Chisholm, 1962, 37; Freeman, 1967, 116 and Hamilton, 1968, 362).

2). The second facet of the industry seems to be that there was an itinerant element in meeting demand which might well be for lower quality bricks. The use of this mode of production seems most appropriate to the lower demand centres where, once initial needs are met, little more is required for a time. The medium-sized demand centres at both Wanborough and Kingscote obtained at least some of their tile from the large industry, and the low demand centres, like Hucclecote used what are possibly the better bricks for jobs involving high loads and stresses. Perhaps, also, important roofs would be better covered with these as the bricks fired at a higher temperature were also less porous.

3). The third facet is at present only hinted at, and may simply be a variation on 2 above. The evidence at present suggests that brickcasters combined in different combinations to make tiles.

The results obtainable from the study of tiles is directly proportional to the quality of the data. The results published here represent a backbone, on to which others are invited to contribute. When whole tile assemblages are available it may be possible to add some real figures and proportions to the economics of the industry. With increased awareness of the importance of bricks and tiles much new data is anticipated over the next few years and then much more detail will be able to be filled in.

It is hoped that the results from the analysis of other groups of stamped tiles from the Cotswolds will soon be ready so that the other levels of production spoken of above can be better elucidated.

ADDENDUM

Subsequent to completing the body of this paper it has been possible to examine, in thin section, samples of the three recently found tiles stamped LHS from Kenchester. One of the three had been fired at a temperature which vitrified the clay, producing a very hard tile. The other two examples are more typical of the output of LHS tiles from Minety and the new microscopic examination shows that they all fall within the range of fabrics produced at Minety. The three LHS tiles from Kenchester are a separate group within the Kenchester assemblage of tiles, and a petrological examination of one RPG stamped tile and one PPG stamped tile emphasised the differences in fabric. This find clearly reinforces the wide spatial distribution of the LHS tiles and the large scale of the industry producing them

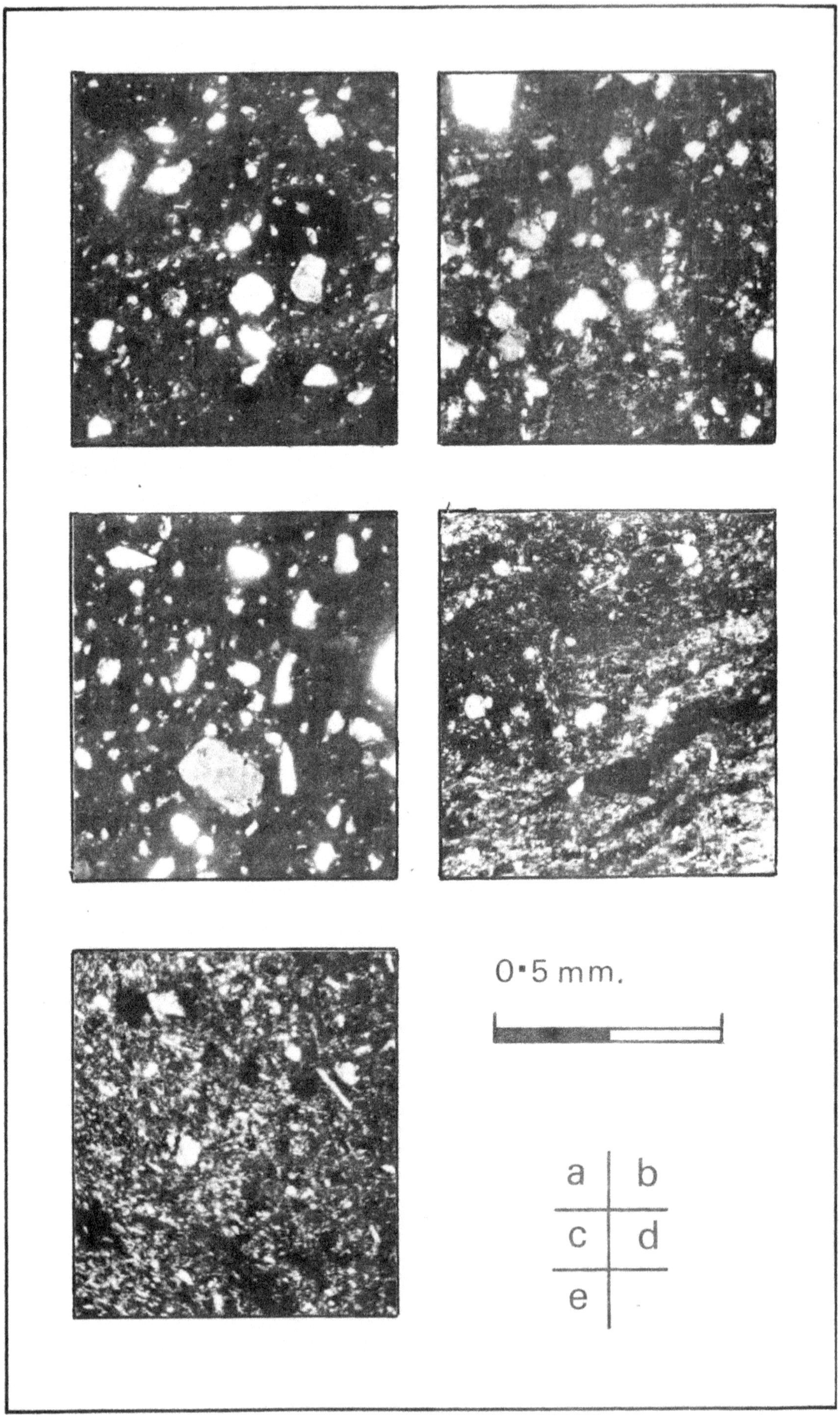
0·5 mm.
a
b
c
d
e

Plate 18.I Photomicrographs of tile sections and clay samples

a. tile sample from Minety (unstamped)
b. clay sample from the top of the most easterly clay pit at Minety
c. TPF stamped tile from Beeches Road, Cirencester, fabric 1
d. TPF stamped tile from Hucclecote Villa, near Gloucester, fabric 2
e. clay sample from eastern Gloucester, Lower Lias clay

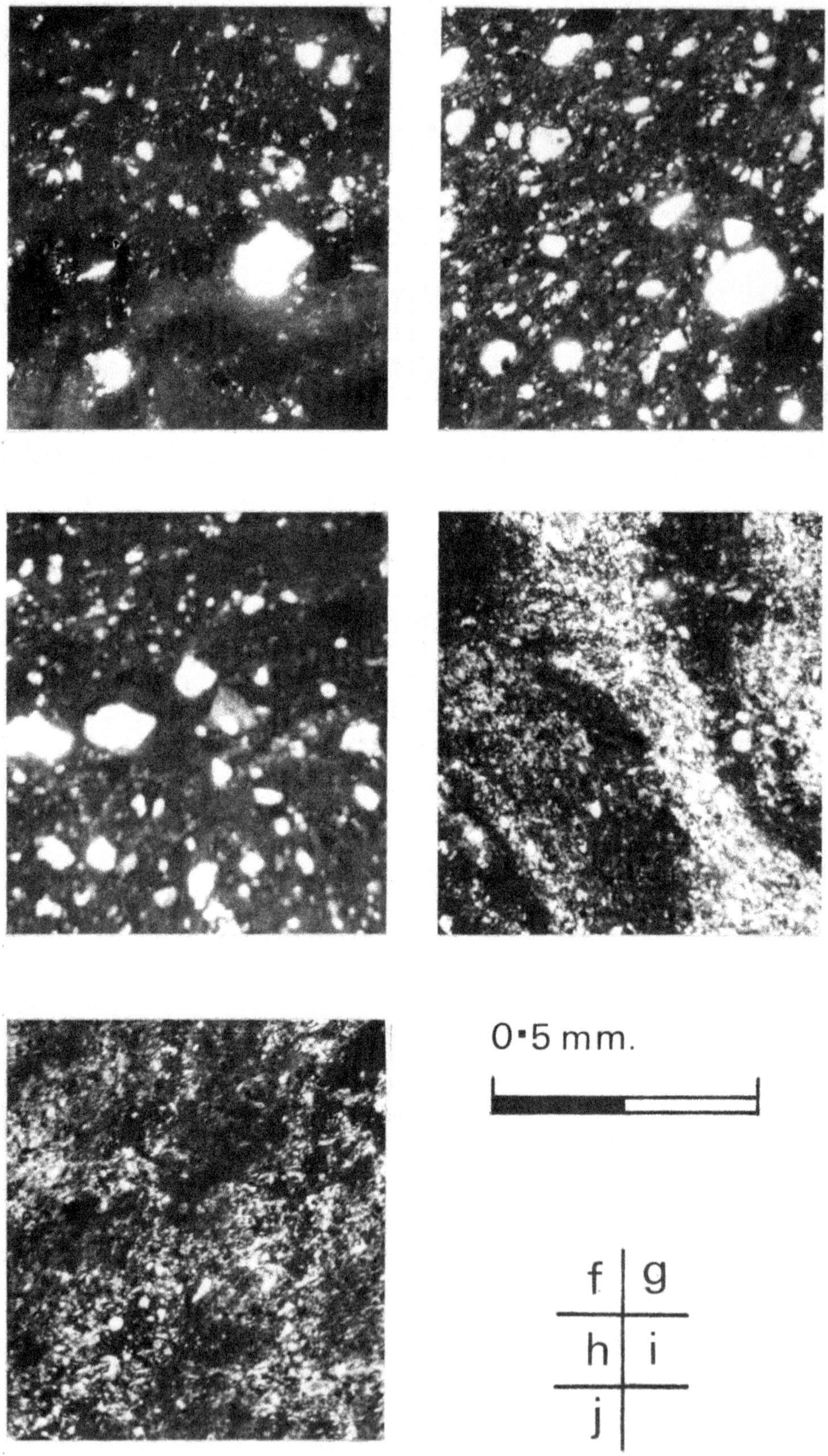
0·5 mm.
f
g
h
i
j

Plate 18.II Photomicrographs of tile sections

f. TPFA tile from Hucclecote Villa
g. TPFP tile from Beeches Road, Cirencester
h. LHS tile from Victoria Road, Cirencester, fabric 1
i. LHS tile from Beeches Road, Cirencester, fabric 2
j. ARVERI tile from Kingscote

APPENDIX I

CATALOGUE OF TILES STAMPED LHS OR IN THE TPF SERIES

NOTE

This catalogue of stamps and petrological groupings was compiled from data available up until June 1979. Some tiles have not yet been assigned to a stamp group, although petrological data may be available, and so these have been listed at the end of the catalogue. This appendix should be used in conjunction with Figs. 3, 7, 8 and 11 in the text which illustrate a representative example of each group. It must be remembered that in some cases stamps are not complete, making classification more difficult. Within each die group there will be variation between examples. It is rare to find stamps evenly impressed and in most cases some letters are deeper than others, and since letters often have a ‾_/‾ profile this will affect the letter width. More important however is the variation caused by clay shrinkage. Stamps are usually applied before the tile is "leather hard", so that shrinkage during drying and firing will occur. The amount of shrinkage will depend on the chemical composition of the clay; fluxes like felspar and mica increase the shrinkage while refractory materials like alumina reduce shrinkage. The grain size of the clay particles and the admixed impurities also affect the rate of shrinkage. Rosenthal suggests an overall shrinkage during drying and firing would be between 8 and 22% (1949, 67). When more tiles are available it may be necessary to reclassify some examples. The groups suggested in this classification are on the basis of differences in the letter shape and spacing. The use of a light table and rubbings on thin paper seems to be the most efficient way of comparing stamps.

For references to the find spots of tiles see McWhirr and Viner 1978.

<u>Abbreviations used in Appendix I</u>

Chelt.	Cheltenham Museum and Art Gallery
CIR	Cirencester excavation site code
C/M	Corinium Museum Cirencester
GCM	Gloucester City Museum
BM	British Museum
KAA	Kingscote Archaeological Association
M	Sample microscopically examined in thin section
V	Macroscopic examination of the sample
NE	Not examined
*	Stamp die is very similar to the others in the group
≠	Sample used in Optical Emission Spectroscopy

Thin section numbers; R000 is the University of Southampton thin Section Library number. S00 is the sample number for sections not in the main collection. See Appendix II for details of sections.

Provenance	Museum Registration	Fabric group	Examination	Thin Section No.
TPF (a)				
no stops, no surround, serifs on T, thick backed F				
Cirencester	CIR 61 D vi 10		NE	
Cirencester	CIR 62 R I	1	V	
Cirencester	CIR 64 H v 43		NE	
Minety	Swindon Museum	1	M	S28 *
TPF (b)				
no stops, no surround, serifs on all letters				
Cirencester	CIR 61 E vi 4	1	V	
TPF (c)				
no stops, frame and well-shaped letters				
Cirencester, Watermoor House	C/M A265i	1	M	R284
Cirencester, Watermoor House	C/M A265ii	1	M	R285
Cirencester, Victoria Road	Chelt. 1922/277/h	1	M	R280
Cirencester, Victoria Road	Chelt. 1922/277/1	1	M	R281
Cirencester, Victoria Road	GCM A 1419	1	M	R282
Hucclecote Villa	Chelt. 1970/82/7	1	M	R297
Kingscote	KAA B5 1977	1	M	R202 *
Kingscote	KAA Ax 2	1	M	R203 *
Kingscote	KAA B5 1977	1	M	R204 *
Kingscote	KAA D 1977	1	M	R205 *
TPF (d)				
stops, frame, and clear serifs on all letters				
Cirencester, Beeches Road	CIR 73/9	1	M	R258
Cirencester, Beeches Road	CIR 73/7	1	M	R257
Cirencester, Beeches Road	CIR 73 M384	1	M	R245
Cirencester, Beeches Road	CIR 73 M342	1	M	R266
Cirencester	C/M B1417 iii	1	V	
Cirencester	C/M B1415	1	V	
TPF (e)				
stops, frame and narrow letters				
Cirencester, Beeches Road	CIR 73 M385	1	M	R254
Cirencester, Beeches Road	CIR 73 M368	1	V	
TPF (f)				
stops and frame, well-cut letters				
Cirencester, Beeches Road	CIR 73 M391	1	M	R214
Cirencester, Beeches Road	CIR 73 M332	1	M	R251

Provenance	Museum Registration	Fabric group	Exami- nation	Thin Section No.	
TPF (f) (contd.)					
Cirencester, Beeches Road	CIR 73 M389	1	M	R231	
Cirencester, Beeches Road	CIR 73 M410	1	M	R256	
Cirencester, Beeches Road	CIR 73 M393	1	M	R242	
Cirencester, Beeches Road	CIR 73 M381	1	M	R253	
Cirencester, Beeches Road	CIR 73 M388	1	M	R255	
Cirencester, Beeches Road	CIR 73 M331	1	M	R264	#
Cirencester, Beeches Road	CIR 72 M311	1	M	R243	#
Cirencester, Beeches Road	CIR 73 M400	1	M	R218	#
Cirencester, Victoria Road	Chelt. 1922/277/B	1	M	R278	
Cirencester, Victoria Road	Chelt. 1922/277/D	1	M	R277	
Cirencester, Victoria Road	Chelt, 1922/277/G	1	M	R279	
Cirencester, Victoria Road	GCM A1492	1	M	R283	
Hucclecote Villa	GCM 5315	2	M	R298	
Hucclecote Villa	Chelt. 1970/82/8	2	M	R296	
TPF (g) <u>stops, frame and serifs on the letters</u>					
Cirencester, Victoria Road	Chelt. 1922/277/E	1	M	R276	
Wanborough	D. vi 14	1	V		
TPFA (a) <u>no stops, no frame, closely set letters</u>					
Cirencester	C/M B1411		NE		
Cirencester	C/M A314	1	V		
Cirencester	CIR 71 Q 266	1	V		#
Rodmarton	BM Lysons Col. 31	1	V		
Rodmarton ?	BM Lysons Col	1	V		
TPFA (b) <u>no stops, no frame, long A foot</u>					
Cirencester	CIR 1971 Q 267ii	1	V		#
Cirencester	CIR 72 X 304	1	M	R217	#
Cirencester	C/M		NE		
Hucclecote Villa	Chelt. 1970/82/3	1	M	R291	
Hucclecote Villa	GCM A 1860	1	V		
TPFA (c) <u>no stops, no frame, uneven letters</u>					
Cirencester, Beeches Road	CIR 73/1	1	M	R234	
Cirencester, Beeches Road	CIR 73/2	1	M	R232	
Cirencester, Beeches Road	CIR 73/8	1	M	R240	
Cirencester	C/M B1413	1	V		
Cirencester	C/M B1414	1	V		

Provenance	Museum Registration		Examination	Thin Section No.	
TPFA (d)					
stops, frame, widely spaced letters					
Hucclecote Villa	Chelt. 1970/82/4	1	M	R290	
Hucclecote Villa	Chelt. 1970/82/5	1	M	R292	
Hucclecote Villa	Chelt. 1970/82/6	1	M	R295	
Hucclecote Villa	GCM A1103	1	V		
Hucclecote Villa	GCM A1861	1	M	R293	
TPFA (e)					
stops, frame and a thin T and wide A					
Hucclecote Villa	GCM A1104	1	V		
Lillyhorn Villa, T2	Private		NE		
TPFB (a)					
no stops, no frame, chunky letters					
Cirencester ?	C/M 1973	1	V		
TPFB (b)					
no stops, no frame, TPF with hand drawn B					
Easton Grey	Swindon Museum	1	V		
TPFC (a)					
no stops; frame closely set letters with out-of-proportion C					
Cirencester	CIR 61 L vi 1	1	V		
Cirencester	C/M B1409 i		NE		
Cirencester	C/M B1409 ii		NE		
Rodmarton ?	BM Lysons Coll.	1	V		
Rodmarton ?	BM Lysons Coll.	1	V		
Rodmarton ?	BM Lysons Coll.	1	V		
TPFC (b)					
no stops; frame, irregularly set letters and large C					
Stanton Fitzwarren 1969	Swindon Museum	1	V		
TPFP (a)					
no stops; frame and hollow based Ps					
Cirencester, Beeches Road	CIR 73/3	1	M	R230	≠ ?
Cirencester, Beeches Road	CIR 73/4	1	M	R261	
Cirencester, Beeches Road	CIR 73/11	1	M	R229	
Cirencester, Beeches Road	CIR 73/12	1	M	R235	
Cirencester, Beeches Road	CIR 73/13	1	M	R236	
Cirencester	CIR 71 M267		NE		

Provenance	Museum Registration	Fabric group	Exami- nation	Thin Section No.
TPFA (a) (contd.)				
Cirencester	C/M A265iii		NE	
Cirencester	C/M C2810c	1	V	
Cirencester	C/M B 1417i	1	V	
Cirencester	C/M C2808i	1	V	
Cirencester	C/M C2808ii		NE	
TPFP (b)				
<u>no stops, frame, full Ps</u>				
Cirencester, Beeches Road	CIR 73/5	1	M	R262
Cirencester, Beeches Road	CIR 73/6	1	M	R263
Cirencester, Beeches Road	CIR 73/10	1	M	R237
Cirencester	C/M B 1397		NE	
Cirencester	C/M B1398	1	V	
Cirencester	C/M B1416i		NE	
Cirencester	C/M C2807ii	1	V	
Cirencester	C/M C2807ii		NE	
Cirencester, Phoenix Way	C/M	1	V	
Lillyhorn Villa	Private possession		NE	
Lillyhorn Villa	Private Possession		NE	
Lillyhorn Villa	Stroud Museum		NE	
TPFP (c)				
<u>no stops, no frame, close letters and full Ps</u>				
Cirencester, Beeches Road	CIR 73/16	1	V	
Hucclecote Villa	GCM A 1850	1	M	R307
TPFP (d)				
<u>no stops, no surround; serifs on letters and distinctive F</u>				
Rodmarton	BM Lysons Coll.	1	V	
Rodmarton ?	BM Lysons Coll.	1	V	
LHS (a)				
<u>no stops; no surround, rough letters distinctive H and S, some variety in size</u>				
Cirencester, Beeches Road	CIR 73 Y 343	1	M	R250 #
Cirencester, Admirals Walk	CIR 74 T 464	1	V	
Cirencester, Victoria Road	Chelt. 1922/277/c	1	M	R272
Cirencester, Victoria Road	Chelt. 1922/277/F	1	M	R271
Cirencester, Victoria Road	GCM 1490	1	M	R273
Cirencester	C/M C2803i		NE	
Cirencester	C/M C2803ii		NE	
Cirencester	C/M C2803iii		NE	
Cirencester	C/M C2803iv		NE	

Provenance	Museum Registration	Fabric group	Examination	Thin Section No.	
LHS (b)					
no stops; no frame; large crude letters with wide H					
Cirencester, Beeches Road	CIR 73/14	2	M	R228	
Cirencester	CIR 65 L iii 66	1?	V		
Cirencester	C/M B1400	2	V		
Minety	Swindon Museum	1	V		*
LHS (c)					
no stops; no frame; narrow thick letters					
Cirencester	C/M B1401ii	1	V		
Silchester	Reading Museum 00131	1	V		*
LHS (d)					
no stops, no frame, short bottom to the L and an uneven S					
Cirencester, Beeches Road	CIR 72 Y 330	1	M	R225	≠≠
LHS (e)					
no stops, no frame, wedge shaped letters					
Cirencester, Beeches Road	CIR 1973 Y 386	2	M	R224	
Cirencester	C/M B1402i	2	V		
LHS (f)					
no stops; frame; widely spaced letters with serifs					
Stratford sub-castle, Old Sarum	Salisbury Museum	1	V		
LHS (g)					
stops, frame; closely set even letters with serifs					
Old Sarum	Salisbury Museum 57 Ba4	1	V		
LHS (i)					
no stops, no frame; small narrow letters					
Kenchester	With post-excavation team at Worcester T6	1	V		
Kenchester	as above, BR 1	1	V		
Kenchester	as above, BR 4	1	V		

OTHER TILES NOT FULLY CLASSIFIED

Minety, three LHS stamps, fabric 1
Lillyhorn Villa, one TPF tile in private possession
Frocester Court Villa, one TPF tile, NE
Wanborough, two TPF tiles and two LHS tiles

APPENDIX II

Fabric analysis statistics from tiles and samples microscopically examined

Section No.	Mean grain size M_Z	Standard deviation 6_I
R202	4.20	0.48
R203	4.10	0.71
R204	4.26	0.54
R205	4.05	0.75
R214	4.27	0.71
R215	4.30	0.95
R217	4.58	0.34
R218	4.17	0.72
R224	5.21	0.27
R225	4.75	0.84
R226	4.29	0.66
R228	5.23	0.29
R229	4.37	0.65
R230	4.58	0.41
R231	4.00	0.82
R232	4.41	0.61
R234	4.04	0.82
R235	4.61	0.42
R236	4.40	0.54
R237	4.50	0.49
R238	4.50	0.46
R240	4.34	0.61
R241	4.00	0.81
R242	4.17	0.74
R243	4.53	0.50
R245	4.30	0.61
R250	4.64	0.38
R253	4.16	0.73
R254	4.01	0.79
R255	4.30	0.62
R257	4.12	0.65
R258	4.29	0.62
R259	4.33	0.70
R261	4.40	0.56
R264	4.46	0.57
R270	4.34	0.69
R271	4.60	0.71
R272	4.75	0.60
R273	4.71	0.68

Mean grain size Section No.	Mean grain size M_Z	Standard deviation 6_I
R276	4.25	0.62
R277	4.29	0.66
R278	4.18	0.65
R279	4.11	0.74
R280	4.23	0.62
R281	4.13	0.81
R282	4.23	0.62
R283	4.39	0.64
R284	4.04	0.80
R285	4.21	0.66
R290	4.57	0.52
R291	3.99	0.84
R292	4.24	0.70
R293	4.12	0.93
R294	4.29	0.80
R296	4.23	0.79
R297	4.56	0.41
R298	4.60	0.42

SAMPLES

No.	Origin	$M_Z(\emptyset)$	6_I
S4	tile sample from Minety	4.70	0.73
S5	tile sample from Minety	4.35	0.95
S11	clay sample from near Gloucester (lower Lias clay)	-	-
S14a	clay sample from Minety (Oxford clay)	4.5	0.92
S14b	clay sample from Minety (Oxford clay)	4.65	0.77
S14c	clay sample from Minety (Oxford clay)	4.41	0.99
S26a	tile sample from Minety	4.06	0.86
S28	TPF stamped tile from Minety	3.95	1.05

APPENDIX III

Correlation of die types in this paper with Clifford, 1955

TPF	a	*	TPFB	a	10
	b	*		b	*
	c	8a	TPFC	a	12a
	d	8b		b	12b
	e	*		c	*
	f	*	LHS	a	16d
	g	*		b	16c
TPFA	a	9a		c	16b
	b	*		d	16a
	c	*		e	*
	d	*		f	*
	e	9b		g	*

* not identified by Mrs. Clifford

BIBLIOGRAPHY

Baddeley, St. Clair, 1906: 'Roman remains at Watercombe, nr. Bisley' TBGAS, 29.

Baddeley, St. Clair, 1921: note in 'Roman Britain in 1921 and 1922,' JRS vol. 11, 1921.

Baddeley, St. Clair, 1922: 'The inscribed brick-tegula found at Cirencester,' TBGAS, 44, 117-120.

Baker, R., 1846: 'Roman Villa, discovered at Bisley, Gloucestershire' Arch. J. II (1836), 42-45.

Britton, D. and Richards, E. E., 1969: 'Optical emission spectroscopy and the study of metallurgy in the European Bronze Age', in Brothwell, D. and Higgs, E. eds., Science in Archaeology, Thames and Hudson, London.

Chisholm, M., 1962: Rural settlement and land use, Hutchinson, London.

Clarke, D. L., 1978: Analytical Archaeology (2nd ed.), Methuen, London.

Clifford, E. M., 1933: 'The Roman Villa at Hucclecote', TBGAS, 55, 323-376.

Clifford, E. M., 1955: 'The stamped tiles found in Gloucestershire', JRS 45, 68-72.

Cornwall, I. W., 1958: Soils for the Archaeologist, Dent, London.

Crawford, O. G. S., 1921: 'Roman tile kilns at Minety', Wilts. Arch. Mag. xli, 424-5.

Folk, R. L. and Ward, W. C., 1957: 'Brazos river bar; a study in the significance of grain size parameters', Journ. of Sedimentary Petrology, 27, 3-26.

Freeman, T. W., 1967: Geography and planning, Hutchinson, London.

Greene, K., 1977: 'Legionary pottery and the significance of Holt', in Dore, J. and Greene, K. eds., Roman Pottery Studies in Britain and Beyond, B.A.R. S30, Oxford.

Grimes, W. F., 1930: 'Holt, Denbighshire; the works department of the Twentieth Legion at Castle Lyons', Y Cymmrodor.

Hamilton, F. E. I., 1968: 'Models of industrial location' in Chorley, R., and Hagget, P., eds., Socio-economic models on Geography, Methuen, London.

Harley, L. S., 1974: 'A typology of brick, with numerical coding of characteristics', Journ. Brit. Arch. Assoc. xxxvii, 63-87.

Hodder, I. R., 1972; 'Locational models and the study of Romano-British settlement', Ch. 23 in D. L. Clarke (ed.) Models in Archaeology, Methuen, London, 887-909.

Hodder, I. R., 1974: 'Some marketing models for Romano-British coarse pottery', Britannia, V 1974, 340-359.

Hodder, I. R. and Orton, C., 1976: Spatial analysis in Archaeology, Cambridge University Press, Cambridge.

Hollestelle, J., 1974: 'Soil-marks of late medieval brick clamps at Wijk bij Duurstede', Berichten van de Rijksdienst voor het oudheidkundig Bodem-onderzok (ROB), 24, 1974, 185-189.

Krumbein, W. C., 1938: 'Size frequency distribution of sediments and the normal phi curve', Journ. of Sedimentary Petrology, 8, 84-90.

McWhirr, A. D., 1978: 'Cirencester, 1973-6: Tenth interim report', Antiq. Journ. lviii pt. I, 61-80.

McWhirr, A. D. and Viner, D. J., 1978: 'The production and distribution of tiles in Roman Britain, with particular reference to the Cirencester region', Britannia ix (1978), 357-377.

Peacock, D. P. S., 1970: 'The scientific analysis of ancient ceramics; a review', World Arch. 1, 375-389.

Peacock, D. P. S., 1971: 'Petrography of the coarse pottery', in Cunliffe, B. W., Excavations at Fishbourne, 1961-69, Soc. of Antiq. Res. Reports, London.

Peacock, D. P. S., 1977: 'Bricks of the Classis Britannica', Britannia viii (1977), 235-248.

Pye, W. D., 1943: 'Rapid methods of making sedimentational analysis of arenaceous sediments', Journ. Sedimentary Petrology, 13, 85-104.

Renfrew, C., 1978: 'Introduction; Production and exchange in Early state societies, the evidence of pottery', in Peacock, D. P. S. ed., Pottery and early Commerce, Academic Press, London.

Richardson, L. and Webb, R. J., 1910: 'Brickcasters, Pottery and brick-making in Gloucestershire', in Proc. of the Cheltenham Natural Science Society, new series vol. 1, 1910, 222-282.

Rosenthal, E., 1949: Pottery and Ceramics, Penguin, London.

Rutton, W. L., 1893: 'Sandgate Castle, A.D. 1539-40', Arch. Cant. xx, 228-257. 257.

Schiffer, M., 1976; Behavioural Archaeology, Academic Press, London.

Swain, E. J., 1978: Excavations at the Chessals Kingscote. Kingscote Archaeological Association.

Tite, M. S., 1972: Methods of physical examination in archaeology, Seminar Press, London.

Williams, J. H., 1971: 'Roman building materials in the south-west', TBGAS (1971), 90, 95-119.

Young, C., 1977: Oxfordshire Roman Pottery, B.A.R. 43, Oxford.

ACKNOWLEDGEMENTS

I would first like to thank Alan McWhirr and David Viner for interesting me in the problems of tile production in the Cotswolds. I would also like to thank Professor Renfrew for allowing me to use the facilities available in the Department of Archaeology at Southampton University for petrological analysis, and David Peacock for his help and encouragement with the work.

I would also like to thank all the museum keepers and excavators who have given samples of tiles for study, or made their collections available for examination; Alan Saville at Cheltenham Museum; Malcolm Watkins at Gloucester Museum; David Viner at Cirencester; Susan Read at Reading Museum; Valery Rigby and Catherine Johns at the British Museum; Ted Swain, Director of the Kingscote excavations; Jennifer Price at Salisbury; and only just in time for this paper, Sebastian Rahtz at Worcester Museum. Mike Stone of Swindon Museum was particularly helpful with information from Wiltshire and his unpublished work at Minety. Jane Timby kindly prepared and fired some of the clay samples, and Simyu Wandibba generously allowed me to draw on the results of some of his experiments on fabric analysis.

The Administrator of the Haverfield Bequest of the University of Oxford kindly gave a grant to cover certain items of expenditure for the study of this group of tiles, and I am also grateful to the Oxford Laboratory for Archaeology and Art History for allowing me to include the results of their analysis of the Beeches Road tiles. Finally I would like to thank all those who have shown me their Roman tiles and discussed them with me.

THE COMPUTER AS AN AID IN THE RECORDING, ANALYSIS AND PUBLICATION OF TILES

John Wilcock

1. Introduction

1.1 The Ubiquitous Computer

It is only necessary to take a trip down the High Street to realise that computers have entered most aspects of our lives. Banks, Building Societies, The Post Office, Local Government, supermarkets, travel agents, most industrial firms and businesses, the Police, libraries and the educational system all use computers. Coal mines and the railways are controlled by computer, industrial processes and production lines run by them, and air traffic control relies almost exclusively on them. Traffic lights are switched on by them and planes flown by them. Each person, from the cradle to the grave is recorded on various computers, as a population statistic, a participant in the educational and medical systems and above all as a consumer. Thus it was only a matter of time before the computer entered archaeology, and now there are several archaeologists and computer scientists in Britain and internationally who have become practitioners of computer archaeology.

1.2 Computers and Man

Psychological reactions to the computer are interesting but mostly mistaken, based on popular myth. The first school regards the computer as an infallible "Electronic Brain" whose workings are so complex that there is no point in trying to understand them. However, computers are rather unintelligent machines which do exactly as they are told, neither more nor less. If told to do the wrong thing they will do it, and if given incorrect information they will treat it as correct, giving wrong answers—or as the Americans have it "garbage in, garbage out".

The second attitude is that a computer is Big Brother who is watching you. Personal information is certainly carried on many computerised data files, but given adequate security and password systems the computer is no threat. Information systems which were prejudicial to human interests could only arise by human design.

Contrary to these myths, the computer is best regarded as an Information Systems Clerk. It is totally obedient, accurate and with a memory beyond human capability, tireless (it can work 24 hours a day, 7 days a week non-stop), and speedy (a typical addition takes one millionth of a second; concordances which were once thought suitable material for a doctorate can now be generated in a few minutes), but it is unintelligent. Its capabilities include data capture, information storage and retrieval, statistical calculations, sorting and listing of all kinds, and graphics (drawing diagrams), indeed anything that we know how to tell it to do.

2. The Microcomputer Revolution

2.1 The History of the Computer

The first significant development in computer design was the technique of the stored program, invented by Charles Babbage about 1820, whereby the program or set of instructions for the computer is stored in the computer memory with consequent speed improvements. But Babbage's machines were mechanical and hand-made by craftsmen. Despite a government grant for their development the lack of a machine-tool industry prevented their success. If this had not happened, we would have had working computers from the 1820s.

War is always a good provider of funds, and it was the second World War which gave the impetus for the development of both the electronic valve and the first working computers: the Colossus was an electronic machine used for decoding German messages at Bletchley Park, Britain; while the Automatic Sequence Controlled Calculator (a mechanical machine, Baggage's Analytical Engine reincarnate) and the Electronic Numerical Integrator and Calculator, both USA machines, were used for defence purposes including the design of the Atomic Bomb. It was for these purposes that the USA poured so much money into the development of computers, and now the USA dominates the computer market, although many of the research teams involved were British, working at Cambridge, Manchester and London Universities.

Electronic valve computers (the "1st Generation") were, however, inherently unreliable, since the chances of getting tens of thousands of valves to work all at the same time were slim. The second significant development in computer design came with the invention in the early 1950s of the transistor, a semiconductor replacement for the electronic valve using the element Germanium. The first reliable computers (the "2nd Generation") used transistors and other discrete components such as diodes, resistors and capacitors, wired together on copper-tracked printed circuit packages. More reliable semiconductors, in particular Silicon were investigated, and soon the engineers were manufacturing ten components on one "chip" of silicon (Small Scale Integration (SSI), c. 1960), then 100 components/chip (Medium Scale Integration (MSI), c. 1965), and 1000 components/chip (Large Scale Integration (LSI), c. 1970). Money was poured into the silicon chip industry in Silicon Valley, California as a result of the Space Race, and "3rd Generation" computers were constructed of LSI chips. By 1975 10,000 components/chip were possible (Very Large Scale Integration (VLSI)),but the circuits were becoming so complex and specialised that their market was dropping. Although the VLSI chip of 1975 cost little more than the transistor of 1950 to manufacture, the development costs had to be recouped over a smaller number of items and their market price consequently rose. To combat this, the electronic engineers hit on the idea of a general purpose circuit capable of fulfilling a large number of different circuit functions, by programming. Thus the microprocessor was born, a child not of the computer industry but of the electronic engineering industry. The microprocessor, a computer processing unit on a wafer of silicon 5 mm square, cost less than £10 in 1979. And Silicon, the basis of the first tool used by man, the stone axe, is also the basis of his latest tool, the microprocessor.

2.2 Consequences of the Microprocessor

Each household will by the mid-1980s have a number of microprocessors with dedicated applications, controlling the cooker, washing machine and TV set. We will without realising it become dependent on the microprocessor, rather after the fashion in which we use small electric motors without thinking, for a variety of purposes (each home probable has 10-15 such motors in clocks, fans, drills, timers, food mixers, etc.). Some applications of microprocessors are already with us—electronic calculators, liquid crystal display watches, and TV games to name but a few—and such complexity at such a small cost was unthinkable a few years ago. In the future car electronics will use a microprocessor, production lines will be automated using robots, and office work will be invaded by "Word Processing" equipments (see below) with profound effects upon employment.

2.3 Teletext and Viewdata

Perhaps one of the most significant developments is the incorporation of the microprocessor in the TV set, and the addition of the telephone, a keypad or full keyboard, and a recorder to make the complete home/office information terminal. The BBC CEEFAX and ITV ORACLE ("Teletext") systems use a few spare scan lines on their normal broadcast frequencies to transmit a continually-circulating set of "pages" of mass-market information such as the new headlines, weather forecast, sports results, travel information, stock market, what's on, etc. Pages are selected at will using a keypad, and appear on the screen instead of the normal TV picture. The use of the system is free, but users need a special TV set incorporating decoding electronics.

The Post Office PRESTEL ("Viewdata") system is much more significant. An unlimited amount of information is inserted by Information Providers, who recoup the cost by charging for the pages—the bill appears along with the telephone account. There are facilities for the user to reply via the keyboard, for business and personal messages to be sent to other users ("electronic mail"), for financial transactions to be completed via the terminal, for educational "teach yourself" packages to be sent and recorded for leisure use, and for computer programs to be sent to the built-in microprocessor ("telesoftware"). The user thus will have a computer terminal in the house, or a stand-alone microcomputer for personal use. This will make it possible for a user to carry out office work and education without leaving home, and to complete via the terminal such complex transactions as:

1. Search the Marketplace information to determine the item to be bought.
2. Search the shop/supermarket files to determine the price of the item. Decide where to buy the item.
3. Initiate the purchase via the terminal. The bank account of the shop will be automatically credited and your bank account debited.

2.4. Microcomputers

Finally, the microprocessor can be configured with memory and some external devices such as video screen, readers, printers, magnetic tapes or discs, keyboard and graphics equipments to make a microcomputer. The chief cost lies in the external devices, the cost of which will not reduce, but the basic microcomputer costs less than a car, and its relative cheapness makes it possible for many people, who could never have afforded large computer use, to have their personal computer which sits on the desk top and is always at their beck and call. Herein lies the true Microcomputer Revolution.

3. Uses of Microcomputers in Archaeology

3.1 Dedicated Uses of a Microprocessor

In the same way as the microprocessor can be made to control a variety of domestic appliances, it will appear in archaeological applications to control:

a) geophysical instruments (resistivity meters, proton gradiometers, etc.) on site, with data logging and production of site maps of where to dig;

b) chemical and physical processes in the laboratory (radiocarbon dating, thermoluminescence, conservation etc.), with data logging and calculation of the results.

3.2 Use of the Microprocessor as a Personal Microcomputer for on-site Processing.

In these uses, a microcomputer is set up in the site office for on-site processing of archaeological data:

a) Data capture via a keyboard, recording data on magnetic tapes or discs for permanent storage, printing labels for attachment to finds;

b) Information retrieval, a Management Information system for use by the Site Director;

c) Routine processing of data: sorting, listing, distribution maps, context sorting and phase diagrams (Harris-Winchester method);

d) Routine and advanced statistics

e) Graphics: plans, maps, diagrams

f) Word Processing (text handling) for publication.

Each of these applications will now be described in detail.

3.2.1 Data Capture: Several computer archaeologists in Britain now take their microcomputer on-site in a motor van which acts as an extra Site Office. The electricity is provided by a petrol-driven generator if necessary. The approach is to have the computer print identity labels for artefacts, but at the same time the information is entered into permanent storage and may be augmented when more is known of the finds. The information is typed into the computer using a keyboard. Summary lists may be printed at any time, providing a daily or cumulative finds catalogue. A common method of recording is the question and answer technique. The finds recorder is presented with a visual display on a TV-like screen and asked a sequence of questions about the find to which he replies using the keyboard. Mistakes may be erased easily, and the fixed sequence of questions prevents omission of data. When a magnetic cassette or disc is full of data it may be sent by post for processing at a computer centre, or even transmitted over a telephone line to a large computer.

3.2.2. Information Retrieval: The advantages of computerised information retrieval are many. First of all there is rapid retrieval of one's own and other people's data (if the recording systems are standardised). Secondly, the retrieval criteria may be as complex as desired, not just limited to a few logical conjunctions (AND) ar disjunctions (OR) as in the use of edge-punched or drilled card systems. Distribution maps are possible as a bonus. There is the possibility of the automatic production of printed lists, sorted by whatever order is desired, then the data may be re-ordered into tables, and illustrated by routine statistical diagrams such as histograms and pie-charts. Finally, the information retrieval may be linked to other programs for advanced statistics and graphics and to Word Processing for publication.

It will be obvious that these benefits cannot flow without standardisation of data structures. This may appear irksome, but in fact it is a blessing in disguise, for the application of explicit and well-defined rules to the description of a collection of artefacts ensures consistency and reliability, and this is true whether the information actually gets onto a computer in the end, or not. The classification of a tile, for example, may then be more easily checked by others, and large collections of tiles may be processed by several different workers without prejudicing the overall analysis. The approach enables fuller access to the cumulative body of tile material, making it easier to test hypotheses and formulate conclusions.

We may note here that computer scientists turned archaeologists are not alone in crying in the wilderness about computer usage and standardisation. The Information Retrieval Group of the Museums Association (IRGMA, renamed in 1977 the Museums Documentation Association) has long encouraged the use of a high-level descriptive language for all sorts of artefacts, now implemented in its GOS computer program (IRGMA, 1975). The Council for British Archaeology also addressed itself to the problem (Royal Commission on Historical Monuments, 1975). Further, the Frere Report (Ancient Monuments Board, 1975) stated that the use of computers for permanent storage of quantifiable information should be encouraged, that records should be centralised eventually, and that the Department of the Environment should study further the use of computerised data for archaeological records (Ancient Monuments Board, Frere (ed.), 1975, paragraphs 3.7, 3.10, appendix II, (v), (vi), (vii)). Since then the Department of the Environment has carried out extensive work using on-site microcomputers as well as remote computer terminals (Jefferies, 1977). Also in 1977 the British Library addressed itself to the problems of information handling in archaeology (The British Library Board, 1977).

It is recommended that the bulk of such computer-produced information for a site should appear in the Level III publication. The use of a computer is particularly appropriate for those handling large quantities of tiles. The material should be classified using carefully-constructed glossaries of allowed words and data categories. Obviously in the case of Roman tiles the formulation of such glossaries and categories will be the work of a sub-committee acceptable to all current workers, but there is no need to delay recording until such a classification is agreed, providing that sufficient detail is given to enable the tiles to be inserted into the eventual format.

As a beginning, the following categories are suggested. These may prove too elaborate in practice, and of course may be extensively modified by the sub-committee, but it is a start. It should be emphasized that categories not known for a given tile may be omitted, and perhaps inserted later when more research has been done or laboratory processes carried out on the tiles. The scope of the retrieval requirement for tiles is yet to be determined, but an attempt must be made at the outset to provide for most contingencies. It is better to err on the side of too much detail than too little: categories may always be "lumped" together later with little difficulty, but the splitting of categories into finer detail (because the original category has proved too imprecise) requires that all previous records be examined and reclassified, and this is wasteful of both time and money. In the complication of a multi-purpose tile classification scheme categories will inevitably appear for which some archaeologists will have little or no use. There is no case for the exclusion of these categories: it is sufficient that someone has asked for their inclusion and intends to use them.

These remarks having been made the suggested prototype classification scheme will now be presented:

IDENTIFICATION

Classified Identification
Architectural function
Authority for classification
Date of classification

PRODUCTION

Body
Texture
Inclusions
Clay provenance
Kiln site
Colour
Firing
Produttion Details
Distribution route

QUANTITIES

Number of fragments
Weight of fragments
Number of tiles represented
Equivalent number of tiles
Displacement
Total surface area

DATING

Period (e.g. CLAUDIAN)
Numerical date

Dating Method (e.g. association, thermoluminescence)
Laboratory ref. no.
Date of age determination

PROVENANCE

Site Name
Roman name of site
Site reference no.
Grid reference
Point on site referenced
Parish
District
County
Country

EXCAVATION DETAILS

Phase
Context no.
Context relationships
Context type
Joined fragments in context...
Context period
Context date
Coordinates
Excavator
Find reference no.

STORAGE

Institution
Accession no.
Storage Location
Date of accession

CONDITION

Condition (EXC./GOOD/FAIR/POOR/URG.CON. etc.)

COMPLETENESS

Completeness

DIMENSIONS

Length
Width
Thickness

MARKS

Maker's mark
Inscription
Graffiti
Military Unit
Position of mark

Method of marking
Comb signature

PART DESCRIPTION

Part } may be repeated
Part description }

PROCESS

Process (e.g. conservation, reproduction, neutron activation analysis, X-ray fluorescence, X-ray diffraction, petrological section, etc.)
Laboratory reference no.
Date of process
Process result

DOCUMENTATION

Associated finds
Documentation type
Author
Document title
Volume/Edition
Year of Publication
Pagination
Page/Figure/Plate Details
Photograph number
Drawing number

RECORDER

Name of recorder
Date of completed record.

Information retrieval, if carried out on site, may form the basis for decisions made by the Site Director, and thus influence the direction and progress of the dig. This may be valuable, particularly in a rescue situation. However, in most cases archaeologists will find remote information retrieval with say a week's delay perfectly acceptable. Much tile information retrieval will probably be of a research nature on the cumulative body of tile data for Britain, in order to test hypotheses.

3.2.3 Routine Processing of Data: The computer is an immense aid in the organisation of data. While necessary for publication, data organisation can be a tedious business when carried out by hand, and so archaeologists tend to do the bare minimum which is necessary; this has also been the cause of many a delayed publication. Data organisation by computer is rapid and accurate. Sorting and listing may be carried out in whatever order or orders are required. Distribution maps may be quickly and accurately produced in publishable form by computer, on-site if required. The unravelling of time sequences for contexts by the Harris-Winchester method (Harris, 1975) has also been successfully undertaken by computer, thus saving much expert archaeologist's time (Bishop and Wilcock, 1976).

3.2.4 Routine and Advanced Statistics: It is not the intention of this paper to give a complete explanation of all statistical techniques which have been used in archaeology. Those readers who are interested in this area should read the many general and specialist books and papers which have formed the bulk of published work in computer archaeology (e.g. Clarke, 1968; Monkhouse and Wilkinson, 1971; Hodson, Kendall and Tautu, 1971; Sneath and Sokal, 1973; Doran and Hodson, 1975; Hodder and Orton, 1975; Wilcock, 1975; Clarke, 1977). A few techniques will be mentioned briefly:

Attention should be paid to numerical quantification of the amount of tiles to be recorded, which has a bearing on later statistical analyses and their meaning. Various measures of quantity have been used—it is up to the individual to decide which method is most relevant, or perhaps a particular method should be made standard. The measures are:

i) Number of tile fragments

ii) Total weight of tile fragments. Weight is a more reliable measure than number, since a large fragment will still weigh the same when broken into several smaller fragments, while some tiles may fragment more than others for a variety of reasons

iii) Number of tiles represented, i.e. the actual number of tiles which have produced fragments found in a context—this measure may be impossible for unmarked tiles

iv) Equivalent number of tiles, i.e. the total obtained by counting a fragment as a fraction of a tile, and adding up the fractions

v) Total volume of fabric, i.e. water displacement of the fragments

vi) Total surface area of the fragments

These last two measures would seem to behave in a very similar manner to total weight of fragments. The number of tiles represented is the ideal, but this may not be attainable as mentioned above.

Some sites produce large quantities of tile which at first sight ought to be the basis for a valid statistical study. When examined closely, however, there may be only a small number of fragments per context, and their analysis may not be valid statistically. The Level III publication should in all cases contain sufficient quantity measurements to enable other workers to perform subsequent statistical analyses where appropriate.

A clear distinction must be made between "statistics", meaning raw data (counts, weights, lengths, etc.) with derived data (percentages, totals, averages, etc.), and "STATISTICS", meaning a way of approaching problems from a mathematical viewpoint, based on the known laws of chance and probability. It is accepted that archaeologists who have not received training in numerical techniques may not wish to apply STATISTICS to their results. We ask these workers to accept that STATISTICS is a scientific technique which others may wish to use, and to present their results at Level III stage so that data may be examined further.

Simple techniques are the incidence matrix, the histogram (or block graph), its equivalent the pie-chart, various bar charts, lenticular (time) diagrams, scattergrams, three-pole plots and cumulative percentage graphs. All these are graphic methods.

Advanced techniques for comparison studies are cluster analysis, principal components analysis and multidimensional scaling. Graphic representation of the results involves the use of dendrograms, skyline plots, Wroclaw diagrams, principal components plots, and scalograms.

Finally, there is a group of spatial analysis techniques (Hodder and Orton, 1975), with complex graphical representations. Raw data may be plotted on a simple distribution map, but note that such maps show only available knowledge. A gap on the map may reflect merely an absence of information, i.e. where archaeologists have not excavated, and these maps may tell us more about archaeologists and their favourite sites than anything else. The problem may be partly overcome by including negative evidence to show, for example, excavations undertaken where none of the defined types of tile has been found, or even where no investigation has taken place at all.

In summary, the most important point for the archaeologist is to understand the assumptions and limitations which underlie statistical techniques. Archaeological samples may not truly be the "random samples" required by statistical theory, and significance tests may not be valid in some archaeological circumstances. Also the methods are not "cure-alls" to make order out of chaotic archaeological data, nor can they derive significance where there is none. When in doubt, consult a statistician, preferably one who has some experience in archaeology.

3.2.5 Graphics: Graphics is the name given in computer terminology to maps and diagrams of all sorts drawn by the computer for publication. It has been said that "one picture is worth a thousand words" and this may well be true for archaeological distribution maps, site plans and statistical diagrams. All these can be drawn by computer, and it is best to use a man-machine interactive technique such as that described by Wilcock (1974). Some computer-drawn statistical diagrams have also been compared by Wilcock(1975).

3.2.6 Word Processing: "Word Processing" is the handling by computer of paperwork of all kinds, such as text, documents, records, summary lists and reports. The basic equipment is a microcomputer with video screen, keyboard, permanent storage (magnetic cassettes or discs) and a high quality printer. Text may be assembled on the screen, either typed in or retrieved from permanent storage, and it may then be edited on the screen by correcting, inserting, deleting or reordering letters, words, sentences and paragraphs. Upper and lower case letters are of course available, together with special features such as underlining, centering of titles, page numbering, right justification of columns of text with automatic hyphenation, headings and footings for pages, and tabulation (left aligned for text, right aligned for integers, aligned by decimal point, or centered). Thus a complete publication can be prepared on the computer in draft form and recorded. Only when the editing is complete is it transferred to paper.

The printer must have a high quality of print it if is to be directly published by photography and offset lithography. A "daisy-wheel" printer is suitable for this, with an exchangeable print head with "petals" which carry the type faces for any number of fonts, and continuously-variable horizontal and vertical spacing of print (which allows right justification, and also diagrams to be produced on the same printer).

An alternative method is "film-setting" where typefaces indistinguishable in appearance from hot metal type are assembled as images on photographic paper by a beam of light shining through character masks on spinning glass discs. This method is now used by several newspapers and journals.

Finally, for bulky publications microfilm is available. For archaeological publications the microfiche form is most suitable (up to 98 full-colour or more than 300 black-and-white page-images per 105 mm x 148 mm fiche). These sheets of film contain rows of page-images in a grid pattern. Because they are easily mass-duplicated, fiche are widely used as a publishing medium for lists, catalogues, research reports and training materials. A microfiche reader is necessary to project the page images to natural size for reading. This inconvenience should not be allowed to outweigh the advantages for fiche publication of archaeological reports: a large Level III publication posted to everybody who requires it using normal letter mail, and at a total cost of less than £1 each copy. Archaeologists should surely look at the advantages of computer-generated microfiche for their publications, and this is the natural conclusion if an archaeological excavation has been using the computer from its inception.

4. Conclusion

This paper has attempted to show the advantages to be gained in archaeological studies from the use of the computer. It is to be hoped that Roman tile studies can join the many other archaeological groups who are actively using the computer, with pleasing results.

Bibliography

Ancient Monuments Board 1975. S. S. Frere (ed.), Principles of Publication in Rescue Archaeology, Department of the Environment, London, 1975.

Bishop, S. and Wilcock, J. D., 1976. 'Archaeological context sorting by computer: the STRATA program', Science and Archaeology 17 (1976), 3-12. See also correspondence N 20/2. Science and Archaeology 20 (1977), 27-29.

British Library Board, 1977. The British Library, Research and Development Department, Problems of information handling in archaeology, London, 1977.

Clarke, D. L., 1968. Analytical Archaeology, London, 1968.

Clarke, D. L., (ed.), 1977. Spatial Archaeology, London, 1977.

Doran, J. E. and Hodson, F. R. 1975. Mathematics and computers in archaeology, Edinburgh, 1975.

Harris, E. C., 1975. 'The stratigraphic sequence: a question of time', World Archaeology 7 (1) (1975), 109-121.

Hodder, I. and Orton, C. R., 1975. Spatial analysis in archaeology, London, 1975.

Hodson, F. R., Kendall, D. G. and Tautu, P., 1971. Mathematics in the archaeological and historical sciences, Edinburgh, 1971.

IRGMA, 1975. Information Retrieval Group of the Museums Association, Archaeology object record card, Cambridge, 1975.

Jefferies, J. S., 1977. Excavations Records: Techniques in use by the Central Excavation Unit, Department of the Environment, London, 1977.

Monkhouse, F. J. and Wilkinson, H. R., 1971. Maps and diagrams: their compilation and construction, London. 1971.

RCHM, 1975. Council for British Archaeology, Report of the Working Party on Archaeological Records, London, 1975.

Sneath, P. H. A. and Sokal, R. R., 1973. Numerical Taxonomy, San Francisco, 1973.

Wilcock, J. D., 1974. 'The facilities of the PLUTARCH System', Science and Archaeology 11 (1974), 16-24.

Wilcock, J. D., 1975. 'Presentation of computer classification results: A comparison of graphical methods', Science and Archaeology 15 (1975), 32-37.

TECHNIQUES FOR STUDYING COMB SIGNATURE DISTRIBUTIONS

Talbot Green

The work to be discussed grew out of the preparation of the report on the Itchingfield Roman tileworks site near Horsham, Sussex, dug in 1964/65 and 1969 (Green 1970). I wanted to show that there was evidence linking the site with the posting station settlement of Alfoldean, 4 km away down the Arun valley. This evidence included the clearly identical comb-markings on some Itchingfield flue tile fragments and on one from Alfoldean, obtained from the Winbolt Collection at Horsham Museum that was subsequently discarded by the then curator. I attempted to make plasticine impressions of the clearest comb-marks and to trace round sections of these to provide comb profiles for publication. (Regretfully, in the published article the wrong scale of reduction was used in making the block fit the page, and so the comb profiles are wrong by a factor of 5: 3 linearly; see Green 1970).

Later, I discovered the advantages of thermo-setting modelling clay for making and comparing comb profiles, and set out to see if I could build up a corpus of them. My initial purpose was to try and identify any other sites in the Arun valley to which Itchingfield tiles had been carried. It was assumed that rafts or punts transported them downstream to Alfoldean, and there seems no reason why they should not have travelled further.

The method evolved for taking comb profiles, which is completely non-destructive, is as follows:

a) The area to be recorded is dusted with French Chalk or talcum powder, to act as a releasing agent. A paint brush is used to spread the powder thinly and evenly over the surface, and the excess is then blown off.

b) Kneaded clay is pressed down firmly into the furrows left by the comb, with the aid of a small flat.

c) The clay is lifted off the tile and eased off the flat. An identification mark is scratched on before storing on a tray until curing.

d) The impressions are cured in an oven, at the temperature indicated on the clay's wrapper. It should not be overheated as the clay blisters.

e) The cured impression is marked on the back in ink and then sectioned with a craft knife and straight-edge. The face must be cut perpendicular to the furrows and at a point where they run straight: not where they curve or swirl, when the comb may be twisted and the gaps between the teeth become artifically narrowed.

f) The cut face of the clay must now be rubbed down on 800-grade paper, to give a smooth surface.

g) The smoothed surface is pressed on a marking-stamp pad, and an impression made on paper.

I should like to consider for a moment what the original combs were like. None are recorded from Roman tileworks, but there is a slightly analogous comb, used to give rilling round the bodies of graphite-coated, Pre-Roman Iron Age pottery, from Manching (Kappel 1969, TAF. 59.2). It is a section of cattle rib with 27 small saw-teeth cut into the concave edge. However, we do have the impression of a wooden comb, bearing the CLBR proprietary mark, from Beauport Park (Wright & Hassell 1973, Pl. XLI A). There is also a flue tile which bears the impression of Itchingfield Comb B stabbed between the arms of a combed saltire (Plate 20.I).[1] From its sharpness a comb like this could have been either of hard wood, such as box, or of bone, like Roman hair- and weaving-combs.

Permission was granted to examine the collection of Roman tiles at Worthing Museum, which covers the Arun basin generally, and to take impressions at Bignor villa (Fig. 20.1). Profiles were collected from the excavations directed by Clair Smith at Binscombe, near Godalming (Smith 1977). As can be seen (Fig. 20.2), there is a variety of comb profiles, but no matches. So far, then, a wider trading pattern remains unproven. (The unnumbered profile shown from Wiggonholt is from one of the two tiles preserved at Worthing that are marked by rows of holes seemingly made using a wooden rake with four or five teeth. From the steep-sided shape, I do not believe that the holes have been made with any sort of roller. The single combed tile from Arundel has an unmeasurable swirl on it.)

As the work of collecting and storing these comb profiles progressed it became obvious that a classification system was necessary. In the first instance the number of teeth is paramount, but then both the width of the teeth and their spacing are clearly of significance. Since clay shrinks on firing, the measured width of the comb impression is no absolute guide, for successive batches of clay may be expected to shrink by different amounts. (From studies made at Silchester, to be published elsewhere, I would suggest that, even within a batch, the standard deviation on linear distance is around 2%). What do appear to remain constant, however, are the relative sizes of teeth marks and gaps.

On the analogy of the stamp collector's perforation gauge, a transparent scale was made to divide the distance across the ends of the teeth into a hundred parts (Fig. 20.3), and to record the relative points at which teeth start and stop (Fig. 20.4). Using an electronic analogy, these readings are Mark/Space readings and comprise my Type I Signature. One has to record the signature twice, in fact, in complementary forms since there is nothing to stop one reading from 'right-to-left' instead of 'left-to-right' when measuring an impression. In order to match a new comb against others in a corpus of Type I signatures, a fairly close match between all the numbers in the signature must be found. How closely the numbers should match is a nice statistical question which need not concern us here, but mathematical models are available. The match will only be a suggested one; visual comparison of impression, if not actual tiles, should always be done before matches are asserted. An obvious way of building up a corpus of Type I signatures is a computer file.

Plate 20.I Flue tile from Itchingfield with comb impressions stabbed between the arms of a combed saltire. (Photo T. K. Green; by courtesy of Barbican House Museum, Lewes.)

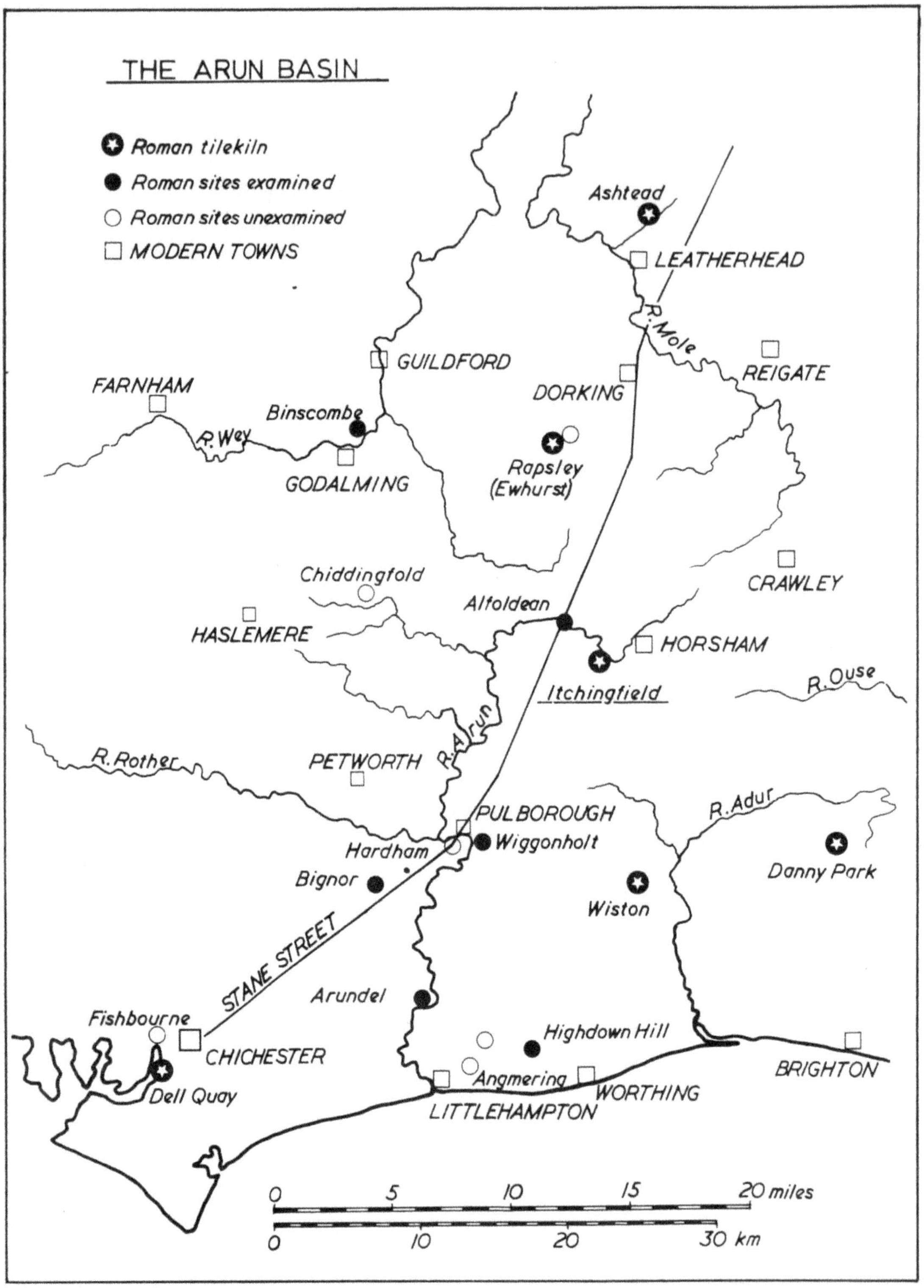
THE ARUN BASIN
Roman tilekiln
Roman sites examined
Roman sites unexamined
MODERN TOWNS
Ashtead
LEATHERHEAD
R. Mole
REIGATE
GUILDFORD
DORKING
FARNHAM
Binscombe
R. Wey
GODALMING
Rapsley
(Ewhurst)
CRAWLEY
Chiddingfold
Alfoldean
HASLEMERE
HORSHAM
Itchingfield
R. Ouse
R. Rother
PETWORTH
R. Arun
PULBOROUGH
R. Adur
Hardham
Wiggonholt
Danny Park
Bignor
Wiston
STANE STREET
Arundel
Fishbourne
Highdown Hill
CHICHESTER
Dell Quay
Angmering
WORTHING
BRIGHTON
LITTLEHAMPTON
0 5 10 15 20 miles
0 10 20 30 km

Fig. 20.1

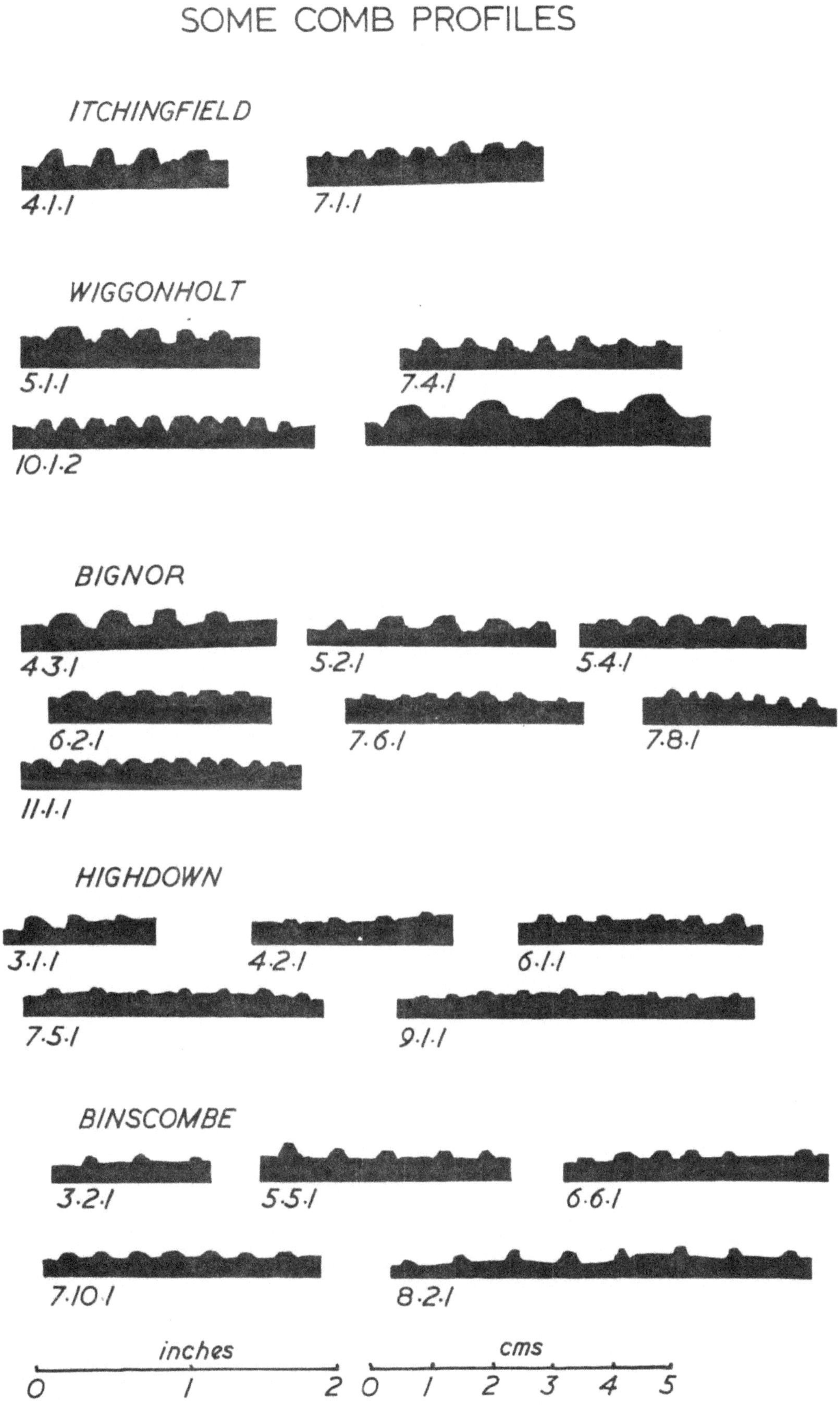
SOME COMB PROFILES
ITCHINGFIELD
4·1·1
7·1·1
WIGGONHOLT
5·1·1
7·4·1
10·1·2
BIGNOR
4·3·1
5·2·1
5·4·1
6·2·1
7·6·1
7·8·1
11·1·1
HIGHDOWN
3·1·1
4·2·1
6·1·1
7·5·1
9·1·1
BINSCOMBE
3·2·1
5·5·1
6·6·1
7·10·1
8·2·1
inches
0
1
2
cms
0
1
2
3
4
5

Fig. 20.2

Leading on from the Mark/Space readings of the Type I signature, a further electronic analogy was drawn with the Pulse Train (Fig. 20.5). This suggests that the information contained in the Type I signature could be stored in a simple Binary String which would be stored in a computer file in a sequence of digits, held in packed binary form. A 'floppy disk' will hold nearly half a million such digits so, even allowing for a fairly elaborate identification number prefacing each signature, the packing density of information could be enormous. The number of teeth are included as the first element of the identification number (see Fig. 20.2) and this will help speed up computer matching. This type of signature is my Type II and it has the advantage that it is of constant length, whereas the Type I has a pair of figures for each tooth. Again, a test for "goodness of match" needs to be devised: measuring the mismatch can be done either using high level languages or at machine level if Exclusive Or instructions can be loaded. The point is that, within a byte, it is merely the number of bits mismatched which counts and not their place in the byte. Any such computer file should be accompanied at the very least by inked impressions of the comb profiles, filed on the same basis as the signatures, so that the identification number which a computer search outputs rapidly allow the profiles to be compared visually.

A final idea, for a Type III signature, should be mentioned. Returning to the comb profile itself it is possible to store a quantified electronic analogy of such a shape using a digitiser and analyse it using a technique known as Fourier Analysis (Brigham 1974). Much work, largely of a defence nature, has been done on Fourier Analysis but it can be said that the result of such an analysis would yield rapidly a very compact signature for the actual shape of the teeth, not just the tips, as the Types I and II merely do. One problem is that the shape varies depending on how deeply the comb was pushed in. Another arises over the misleading shape on the lip of the furrows. There is, first, a bulge due to the release of compression forces in the clay alongside the furrow as it is made, and secondly there is a trail of clay swarf that sometimes gets left on the lip. These vary from point to point along a furrow and introduce an unmatchable element into the signature. The solution is to record the profile only to a constant depth (Fig. 20.6): this will be impossible with lightly impressed combs and so the other type signatures will have wider applications. But an intriguing possibility of Type III signatures is that changes of shape of an individual comb might be identified by Fourier Analysis: it is sensitive enough for this to be possible. One might even be able to detect wear on a comb.[2]

NOTES

1. Lent for the exhibition at this Conference, by kind permission of Fiona Marsden, curator of Barbican House Museum, Lewes.

2. I should like to acknowledge with gratitude the help given in these researches and the preparation of this paper by: Fiona Marsden and Norman Norris, the present and former curators of Barbican House Museum, Lewes; L. M. Bickerton, curator of Worthing Museum and Art Gallery; Miss Jane K. Evans; Mrs. Sylvia Standing; Miss Claire Smith; Mr. J. Morgan,

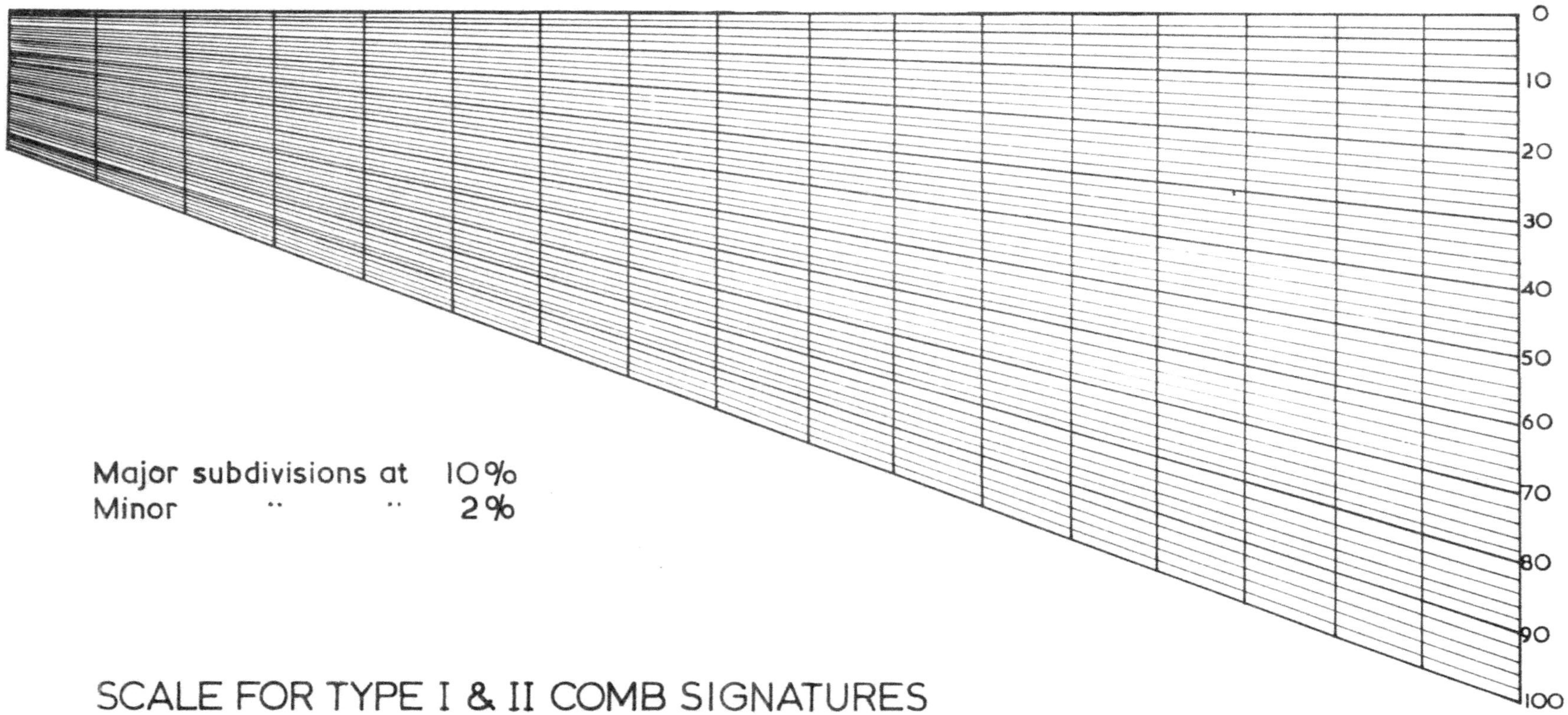
0
10
20
30
40
50
60
70
80
90
100
Major subdivisions at 10%
Minor 2%
SCALE FOR TYPE I & II COMB SIGNATURES

Fig. 20.3

THE TYPE I COMB SIGNATURE

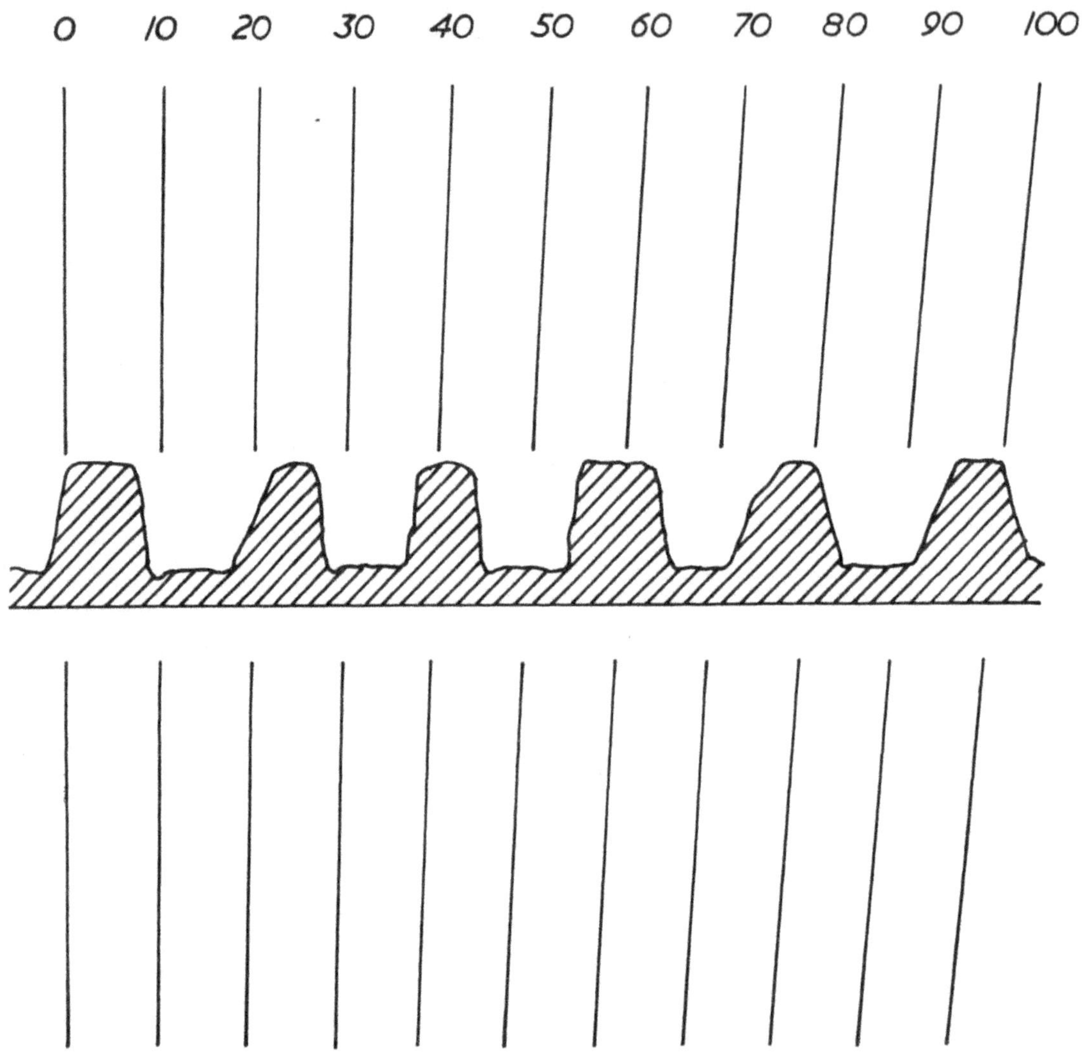

TOOTH

1	2	3	4	5	6
0 -7	22 -27	38 -44	55 -63	76 -80	95 -100

TYPE I SIGNATURE

Fig. 20.4

curator of Bignor Villa Museum, and Bignor Estates; Dr. M. G. Fulford, and the staff of Reading University Library, for help with slides; and to John Wood and Vic Fairchild, for useful and critical discussions of the ideas here put forward.

BIBLIOGRAPHY

Brigham 1974: E. Cran Brigham, The Fast Fourier Transform, Prentice Hall, N.J., 1974.

Green 1970: T. K. Green, 'Roman Tileworks at Itchingfield', Sussex A.C. 108 (1970), 23-38.

Hassell & Wright 1973: M. C. Hassell & R. P. Wright, 'Roman Britain in 1972', Britannia 4 (1973), 333 no. 32 and PL.XLI A.

Kappel 1969: Irene Kappel, 'Die Graphittonkeramik von Manching', Die Ausgrabungen von Manching, Band 2, R.G.K., Wiesbaden, 1969.

Smith 1977: C. Smith, 'A Romano-British site at Binscombe, Godalming', Surrey A.C. 71 (1977), 13-42.

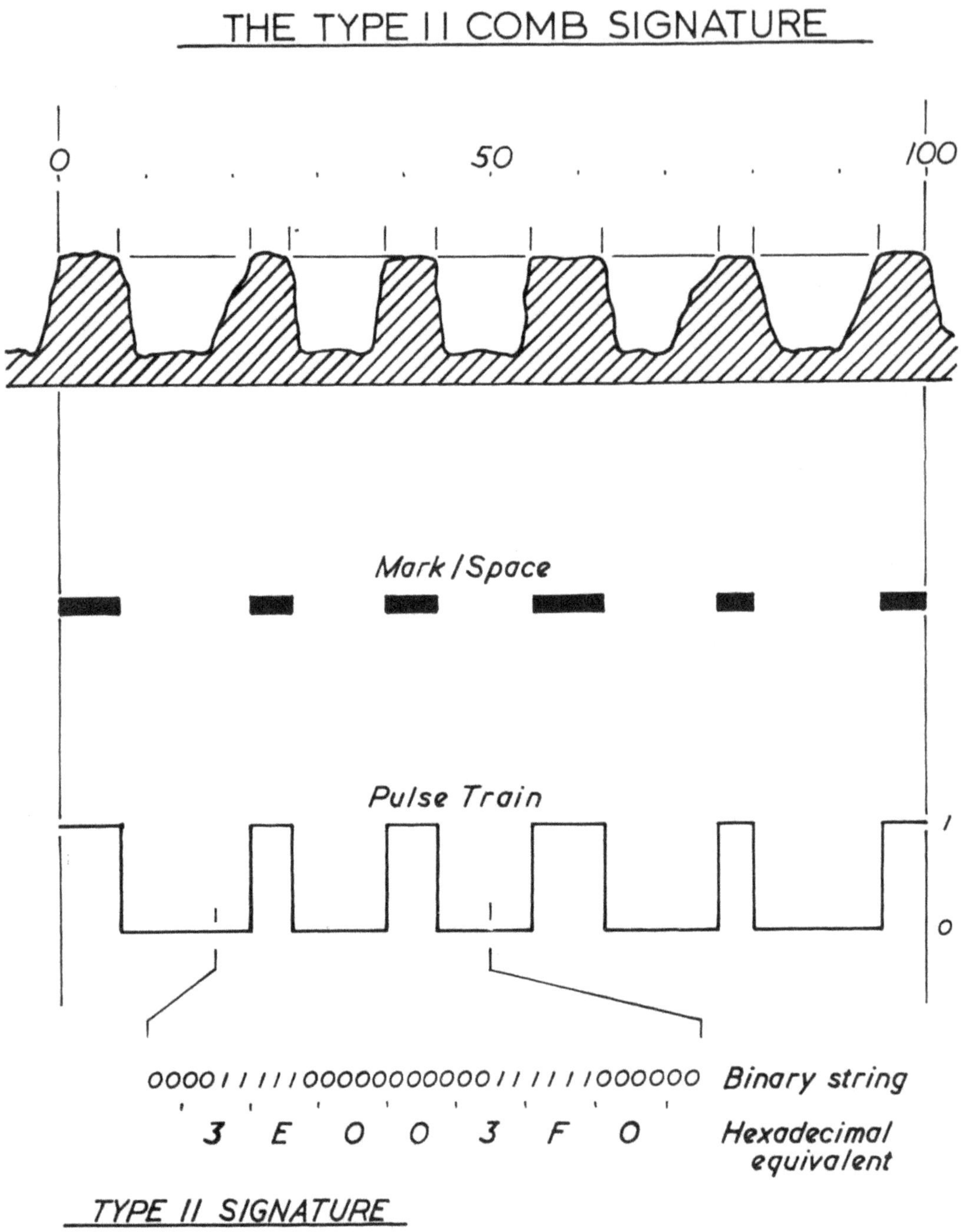
THE TYPE II COMB SIGNATURE
0
50
100
Mark / Space
Pulse Train
1
0
00001111100000000000011111000000 Binary string
3 E 0 0 3 F 0 Hexadecimal equivalent
TYPE II SIGNATURE
FE0003E003F001FF000F0001F

Fig. 20.5

THE TYPE III COMB SIGNATURE

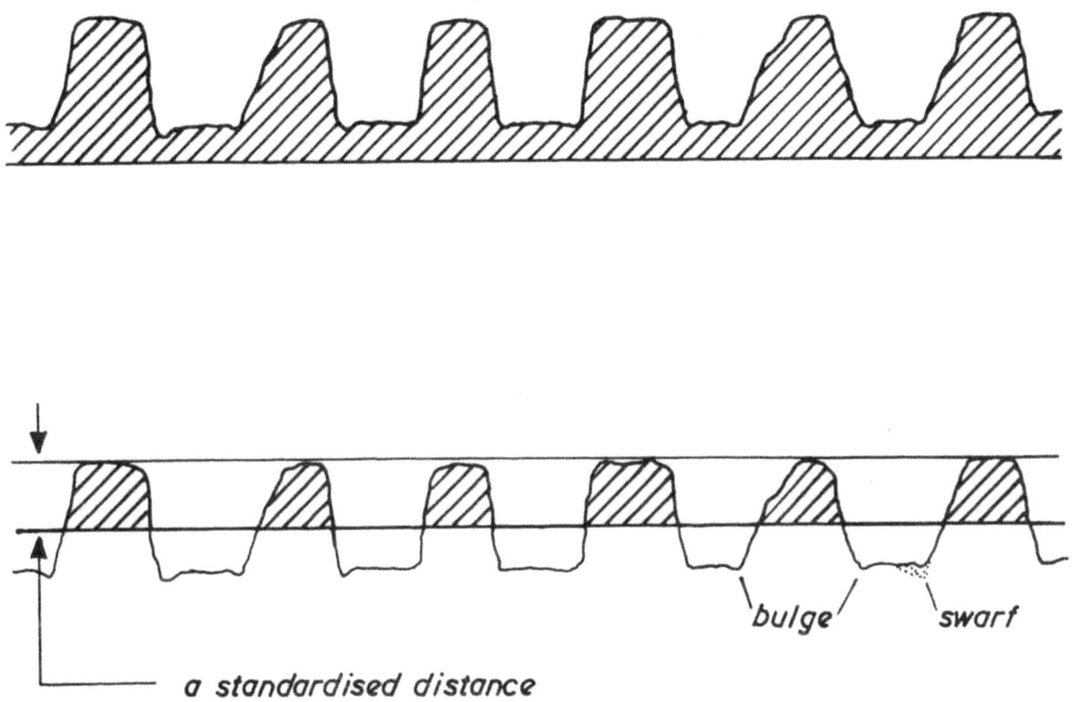

FOURIER SERIES			TYPE III SIGNATURE				
Harmonic	0	1	2	3	4	. . .	n
Term	x_0	x_1	x_2	x_3	x_4	. .	x_n

Fig. 20.6

RELIEF-PATTERNED TILES: A REAPPRAISAL

David Johnston and David Williams

It is now over thirty years since the late A. W. G. Lowther published his valuable Study of the Patterns on Roman Flue-tiles and their Distribution (1948) and more than fifty since he began the research for it. Until his death he was frequently consulted on newly discovered specimens, and his last published identification was of some Shakenoak specimens (Brodribb et al. 1968, 39-40). In 1948 he listed 130 occurrences of these tiles: since then more than a third as many new discoveries have been recorded, and many more probably remain to be traced, identified and listed. In recent years, two important discussions have advanced our thinking on their significance: in 1978 W. Rodwell reviewed the evidence and used it in his study of villas in northern Essex, and in the forthcoming report on the mansio at Chelmsford P. Drury considers an important group of material in the light of Rodwell's hypotheses. Both are relatively localised studies, and it has seemed to us that a general review is required, using techniques of petrological analysis that were not available to Lowther and have been little used in the study of the subject so far. In this paper we include much new material that has kindly been made available to us by excavators and museums; the visual examination has been carried out by Mr. Johnston and the petrological analysis by Dr. Williams. The first fruits, incidentally, of this collaboration were the discovery of a false provenance for a tile formerly displayed in Kendal Museum. Allegedly from the Roman fort at Watercrook, it was shown to be of non-local fabric, and energetic research by the Curator, Mr. J. Anstee, revealed that it belonged to a collection of specimens and replicas from Ashtead that had somehow become detached from the main collection now in Guildford Museum.

The nature of the problem

One of the important aspects of the Roman tile and brick industry that is at present imperfectly understood is the nature of its organisation. What is lacking is a knowledge of the method of production employed and of the distribution of the final product to the building-sites. In an illuminating ethnoarchaeological approach to this subject, Dr. D. P. S. Peacock (Chapter 1) describes five modes of modern brick and tile manufacture, one or more of which might be expected to fit the situation in Roman times: 1) household production, 2) small rural brickyards, 3) a nucleated brickyard complex, 4) estate brickworks and 5) municipal brickworks. The presence of name-stamps and relief-patterning on some tiles suggests a degree of specialist control beyond the range of simple household production. Studies of decorated flue-tiles (Lowther 1948, Rodwell 1978) and tile name-stamps (McWhirr and Viner 1978) have tended to indicate a fairly wide, but on the whole regional, distribution for certain types of stamps. It is not clear, however, whether the decorated tiles, or indeed the more numerous plain tiles, were made at a central point for simultaneous distribution to villa and town sites, or if a team of tile-makers travelled to each building-site, taking their die-stamps with them, to produce the tiles locally.

The patterning was apparently impressed with a roller that was normally made of wood, though Lowther suggested the occasional use of thin bronze strips fixed to a wooden core. However, it is hard to see how delicate bronze pieces could have been secured without visible nails. The cylindrical dies were normally 3-4 in. long and about 2-2 1/2 in. in diameter. Each face of a flue-tile had to be rolled lengthways twice, and the frequent overlapping makes identification from small fragments difficult at times. The effects of wear and tear can sometimes be recognised (Pl. 21. I) and occasionally the re-cutting of a die is noted. It is clear, from the modern replicas that turned up in Kendal Museum, that Lowther put his theories to the practical test of making and using a replica die; the results look convincing, but his conclusions from the experiment are, as far as we know, unrecorded.

As Lowther observed, the patterning is confined to flue-tiles, though he noted 3 examples of 'brick' and one exceptional piece, a 'curved tile' that he later suggested was 'apparently for building ornamental plastered half-columns' (Brodribb et al. 1968, 39). No examples of this have appeared subsequently, though 8 pieces and one nearly complete specimen of rectangular brick (possibly for wall-facings?) have been found at Shakenoak (Brodribb et al. 1973, 41:1978,37). The flue-tiles are normally of the box-flue variety, though Lowther found patterning on 3 cases of 'voussoir' tiles.

The main body of Lowther's work was the classification of 46 patterns or dies, which he reduced to 9 'Groups'. These groups may one day need revision, but they were as follows:

1. W-chevron (5 variant dies, one recut)
2. Dog and Stag (1 die, recut)
3. Florid (2 dies)
4. Compass-drawn (2 dies)
5. Diamond and Lattice (15 dies, one recut)
6. Billet (2 dies)
7. Rosette (2 or 3 dies)
8. Addenda (7 dies)
9. Plain chevron (7 dies)

Lowther did not jump to the easy conclusion that 46 rollers represented 46 tilemakers, as we do not know how many rollers each man owned, or whether they passed from hand to hand, or even how long they lasted before being thrown away. But the first conclusion to be drawn was that they were personal—in the nature of a trade-mark or signature, and useful for calculating individual output in the tileyard (as the better known literate graffiti sometimes were). One die even bears the personal name CABRIABANUS.

The next conclusion was not drawn by Lowther, though it was implied and has been assumed subsequently; this is that each Group represents a team, or one master and his assistants, perhaps even a 'firm' whose products are thus characterised.

Plate 21.I Die 46: North Leigh (left) and Chelmsford (right) showing splintering of roller in use. (Note, the impressions are from different parts of the die.) Photo: Southampton University

Lowther's third conclusion was drawn from his distribution maps. The distribution patterns varied, but all were very extensive, the find-spots being as far afield as Kenchester (Herefordshire) in the west and Castle Hill (Margidunum) in the north, the main concentration being in London and the south-east. The distribution maps reflect either the distribution of the products, or the movements of itinerant tilemakers. Lowther considered the fragility of some Ashtead products and some evident fabric differences, and settled for the latter hypothesis.

Why, then, was the relief-patterning confined to flue-tiles, and then only to certain tiles, the rest being combed or marked with a knife? If the patterning was a key for wall-plaster, then the range of tiles that needed this was limited. If it was a trade-mark or signature, it would surely have been used on other products. It was clearly personal, as there is no case of two dies being used on one tile—though there is at least one instance of two combed designs sharing the same tile (Shyde, I. O. W., unpublished). With the single exception of a Chelmsford tile bearing both roller-patterning and knife-cut cross-hatching (Drury, forthcoming) the uses of the roller and the comb or knife seem to have been distinct; this might imply either some specialisation within the industry, or a difference of status in the tileyard. Rodwell (1978, 24) dismissed the notion that flue-tiles were a specialist job and equally briskly rejected the hypothesis of itinerant tile-makers. He modified the latter, by postulating a limited number of building contractors operating on a wide scale and employing their own tile-makers who set up temporary tileries specifically for supplying the jobs in hand. He reminds us of those dies with initials probably representing the tria nomina of proprietors, in one case apparently accompanied by the slave's initials and FE (CIT). Moreover, he points to a 'near-total bias towards villas and substantial towns' and wonders whether these firms specialised not in particular types of building but in public and official contracts. One important piece of evidence that has so far gone unremarked suggests that these men were not only tile-makers but builders as well, experienced in the use of clay. For at Silchester, Verulamium and Wroxeter the rollers were used to key daub walls for wallplaster. (Lowther 1948, 9, where the footnote should refer to die 38 and Archaeologia 58, 25 fig. 1.) These walls were of late first or early second-century date, and this implies that either one individual or more probably his firm was called in to build shops and houses as part of an officially sponsored programme of urban development in each of these towns.

A puzzling aspect of the subject is the small quantities of these tiles often found on sites where plain brick is abundant. This may perhaps be explained partly by the relative expense of tile and brick in the earliest phase of a site, and partly by the restriction of the patterning to flue-tiles. But this will hardly account for the discovery of only one or two fragments on some excavated sites, and one is tempted to regard their presence as accidental, the result of moving hardcore or builder's rubble from one site to another in Roman times. Some light has been shed on this problem at Chelmsford (Drury, forthcoming) where the relief-patterned tiles were varied and plentiful. All the dies used except one (44) appeared to have used the same clay, and are presumed to have come from the same tilery. Of these, die 16 was dominant (65.5% of total), with the other dies (4, 6, 8, 16a, 46) in insignificant quantities as low as 0.9% and 1.8%. Drury explains these 'oddments' as evidence for a contractor purchasing tiles largely impressed with a die in current use, but including a few tiles from the mixed stock of the tilery concerned. He also notes the resemblance of the fabric to those at Ashtead, Surrey, and wonders if there may have been fewer production centres than we usually suppose. He suggests four major concentrations of sites: in London; around Ashtead in Surrey; Verulamium in Hertfordshire; and Angerming in Sussex. Die 44 on the other hand (9.7% of total) was used on tiles of a different fabric, indicating a different source; the design, moreover, was relatively crude, and these tiles may be slightly later in date, a supplementary delivery of material perhaps from a more local source. This leads Drury to a further suggestion—'that the idea was copied by small local tileries, perhaps resulting in some of the less well executed and very localised designs, for example die 44, which is known only from Chelmsford (Lowther 1948, 15), and the similar dies 41-3, which are found only in Canterbury.' The Chelmsford evidence also confirms what had already been suspected, that the tileries which produced the bulk of relief-patterned tiles also produced some of the more common cross-hatched and combed types.

It is clear from this summary that the crucial, unresolved, question is this: did the tilemakers travel extensively, in teams or individually, making tiles wherever there was a demand, from local materials? Or did they work in a limited number of major production areas, their techniques being copied locally in more ephemeral centres? Our approach has been to start with a kiln site (Long Hanborough, Oxfordshire) where the relief-patterned tiles might have been made, and to seek to trace its products in a limited region, that of North Oxfordshire. Our progress with this is described below.

Finally, there remains the problem of the dating of the relief-patterned tiles, as Lowther felt that 'they may have considerable value in dating structural remains, particularly when stratified coins or pottery happen to be absent from the finds.' Subsequent events have done little to fulfill his hopes, in spite of the fairly narrow date-range of A.D. 80-150 that he proposed; a review of the dating evidence in the light of new finds has seemed overdue to us.

The results of visual examination

To date we have recorded 189 examples (compared to Lowther's 130)—that is, 4 cases of patterned daub already published and 55 new examples. (By 'examples' we mean one or more specimens of the same pattern from the same site.) Of these, some have been published or are due for publication, but most are unrecorded and unpublished. 13 have been thin-sectioned for analysis.

Identification of specimens is fraught with problems, particularly if drawings are used for recording and publication. Where practicable, it is now our practice to record by means of plaster casts and photographs where appropriate. Even so, specimens are often very small, the impressions uneven, overlapping and damaged; uneven pressure may have distorted the impression, and clay can sometimes have clogged the roller. It could even be said that no two impressions are exactly alike, and for this reason considerable conservatism has been exercised in the recognition of new dies. Nevertheless, the discovery of 11 new dies represents a considerable advance in our knowledge. Appropriately, most of the new dies fall into Group 5 ('Diamond and Lattice') which is the largest anyway.

The position by September 1979 was as follows:

New dies:	Group 1	1	
	5	5	
	6	2	
	7	1	
	8	2	total: 11
Old dies:		24	

(new examples, or further examples of new dies)

Not examined/identified	20
Falsa	1

The new dies are described in Appendix II and unpublished examples illustrated in Figs. 21.2-21.5.

Petrological examination

In an attempt to shed a little light on the vexed question of 'centralised' tilemaking as opposed to 'on-site' tilemaking, a small programme of petrological analysis was undertaken on samples of tiles from a number of sites in the North Oxford region (see Fig. 21.1 and the Ordnance Survey Map of Roman Britain 3rd edition fig. 4). This region includes a suspected tile-kiln at Long Hanborough, a walled town (Alchester) and a high concentration of villas and suspected buildings. A sample of a relief-patterned tile from Chelmsford, Essex, has also been included, as it is almost certainly from the same die-stamp (no. 46) as one from the North Leigh villa (Plate 21.I). It is from a different part of the impression, and shows considerable splintering of the roller—in other words, if the identification is correct it implies movement from one site to another and allows the fabrics to be compared. The majority of the 27 tile samples in Appendix I are from plain tiles, selected on a fairly random basis where tiles were in quantity at certain sites. Also included are a small number of relief-patterned samples, the only ones available for analysis in the designated area. No attention has been paid to the dating of the sites from which the samples were taken. Indeed, in many cases this information was not available. Bearing these points in mind, it is clear that the tile samples in Appendix I are far from being a controlled group and the results should therefore be regarded as tentative in advance of further detailed work in this area.

All the samples in Appendix I were thin-sectioned and studied under the petrological microscope. In each case the dominant inclusions were found to be subangular quartz grains, with usually lesser or greater amounts of mica. Due to the common nature of these minerals it is not possible to be specific about geological origins on this information alone. However, a textural analysis of the slides, taking into account the size, shape and frequency of

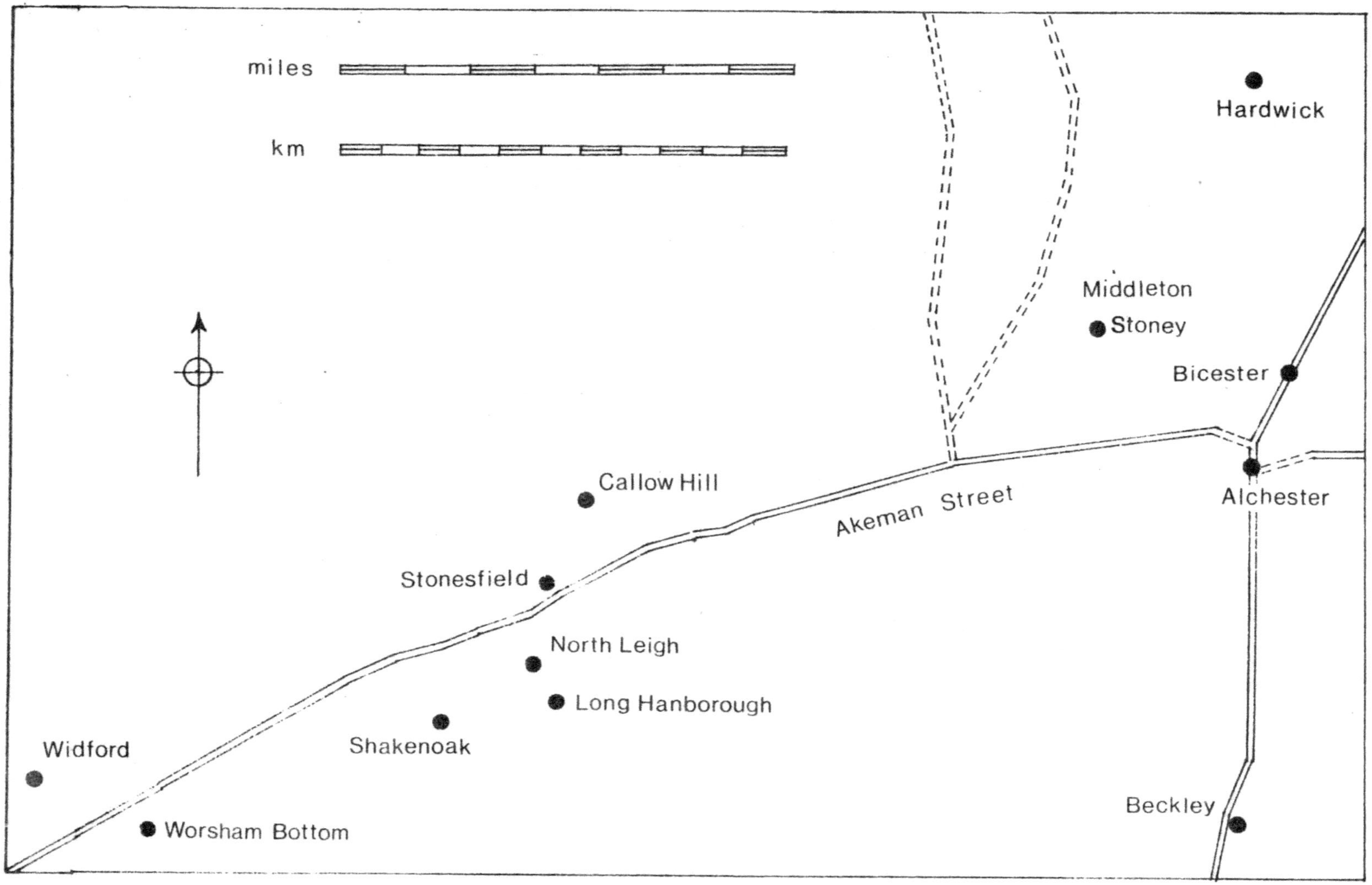

Fig. 21.1 **Location** of sites mentioned in Appendix I

the mineral grains, does allow similarity tempered fabrics to be characterised and comparisons made. The area of North Oxfordshire under discussion contains a variety of Jurassic deposits as well as a spread of boulder clays. It should be possible, therefore, to suggest if a series of tiles was made at the same or different localities.

Appendix I shows the range of inclusions present in all the samples, a rough guide to their frequency and, in the case of quartz, a breakdown of average grain-size. From this information it is clear that the two samples from the suspected kiln at Long Hanborough and the two from the nearby North Leigh villa (one plain, one relief-patterned) form a fairly tight textural group containing a scatter of smallish quartz grains and fragments of iron ore. This fabric is very similar to that of two relief-patterned samples from Hardwick and Stonesfield (19), where the die is considered to be identical to that of the North Leigh sample (die 46). It seems quite possible that all three tiles were made in the same location. However, the fabric of the relief-patterned Chelmsford tile, also considered to be from die 46, is quite different from this group. If both stamps are from the same die, the petrological results in this instance suggest a movement of tilemakers rather than tiles. Samples of plain tile from Stonesfield and Shakenoak are also similar in fabric to the North Leigh group and might also imply an origin at Long Hanborough, situated approximately halfway between the two villas.

Two other groups of tiles appear to have fairly similar fabrics. Those from Worsham Bottom, Widford and Shakenoak (14 and 15) form one group, and those from Alchester (1) and Bicester form another. The sites in both groups are fairly close to each other and the geology in each case is similar, so that it is difficult at this stage to say if both groups imply the same origins for their respective tiles, though this is a possibility.

None of the remainder of the tile samples in Appendix I are sufficiently similar to tiles from other sites to suspect the same origin. It is, however, interesting to note that there are differences in fabric in the samples of tiles from Alchester, Middleton Stoney, Stonesfield and Shakenoak, respectively. As mentioned above, it is not certain that the tiles from these sites are contemporary, and the differences in fabric may merely reflect different periods of building. Alternatively, the petrology may imply that the tiles were coming from different sources to individual sites—perhaps some were produced locally while others were brought in from some distance away.

Many of the views expressed here are of course highly speculative. However, taking the petrological results at their face-value, the following tentative conclusions may be stated:

1) The suspected tile-kiln at Long Hanborough was probably producing plain and relief-stamped tiles for local villas at Stonesfield and Shakenoak, as well as for Hardwick some 14 miles distant.

2) If the dies on the North Leigh and Chelmsford samples really are identical, then the tilemaker, or his die, was moving around the country, and not the finished product.

3) The tile industry may have been composed of several elements. The differences in fabrics of tiles from the same site may imply both locally produced and imported tiles. Perhaps based on Peacock's estate brickwork supplemented by some importation?

Discussion

a) Dating: Since thermoluminescence cannot yet give us the precision we require, we are still dependent on conventional methods—that is, stratigraphy and association. Lowther (1948, 10) summarised the evidence and proposed a date-range of A. D. 80-150 for the series. Subsequent discoveries, mostly with rather shaky dating evidence, have tended to confirm this. More precise and secure dates are available for a few dies at both ends of the date bracket. But since it is impossible at present to define phases within it, and to allocate individual dies or whole Groups to those phases, we must wait until a sufficiently large number of dated dies has accumulated. Rodwell's suggestion (1978, 23) that each Group represents a separate firm of contractors would imply a different building phase for each Group represented on any site; and this was not borne out at Chelmsford (Drury, forthcoming), where the 'main group' was dated to before 125, and probably 60-80, while only die 44 may be slightly later. It does not, in fact, appear on the site before Phase VI (200 +); but this may be misleadingly late and caused by re-use, as dies 4, 6 and 8 also appear only then and of these 4 and 6 are almost certainly Flavian.

The lack of a proper chronology for what must be the work of at least two generations of craftsmen prevents us tracing the origins, development and decline of this branch of the industry. As far as we know, the basic forms of bricks and tiles in Britain underwent little change over four centuries. Why then do we find this attractive manufacturing technique appearing in the Flavian period and lasting for a mere 70 years? The answer will probably appear within the industry when we understand its organisation and chronology better. Its origins are probably to be found in a firm or firms employed in Flavian times on official contracts such as the Fishbourne Palace; the early demise of the technique could mean the extinction of firms using these designs as a trade-mark in consignments that included combed and plain tiles. Alternatively, it could mean the retirement or death of, say, a second generation of specialists using what were, by then, old-fashioned rollers. Dies appearing later in the second century could even imply subsequent pockets of conservatism in remote rural brickyards.

b) Organisation: The lack of a chronology not only deprives us of an evolution for this branch of the industry but also leaves the particular problem of the itinerant tilemarkers unresolved. The starting-point for an answer to this remains Lowther's Groups, which are still valid (though Rodwell has subdivided his Group 5) and his distribution maps (which are not out of date). With so much material still unexamined, we have not attempted at this stage to produce a definitive revision of these maps, and the reader is referred to Rodwell's revision (1978, fig. 6) which incorporates some new material and is an improvement on the originals (which confusingly amalgamated several of the Groups). Rodwell's suggestion that each Group represented a distinct firm is hard to sustain in view of the Chelmsford evidence and the occurrence of more than one Group on some sites. From the maps it is preferable to follow Drury and recognise four major production centres, in London, around Ashtead in Surrey, Verulamium in Hertfordshire and Angmering in Sussex, with outliers. Naturally, these served sites in their immediate region; but Drury is convinced that Chelmsford derived most of its tiles from the Ashtead area, and provides the first evidence for the transport of tiles over long distances in Britain. If this practice was indeed widespread the logistics of it must have been formidable, especially in view of the steep gradients on some Roman roads. Nevertheless, this was achieved for lead ingots by a judicious combination of road and river (Johnston 1979, 79 and 116); and loads of raw clay, which was even heavier than bricks, were transported cross-country from claypit to kiln as a matter of course.

There is, however, some good evidence to support the notion of itinerant tilemakers. It is reasonably certain that the Chelmsford and North Leigh specimens are from the same die, and a comparison of the impressions (Pl. 21. I) shows very clearly how the roller was subsequently damaged by splintering. The Chelmsford specimen is believed to derive from the Ashtead area, the North Leigh one from the Long Hanborough kilns. So one tilemaker, at least, moved his centre of operations and took his roller with him. This might have happened, on Chelmsford evidence, at the end of the first century, certainly before 125. By 120 we find the distinctive 'billet' dies used in this same Oxfordshire area but in a different fabric. Either Long Hanborough was drawing its clay from a fresh source or a new tilemaker was now in business at a new kiln site in the same area. The last line in this game of consequences—the die link with Charterhouse-on-Mendip—has yet to be checked petrologically,

The second piece of evidence for the itinerant tilemakers is the important use of the roller on walls at Silchester, Verulamium and Wroxeter. Were these done by the same man? The dating is not precise enough to help, and the dies cannot be confidently identified with those used on tiles, with the possible exception of one Wroxeter impression that might be from die 38. Provisionally, however, three dies seem to have been used. There is a possible link between Wroxeter and Verulamium and between Wroxeter and Silchester (W-chevrons in each case), so we may be able to imagine the same team, or firm, at work.

Nobody has yet excavated a tile-kiln that can be proved to have made the relief-patterned tiles. Even at Ashtead the connection is assumed, though at Long Hanborough the petrological evidence looks hopeful and awaits proof that the surface indications are indeed those of a kiln or kilns. At Minety, Wilts., waster evidence suggests another source (Mr. N. P. Thompson, in pers. comm.) and the supposed kiln site at Little London, near Silchester, is another strong possibility (Boon 1974, 278 and Mrs. J. Greenaway in pers. comm.). Moreover, the ephemeral use of clamps must not be forgotten.

To sum up, then: firm conclusions are premature at this stage, in view of the conflicting evidence for mobility of tilemakers and centralised production. This is, provisionally, best reconciled by supposing four major production areas, each consisting of scattered kiln sites and clamps rather like some of the large pottery-producing areas. Whether they were owned by individual industrialists (such as the owner of the Ashtead villa?) or worked for government contracts (such as the Fishbourne Palace) is open to speculation. What is now clear is that they were prepared to move their products to distant sites. Their products were plain, combed, knife-scored and roller-patterned, the last being the mark of the flue-tile specialists. Unlikely as it may seem, these men may even have invented a name for themselves—at least (if the reading is correct) Cabriabanus described himself on die 31 as PARIETAIEUS—an interesting combination of Latin and Greek elements (paries meaning 'wall'), Such men were also builders (at least of wattle-and-daub walls) and were prepared to travel to other parts of Britain to practise their speciality, to found new tileries and to inspire others to copy their prestigious trade-marks. One is tempted to compare this with another case of industrial mobility in Roman Britain—that of the four 'schools' of mosaic-making and the 'Central Southern Group' that may be an offshoot (Smith 1969, Johnston 1977). There is, however, one important difference; the demand for mosaics was finite in any area, and soon exhausted, while the demand for tile-makers and central heating specialists was (and still is) virtually inexhaustible.

c) Future work: This study has been based partly on other people's work, partly on material submitted to us for examination, and on a small research project. We are aware that there must be many unrecorded specimens that we have not seen, and naturally would be happy to examine and comment on any new material. It is clear that petrological analysis holds the key to the questions outlined above, and this is to be the direction of our future work. We shall seek to identify the products of particular kilns and to trace their distribution as far as possible.

Acknowledgements

We are particularly grateful to Mr. Paul Drury not only for providing the important specimen from Chelmsford, but also for allowing us to use his forthcoming report on the mansio excavations in advance of publication. We have also received invaluable help from the following museums: the Ashmolean (Mr. D. Brown), Bedford (Miss J. Hassall), Chelmsford (Mr. D. Jones) and Kendal (Miss M. D. Burkett and Mr. J. Anstee). Numerous individuals have made this study possible, by providing specimens and casts for analysis and for the reference collection now being formed in Southampton University. Our particular thanks go to Mr. A. C. C. Brodribb, Mr. G. Brodribb, Mr. C. Chambers, Mr. G. H. Cole, Mrs. A. Fairey, Mr. S. D. Ford, Mr. C. Going, Miss J. Greenaway, Dr. A. R. Hands, Mr. A. Rook, Mr. A J. Scammell, Mr. N. P. Thompson and Mr. D. Walker. The Departments of Archaeology and Adult Education of Southampton University have provided the opportunities and facilities for research; and above all we would thank the Editor of this volume, Mr. A. D. McWhirr, for his encouragement and patience during the preparation of this contribution to it.

APPENDIX II

NEW DIES

Group 1 'W-chevron'

Die 47 Roake Farm Broughton, Hants. (SU 316318). A good, clear impression from the edge of a box-flue tile. The impression disappears at the edge of the tile, and at one point the impression is marred by clay clogging the roller. At first site it resembles Lowther's die 5a (a re-cut version of die 5); however, the horizontal recessed (i. e. black) bars are thicker and the angle and position of the vertical relief line cutting them is quite different. It is therefore best regarded as a new die. (Fig. 21. 2)

Group 5 'Diamond and Lattice'

Die 16a Chelmsford, Essex. Identified by P. Drury as a version of die 16 (publication with illustration forthcoming).

Die 48 Fishbourne, Sussex. This large die is unusual in two respects: the adjacent rows of diamonds have alternately open and filled centres, and the left-hand edge (at least) is accompanied by a row of single diamonds. N. B. the illustration in Cunliffe 1971 fig. 25 is in negative; cf. ibid. pl. xii.

Die 49 Gadebridge Park, Herts. A crudely executed die with a single series of concentric diamonds, the remaining space filled with random lines and triangles. (Neal 1974, fig. 86 no. 716)

Die 50 Shakenoak, Oxfordshire. Two fragments, apparently from different parts of impressions of the one roller, have already been published separately (Brodribb et al. 1973, fig. 21 no. 74 and 1968, fig. 11 no. 16). Both were studied in the Ashmolean Museum in 1979 and the latter was found to have been joined to a further (unpublished) fragment. The possible relationship of the fragments is shown in Fig. 21. 4. One relief bar on the lower portion seems to have a cut end or edge; perhaps the roller had plain, raised borders. This would be unique, though dies 14 and 30 were framed by relief lines. The impressions are defective at the edges, but the roller seems to have been unusually wide, at least 6 in.

Die 51 Shakenoak, Oxfordshire. A small fragment, from which it is difficult to draw firm conclusions. But the change of thickness where the incised lines (on the roller) cross cannot be paralleled in any known die, and it is best regarded as a new die. (Brodribb et al. 1968, fig. 11 no. 17)

Fig. 21. 2 Die 47: Broughton Farm. 1:1

Fig. 21. 3 Die 32a: Sandy. 1:1

Group 6 'Billet'

Die 52 Shakenoak, Oxfordshire. The two specimens illustrated in Brodribb et al. 1968, fig. 11 no. 15 and 1977, fig. 11 no. 79 are best regarded, in spite of superficial differences, as from the same die. The statement attributed to Lowther (ibid. 1968, 40) is almost certainly incorrect, if it identifies the first of these two fragments with his die 25. The Shakenoak pattern is in fact considerably larger and seems to be one of two new dies; he may of course have realised this when he added 'a different die had been employed... '. This die, like the next, was used for patterning bricks, and one of the specimens illustrated (1977, no. 79) is almost complete.

Die 53 Shakenoak, Oxfordshire. Illustrated in Brodribb et al. 1973, fig. 21 no. 73 this large specimen appears to be the same as the previous impression. Close inspection, however, shows that small but repeated cutting errors in both establish this as a new die. Both these patterns are of the large variety, and Mr. C. Brodribb has provided two small fragments from Worsham Bottom (for location, see fig. 21. 1) which are also from one or the other.

Note: a third 'billet' die was also used in the Shakenoak region. This is the smaller one, recorded by Lowther (die 25) from Worsham Bottom and Charterhouse-on-Mendip. The former cannot now be traced in the Ashmolean Museum. Mr. Brodribb has provided a third specimen found at Widford, also in the Shakenoak region (see Fig. 21. 1).

Group 7 'Rosette'

Die 32a Sandy, Bedfordshire. A single specimen, of uncertain provenance and unpublished, in the Bedford Museum. It is from a box-flue tile with two slightly overlapping impressions in parallel. Fig. 21. 3 shows the design reconstructed as far as possible. The impression is unclear at top and bottom, and is made less clear by the traces of mortar on the surface of the tile. There is therefore some uncertainty about the form of the diamonds at the bottom and on the left of the drawing.

The distinctive feature is the linking of the bars of the diamond at two points, though at each point the impression is unclear and there is no absolute certainty on this. This feature, and the rest of the impression, show that the diamond is identical with that of die 32 as found at Boxmoor (Hertfordshire, Woodperry (Oxfordshire) and Canterbury (Kent).

There is, however, one important difference. The right-hand impression, later and deeper than the other, shows clearly the left-hand edge of the roller. This is considerably narrower than die 32 and can only be the result of shortening a damaged roller—akin to the re-cutting of the die that Lowther noted in other instances. The full design of die 32 includes a rosette, and the shortening of the roller would leave this feature intact. It is proposed, therefore, to follow Lowther's practice with recut dies, and give it the die number 32a.

Fig. 21.4 Die 50: Shakenoak. 1:1

Group 8 'Addenda'

Die 54 Shaw, Newbury, Berkshire (SU 472684). Drawings and casts of two specimens were submitted by Mr. S. D. Ford, from the fill of a pottery kiln (Fig. 21. 5). Another kiln in the group contained 1st century pottery. Both impressions are presumed to be from the same roller, though there is no overlap of the two impressions. It is impossible to relate them as parts of one complete design, though the angles of each relative to the movement of the roller can be deduced from the impressions of the wood grain.

Compared to other dies, this is larger and cruder. The impressions are shallow, and not very distinct. The raised parts of the die were evidently unworn, and the recessed parts of it probably roughly cut to no great depth. The leaf-like elements are most appropriate to the use of a gouge rather than the square-ended chisel normally used.

The leaf-like elements are clearly deliberate, a contrast to the straight bars of the chevrons. The design is clearly inspired by the W-chevrons of Group I, but in view of the leaf-like elements it is preferable to allocate it to Group 8 as a new die. Its closest relative is die 26, from Alresford, Essex, a site that significantly produced fragments of a truly 'florid' design, no. 8.

Die 55 Fishbourne, Sussex. (Cunliffe 1971 fig. 25 E, where the illustration is drawn in negative.) An incomprehensible fragment of a geometrical design that has some slight affinity to die 28, from London.

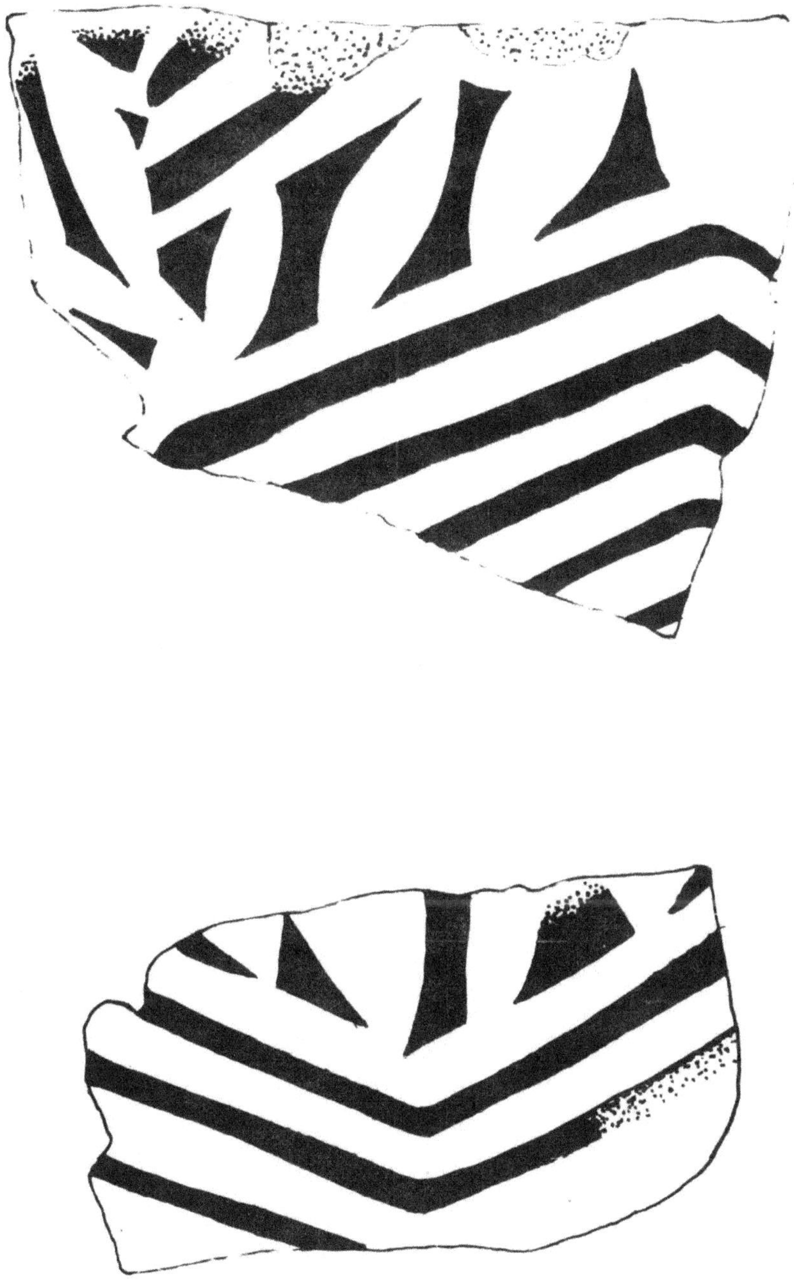

Fig. 21.5 Die 54: Shaw. 1:1

REFERENCES

Boon, G. C. 1974. Silchester: the Roman Town of Calleva. David and Charles.

Brodribb, A. A. C., et al. 1968-1978. Excavations at Shakenoak I, 1968: II, 1971: III, 1972: IV, 1973: V, 1978. B.A.R.

Cunliffe, B. W. 1971. Excavations at Fishbourne, 1961-9, Society of Antiquaries.

Johnston, D. E. 1977. 'The Central Southern Group of Romano-British Mosaics' in J. Munby and M. Henig, Roman Life and Art in Britain, B. A. R. No. 41.

Johnston, D. E. 1979. Roman Roads in Britain, Spurbooks.

Lowther, A. W. G. 1948. A Study of the Patterns on Roman Flue-tiles and their Distribution. Res. Papers of the Surrey Archaeol. Soc. no. 1.

McWhirr, A. D. & Viner, D. 1978. 'The Production and Distribution of Tiles in Roman Britain with particular reference to the Cirencester region' Britannia 9 (1978), 359-376.

Neal, D. S. 1974. The Excavation of the Roman Villa in Gadebridge Park, Hemel Hempstead, 1963-8, Society of Antiquaries.

Rodwell, W. 1978. 'Rivenhall and the emergence of first-century villas in northern Essex', in M. Todd (ed.) Studies in the Romano-British Villa, Leicester University Press, 11-32.

Smith, D. J., 1969. 'The Mosaic Pavements', in A. L. F. Rivet, (ed.) The Roman Villa in Britain, Routledge, 71-125.

APPENDIX I Petrological Analysis of Selected Roman Tiles from North Oxfordshire

No.			Quartz grain size (in mm) c. .05	c. 10	c. 20	c. 30	c. 40	c. 60	mica	iron ore	sandstone
1.	Alchester	P. T.		M	M	M	M				
2.	Alcester	P. T.	M	M					A		
3.	Beckley	P. T.		M	M	M		S			
4.	Bicester	P. T.	M		M	M			A		
5.	Callow Hill	P. T.		M	M				M		
6.	Hardwick	R. P.	S	S	S				M		
7.	Middleton Stoney	P. T.		A	A		M				S
8.	Middleton Stoney	P. T.			S	S	S	S			
9.	Middleton Stoney	P. T.		A	A		M				S
10.	Long Hanborough	P. T.	S	S	S					M	
11.	Long Hanborough	P. T.	S	S	S					M	
12.	North Leigh	P. T.	S	S	S					M	
13.	North Leigh	R. P.	S	S	S					M	
14.	Shakenoak	P. T.		A					A		
15.	Shakenoak	P. T.		A					A		
16.	Shakenoak	P. T.	M	S	S						
17.	Shakenoak	P. T.	M	S	S						
18.	Shakenoak	P. T.	M	S	S						
19.	Stonesfield	R. P.	S	S	S						
20.	Stonesfield	P. T.	S	S							
22.	Stonesfield	P. T.	S	S	S						
23.	Stonesfield	P. T.		M	M						
24.	Stonesfield	P. T.		M	M						
25.	Worsham Bottom	R. P.		A					A		S
26.	Chelmsford	R. P.		M	M	M	M	M	M		
27.	Widford	P. T.		A					A		

P. T. = Plain tile; R. P. = Relief-patterned; A = Abundant; M = Moderate; S = Scatter

EXPERIMENTS IN MAKING AND FIRING BOX-FLUE TILES

Graham Morgan

This study was carried out following ideas published by McWhirr and Viner (1978, 362) concerning techniques used in the manufacture of box-flue tiles.

It is well known that broken-pottery sections, when polished, often show flow lines caused by movement of the wet clay and air bubbles or inclusion orientation. Consequently it was decided to cut up samples of tile and to examine the sections. These sections showed flow lines and joins the latter were confirmed by surface indications on larger tile samples.

The usual features which were found on the samples were variation in oxidation colour (particularly at joins and cracks), inclusion orientation, air-bubble distortion and orientation. Tiles of very fine clay or uniformly fired, with particularly good joins proved very difficult to analyse. In order to have a standard with which to compare the sections a series of three different types of joins, butt, lap and scarf (fig. 22.1) were made. These were fired at about 800°C in a small electric kiln, cut with a diamond saw and polished with carborundum paper.

Comparison of these experimental joins with those found in the Roman samples suggested that the distortions found in the lap or scarf joins are most similar and strongest, Butt joins tended to break of firing.

Examination of Roman box-flue tile sections shows that there was only one join, usually on the wide side of the tile and near the middle (vousoirs seem to be made differently). This suggested that tiles were made by wrapping a single piece of clay around a suitable former. A number of difficulties have been raised about this method particularly concerning the design of the former, including problems of its removal, the use of split-piece formers etc. Ignoring these theoretical problems a tile was made by wrapping a sheet of sanded clay around a parallel-sided wooden former and a lap joint made in the centre of one wide face. When the tile was turned over on to a flat surface to smooth the corners it was found that the clay stretched sufficiently to allow the former to be removed easily leaving the finished wet tile standing vertically.

During the preparation of the clay an oblong sheet of clay was rolled out between wooden guides. However it is more likely that the wet clay slab would have been formed originally in an open mould as with roof tiles, allowed to dry somewhat and then wrapped around the former.

The tiles were fired in an oblong kiln, cut into the ground (clay), the chamber of which was about 1 m by 0.75 m and 0.75 m deep. The fire tunnel was 0.5 m high with five brick arches and no other floor. Twenty tiles were

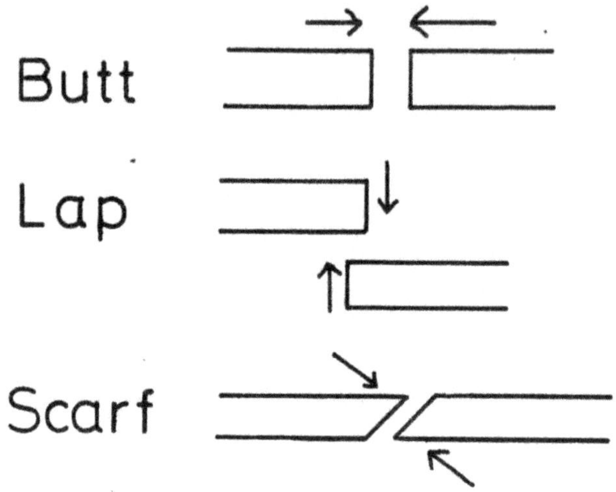

Fig. 22.1 Different types of joints used in making box-flue tiles

placed on end in the kiln in a single layer and covered with flat tiles and bricks. This probably resulted in too great a restriction of heat flow. Using in all about 150 kg (3 cwt) of split wood and twigs, the temperature was slowly raised over four hours to a maximum of around 800°C. This was not high enough for a completely satisfactory firing of the tile, which consisted of particularly sandy clay. Some small slip-cast lamps made of the same clay but with the sand removed were fired with the tiles and became quite hard. It seems that the tile and brick roof of the experimental kiln caused too much restriction of the fire and kept the gas flow and temperature down.

The resulting tiles were cut with a diamond saw and polished with carborundum paper as with the original samples. The sections of the tiles, being sandy, were not very clear, but the joins compared favourably with the initial test specimens. The possibility of making tiles out of four slabs of clay as suggested by McWhirr and Viner (1978, 362) was also investigated. Four pieces of clay were rolled out and cut so as to be slightly larger than the walls of the tile. The edges were moistened and the pieces pinched together around the wooden former. The corners thus formed were smoothed down with a wet hand. The tiles produced in this way stood up to firing as well as the one-piece tiles, but took about three times as long to make. The joins made by pinching together and smoothing were quite difficult to see in the polished sections.

Reference

McWhirr and Viner 1978. A. McWhirr and D. Viner. 'The Production and Distribution of Tiles in Roman Britain with particular reference to the Cirencester Region'. Britannia, 9 (1978), 359-377.

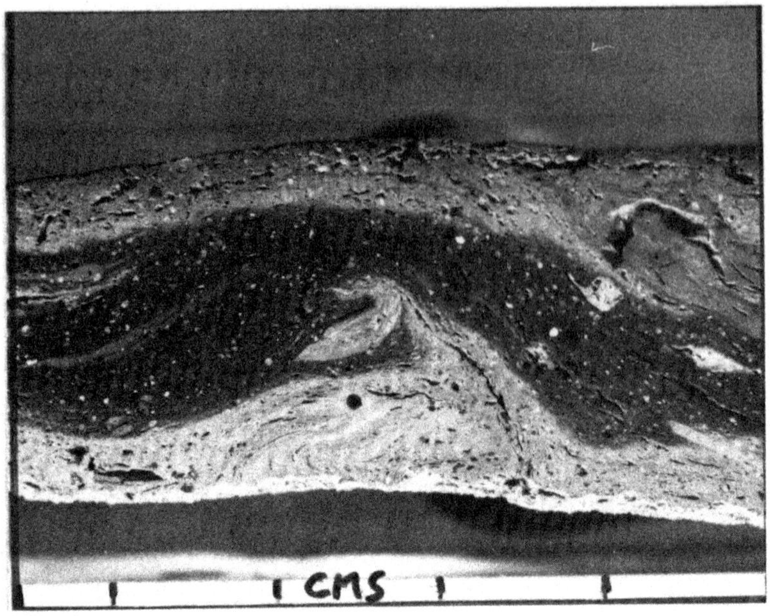

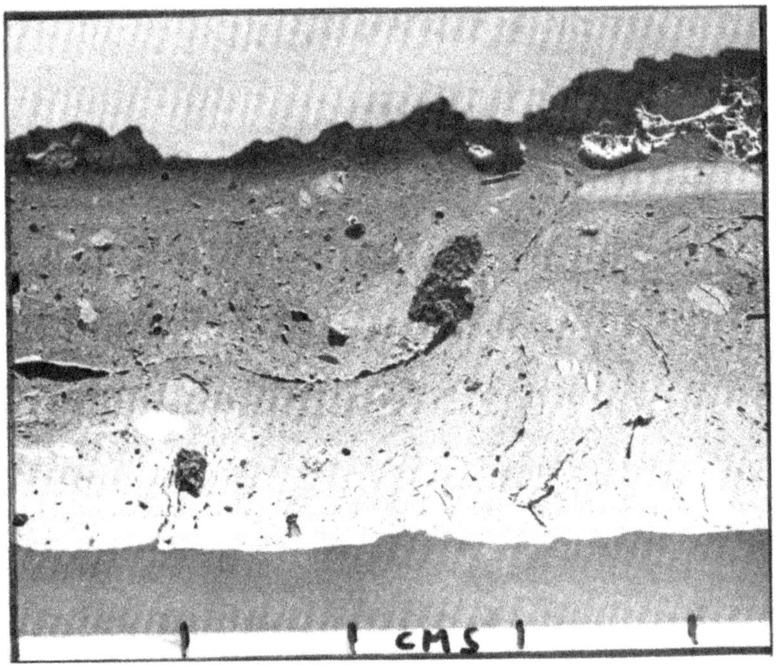

Plate 22. I a. Fragment of a Roman box flue tile from Heckington showing air-bubble flow lines and inclusion orientation. The dark central area is reduced fired and the outer part oxidised

b. Section of a join on a Roman box flue tile from Cirencester. Comb impression on the surface of the tile is also visible

Plate 22. II a. Experimental tile section with a lap joint showing the two pieces well bonded together. The section is at the top, the surface at the bottom of the picture

b. Experimental tile section with a butt joint showing the two pieces poorly bonded together. The top of the picture shows the section, and the bottom the surface, of the tile

THE PROCESSING OF ROMAN TILE[1]

Christopher Young

The various contributions to the conference more than adequately demonstrated the potential of tiles as a source of information for Roman Britain on methods of building, trade patterns and the general economy and society. However, much concern was expressed over the problems of extracting this information economically and expeditiously from the often vast quantities of ceramic building materials found on Roman sites. This paper offers tentative suggestions as to how these problems of scale can be overcome.

In all discussions of the processing and publication of material from excavations it is necessary first to define the questions that can or may be answered by the study of that particular class of artefact and the ways in which those questions can be approached. Undifferentiated recording of every conceivable aspect of the material can only result in a vast mass of undigested and indigestible data. Therefore this paper discusses first the various types of information tiles can provide and the methods by which these can be extracted and then how this methodology can be applied to large quantities of Roman tile.

As David Peacock stressed at the conference the tile industry was probably more important within the Romano-British economy than pottery production. It must have been among the major non-agricultural industries. He, and others, also demonstrated the wide variety of possible organisations for tile production, and a wider understanding of which organisations did occur is essential for a closer knowledge of the tile industry and also of Romano-British society and economy in general. Information is needed on where tiles were made, who made them, whether the same tilers worked in more than one place, whether and in what proportions different sorts of tile were made at the same tilery, quantities produced, and size ranges and standardisation of sizes of particular tile-types.

To attempt to answer these points fabrics must be identified to locate sources and areas of production, types of tiles present at any site must be classified according to fabrics and the sizes of tiles must be recorded where possible. Tilers' stamps, batch-marks, graffiti referring to quantities of tiles, or individual tiles, and tilers' signatures are all essential and must be recorded in relation to fabrics as well as in their own right, since this may show up itinerant tilers (cf. Darvill in this volume). Features such as combing or roller-stamping may also be of some use for identification of particular tilers or workshops.

Secondly and related to manufacturing methods and organisation is the evidence for trade. It is necessary to know from what sources different buildings drew their tiles, whether there was any specialisation in supply and the extent of the marketing patterns of different tileries or tilers. Here again fabric identification must be the major source of information, linked with the evidence of stamps and other identifiers of particular tilers to answer more specific questions of origin and organisation.

Thirdly tilers can provide much information on the nature of buildings under excavation. The presence of different types of tiles implies different types of structure or building methods. For example, the presence of box flue-tiles must imply a heated structure, while that of roofing tiles is evidence for the existence of walls strong enough to carry a roof of such weight. It is essential to know the type and size range of the tiles present on a site. It is also essential that they should not be considered in isolation from other types of building material.

It is essential for all the aspects so far discussed that the information recorded should be quantified, preferably on a uniform basis across the country for ease of comparative study. Quantification is the only way in which the relative importance of various fabrics, tile types, size ranges or other attributes can be gauged. The appropriate method of quantification will obviously vary according to the factor being recorded. Weight is probably the most appropriate for recording fabrics and tile-types present, but fragment count is also possible.

Apart from the major aspects discussed above, there are more incidental areas of study related to tiles. These include evidence for literacy and personal names as shown by graffiti and for flora and fauna as shown by impressions on the tiles. Here too all relevant information must be recorded.

It is thus possible to make a list of the categories of evidence which must be recorded in the study of tiles. Firstly they should be classified according to fabric, using the normal criteria of colour, hardness, feel and inclusions. Most of this can be done by macroscopic examination of the hand specimen, although more detailed petrological work will sometimes be needed.

Secondly tiles should be classified according to type, such as imbrex, tegula, hypocaust brick or box flue-tile. Any evidence for size and standardisation must be recorded, though tiles having at least one side complete appear to be comparatively rare. Other identifiers such as name-stamps, signatures, batch-marks, roller-stamps, graffiti and the presence of combing should be recorded also. All this information should be appropriately quantified and could most conveniently be recorded on a proforma.

If all these variables are recorded adequately, the excavator or student will have the basic evidence with which to approach the types of problems outlined above and discussed more fully by the contributors to the conference. In fact, both the problems to be tackled and the methods for so-doing are similar to those used for pottery which are now relatively well-established (cf. Peacock 1977, Young in press). The major differences are the smaller number of typological variants of form and the frequently much greater quantities of tile found.

This last seems to be the most daunting aspect of tile studies for the excavator, and for the museum in which it will be deposited. If tile is present in large quantities it will frequently be impossible to retain and store all of it, and the need to mark, box and rack such quantities can impose considerable strains on the logistics of an excavation or the store of a museum. It is doubtful whether all tile should in fact be retained. Much of it will be fragmentary and repetitious, and will not be required for future study. It is also more than possible that large quantities of tile from an excavation, once stored, will never be properly studied due to lack of resources or interest.

It is probably most economical to do the bulk of tile processing on site during an excavation, so that the basic data on fabrics, types and quantities is obtained at this stage, following which most of the tile can be dumped, only pieces of interest being retained for the future. Obviously this will impose a strain on the organisation of the finds shed, since at least one person will have to work full-time on this. On the other hand, time will be saved, because it will not be necessary to mark or box large quantities of tile, and fewer resources would be needed for post-excavation work on the report.

It is suggested, therefore, that all tile should be washed on site and sorted according to fabric and type. It should then be quantified within these categories by weight, and perhaps, by fragment count, and any useful dimensions (such as complete sides) should be recorded. All tile should be examined carefully for evidence of stamps, batch-marks, signatures, graffiti, roller-stamps, combing or any other oddities. This should be recorded and all pieces bearing such traces should be retained, as should any complete or near-complete tiles and an adequate sample of all the different fabrics. Remaining fragments should be discarded.

This procedure should produce an adequate archive on which a published report or future research can be based. The retained material should also be a sufficient sample of the original material to allow for meaningful re-examination. It seems also to provide the most economic method of obtaining the necessary information. It is surely not unreasonable to expect that the appropriate resources should be costed into an excavation project when it is expected that large quantities of tile will be found, since so much information on the economy and society of Roman Britain can be obtained from this material.

Bibliography

Peacock, D. P., 1977. 'Ceramics in Roman and Medieval Archaeology' in Peacock, D. P. (ed.), Pottery and Early Commerce, London.

Young, C. J. (ed.), in press. The Processing and Publication of Roman Pottery, DOE Occasional Paper.

1. The views expressed in this paper are those of the author and not necessarily those of the Department of the Environment.

BRIXWORTH CHURCH—ARE THE BRICKS REALLY ROMAN?

Paul Everson and David Parsons

This final paper of the conference differs from most of the rest in a way that may be appropriate in not presenting the results of a project, but being an interim statement that amounts to little more than an exposition of a problem and possible methods of tackling it. The paper is presented on behalf of the Brixworth Archaeological Research Committee.

The church of All Saints, Brixworth in Northamptonshire is known to a wide scholarly and general public for the sheer size, survival and evident structural complexity of its pre-Conquest fabric. Though it has often been assumed that parts of the fabric date from the documented seventh-century foundation of a monastery here as a daughter of Medeshamstede (Peterborough), recent scholarship has concentrated upon the architectural stratification of the building and the analogies of plan-form of parts or the whole, with the result that a wider range of possible developments can be envisaged, which might include anything from the reuse of a standing Roman building to an eighth-century or later initial construction (Parsons 1972; Cherry 1976, 170-2). Nevertheless, the very extensive use of brick in the fabric combines with the scale and details of its architectural conception to emphasise the building's essential Romanitas; and the context for this, whether in its date, stylistic influences, availability of materials, liturgical needs, imagination of an individual (Wilfrid, bishop and saint, has been suggested), or some combination of them all, needs to be sought. It is far from certain, of course, that the use of brick belongs exclusively to one phase, or indeed that its dominant use in the nave arcades and clerestory is part of the original conception.

If the bricks at Brixworth were recovered from a Roman site, their source is an important consideration. Indeed the two questions are so closely connected that a plausible answer for the source has been seen as strong support for the identification as Roman. Until recently the suggested source had been a Roman villa lying about half a mile north of the church. This at least is untenable, since excavation has shown the villa to be a modest structure, not capable of supplying the large numbers of large flat bricks used in the church, and itself furnished with tile of a distinct fabric. Another local source might have been the substantial Roman remains lying immediately to the north of the churchyard which were evidently destroyed by ironstone extraction last century. It is not possible now to gauge the nature of these remains accurately and little weight can be placed on them until other possibilities have been investigated; but Blagg's (1978, 110-12) assessment of the late second- or third-century limestone cornice in Brixworth church at least holds out the possibility of a rural temple or large mausoleum sited locally, that might have contributed an adequate source for the brick. Certainly we appear to be in need of a major public (and probably urban) building from which to derive the brick, if reused.

For the Brixworth Archaeological Research Committee, the date of manufacture, source and date or dates of use of the brick constitute basic questions within a programme of comprehensive investigation of the church and its setting in village and parish. A stimulus for preliminary work was provided by excavations at the west end of the church in 1971 by Hall and in the vicarage garden west of the church in 1972 by Everson (Hall 1977, Everson 1977). The latter excavation in particular produced several hundred brick fragments, all arguably derived from the church or associated buildings and several stratified in the fill of a major ditch interpreted as the vallum monasterii. They were manufactured like those in the church from local Chalky Boulder Clay using methods known for Roman bricks—Harley's 'rolling pin' method or in moulds (Harley 1974, 64-66; Everson 1977, 97-102). Chemical analysis by X-ray fluorescence on five excavated samples produced the sort of variation of elements to be expected in the local Boulder Clay, and the principal interest lay in the fact that one of the samples was one of several raw clay lumps from the vallum, that could in respect of chemical composition be residue from a manufacturing process. Since thermoluminescent dating offers exactly the information -namely date of manufacture rather than of use—that is a primary question at Brixworth, one excavated sample was submitted to the Oxford University Research Laboratory for Archaeology and the History of Art. When this gave a 'Saxon' result, Dr Aitken agreed to take further samples from the church fabric, under more controlled conditions. From these eight samples, four available results showed one broadly late Roman date (A.D. 200-600), two later Saxon (700-1000 and 750-1050) and one late medieval (1400-1600, perhaps resulting from resetting of the fabric).

The BARC is now proposing to follow up this preliminary work. One principal line of approach already well advanced is the stone-by-stone drawing of the external elevations of the church supervised by David Parsons and the linked petrological identification by Dr Diana Sutherland (Parsons 1979, 69-70). This will allow for the first time an accurate assessment of the quantity of brick, and appreciation of its distribution in the fabric and association with other categories of building material. There are certainly signs of reuse of materials. In the south arcade and the ambulatory primary mortar or plaster of opus signinum type found adhering to bricks has given positive evidence for the reuse of some brick. The random distribution of reddened limestone now identified in limited quantities in the fabric indicates reused material, and the tufa also present is best known as a stone quarried for Roman use. Dr Sutherland has made it clear to BARC seminars that in her best assessment the striking assemblage of Leicestershire igneous rocks in the lowest part of the nave and tower is very unlikely to be derived either from the local drift in Northamptonshire or (because of the variety) by direct supply from quarry sources in the Charnwood area. It may indeed be possible to view brick as one element in any given reused fabric assemblage that has significance in point of date and source of materials. With this in mind, Dr Sutherland has recently pointed out the close similarity of the assemblages, comprising igneous rocks plus brick, in parts of the Jewry Wall in Leicester and the relevant sections at Brixworth. This line of approach has to go hand in hand with architectural

Plate 24.I Brixworth Church from the south-east. The nave arcade arch on the extreme left and three clerestory windows have heads consisting mainly of brick (photo: J. Jeffs)

analysis, since it must for example take into account the possibility suggested by Parsons (1977, 180-1) that the massive brick-lined arcades were secondary insertions.

Detailed measurements of the sizes and shape of bricks, based on as large a sample as possible, are obviously necessary for statistical comparison with Roman examples. A start has been made with this task in summer 1979 by Everson and John Williams. Measurements of just over a thousand bricks have been taken on the exterior north wall of the nave and choir while the mortar joints were raked out in preparation for repointing. From this total sample, 327 (32.5%) complete long dimensions were recorded; and of these complete dimensions no less than 90% fell within one or other of two ranges, 260-320 mm (mean 290 mm ± 10%) and 375-440 mm (mean 407.5 mm ± 8%). Even if dimensions that were classified as 'possibly complete' and 'uncertain' were to be included, giving a sub-total of 456 or 45.3% of the total sample, nearly 83% fell within the two ranges. Two examples in the exterior north elevation where both long dimensions were intact and measurable, and the frequent recurrence of measurements approximating to 290 x 410 mm elsewhere in the building where both long dimensions are accessible and bricks may be presumed complete, confirm that a standard large brick is used in overwhelming numbers. From preliminary observations, bricks of the same size are used in the Jewry Wall at Leicester. The measurements have also identified a second standard size in the sample, measuring approximately 210 x 210 mm and making up no more than 10% of the sub-total of complete/possibly complete/uncertain measurements. This size accords with pilae tiles from hypocaust systems.

Work on size and shape of the bricks has also drawn attention to their detailed structural use both in wall courses and in turning arches. With further study this may offer an additional point of comparison with Roman structural technique.

Any extension of the TL programme presents problems. The earlier results were obtained using the fine grain technique, for which the Brixworth samples were reported as having poor TL characteristics. The material was taken up again recently and used with the pre-dose technique. This technique is generally limited to material of less than 1000 years BP, but may be susceptible of stretching to encompass Saxon material. In this case, however, the work threw up a major problem, that requires further basic technical research. We hope that the existing Brixworth samples will form part of the research into the pre-dose method being undertaken at Durham University, and that they may yield further dates by this method. Preliminary tests at Durham appear to confirm the earlier Oxford dates. Even with poor TL characteristics, an accuracy of 10-15% may be possible, and this would be sufficient to distinguish Roman from Saxon manufacture. This limited accuracy to a broad date band should be distinguished from the reliability of the results, which can be very good. Without information from TL, we should not be in a position to assess the possibility of Saxon revival and copying of Roman techniques and standards.

The other main approach to the material can be through petrological study, in particular using thin-sectioning, that has been successfully applied to similar archaeological problems, particularly of Roman and early post-Roman pottery

Plate 24. II a. Brixworth Church: South door of tower. A good example of a brick arch head (photo: J. Jeffs)

b. Brixworth Church: Detail of tower arch seen from the nave (photo: J. Jeffs) (see Darvill's paper in this volume). There appear to be two complementary fields of study that might be considered:

a) analysis of suitable clay deposits within a given radius of Brixworth. This would involve essentially a geological study, such as has not hitherto been attempted, of the Boulder Clay of central Northamptonshire and southern Leicestershire, with many ancillary benefits; but there is a danger that the extremely localised variability of the Boulder Clay might lead to no firm conclusions on the Brixworth questions;

b) analysis of brick and hypocaust tiles from Roman sites within perhaps a wider area, with particular attention to the towns—Irchester, Bannaventa, Towcester, Duston, and perhaps especially Leicester. A reasonable number of samples and sites would need to be available to make the project worthwhile, but this approach might be found to have implications as regards chronology as well as source.

The use of brick in Brixworth church, whether it is reused Roman or of Saxon manufacture or some combination, is remarkable, not least because it occurs in an area of easily won and easily worked stone. It would be wrong, however, to view Brixworth as an isolated phenomenon. What is at issue is the whole question of the reuse of Roman building materials, indeed of Roman buildings themselves. Recent papers by Cramp (1974, 33-34) and Higgit (1973, 1-6) have focused attention on the survival of Roman buildings, in some cases into the twelfth century and later, and Higgit brings out their availability as quarries for building material (1973, 4). The Roman buildings known archaeologically to have continued in use or to have been revived in post-Roman times, such as St Martin's Canterbury or the ruined church at Stone-by-Faversham, are easily listed, but the catalogue of medieval buildings incorporating Roman materials, either in the initial structure or by way of repair, would be enormous and would include such well-documented examples as the now almost illegible legionary stone in the wall of Escomb church. Indeed, the widespread habit of drawing on disused Roman buildings for materials in the construction of churches during the early Anglo-Saxon period leads one to wonder whether Bede's famous reference to building more Romanorum might not mean specifically '(re)using Roman materials' rather than simply 'of stone' or generally 'in a Roman technique' (cf. Cramp 1974, 34).

As far as brick is concerned, buildings that spring immediately to mind are St Pancras Canterbury in the seventh century, Holy Trinity Colchester later in the Saxon period, and St Albans cathedral after the Norman Conquest. In each case there is such an obvious and, one would have thought, abundant source of used brick in the vicinity that the common assumption of a wholesale reuse of Roman materials in the construction of these churches seems inescapable. Yet it is proper to ask whether perhaps the church builders merely supplemented supplies of bricks which they made themselves by robbing the adjacent ruins—or even vice versa. Such 'native' brick-making might, in the light of the articles referred to in the previous paragraph, be regarded as a continuation or—more likely—a revival either of the Roman technology as such, or more generally of Roman structural concepts. The Roman remains surrounding the church builders, in a less ruinous state perhaps than we tend to imagine, may have served more as a source of inspiration and an ever-present model of a type of building construction than as an actual quarry for the building material itself.

Whether the bricks in a given building are Roman or Saxon or both is not simply a question of scientific fact. It has also to do with cultural history: were the early medieval builders led by the idea of Romanitas to imitate the structural forms and masonry types of their predecessors on such sites, or were they attracted merely by the legacy of convenient, ready-made and cheap building materials? Such questions can be answered only by schemes of investigation such as that outlined here. Brixworth is a key site, but others are equally worthy of attention, in particular ruinous buildings like St Pancras Canterbury, where large-scale systematic sampling presents fewer problems than it does in the case of standing structures still in use. The Brixworth programme is being carried out by people whose main interest is in the post-Roman period, but it is essential for the full understanding of the subject that specialists in Romano-British ceramics should regard the reuse of Roman brick as an integral part of their area of study.

BIBLIOGRAPHY

Blagg 1978: T. F. C. Blagg, 'A Decorated Roman Cornice from Brixworth, Northamptonshire', J. Br. Archaeol. Assoc., 131 (1978), 110-12.

Cherry 1976: B. Cherry, 'Ecclesiastical Architecture', in D. M. Wilson (ed.), The Archaeology of Anglo-Saxon England, London, 1976, 151-200.

Cramp 1974: R. Cramp, 'The Anglo-Saxons and Rome', Trans. Archit. Archaeol. Soc. Durham Northumberland, new ser., 3 (1974), 27-37.

Everson 1977: P. Everson, 'Excavations in the Vicarage Garden at Brixworth 1972', J. Br. Archaeol. Assoc., 130 (1977), 55-122.

Hall 1977: D. N. Hall, 'Excavations in Brixworth Churchyard 1971', J. Br. Archaeol. Assoc., 130 (1977), 123-32.

Harley 1974: L. S. Harley, 'A Typology of Brick: with numerical coding of brick characteristics', J. Br. Archaeol. Assoc., ser. 3, 37 (1974), 63-87.

Higgit 1973: J. C. Higgit, 'The Roman background to medieval England', J. Br. Archaeol. Assoc., ser 3, 36 (1973), 1-15.

Parsons 1972: D. Parsons, 'Brixworth, Church of All Saints', Bulletin of the Northamptonshire Federation of Archaeological Societies, 7 (1972), 39-40.

Parsons 1977: D. Parsons, 'Brixworth and its monastery church', in A. Dornier (ed.), Mercian Studies, Leicester, 1977, 173-90.

Parsons 1979: D. Parsons, 'Past History and Present Research at All Saints' Church, Brixworth', Northamptonshire Past and Present, vol. 4 no. 2 (1979), 61-71.

www.ingramcontent.com/pod-product-compliance
Ingram Content Group UK Ltd.
Pitfield, Milton Keynes, MK11 3LW, UK
UKHW061843041026
15591UKWH00015B/894